Mapping, Monitoring and Assessing Disasters

Mapping, Monitoring and Assessing Disasters

Editors

Spyridon Mavroulis
Efthymios Lekkas

MDPI • Basel • Beijing • Wuhan • Barcelona • Belgrade • Manchester • Tokyo • Cluj • Tianjin

Editors

Spyridon Mavroulis
National and Kapodistrian
University of Athens
Greece

Efthymios Lekkas
National and Kapodistrian
University of Athens
Greece

Editorial Office
MDPI
St. Alban-Anlage 66
4052 Basel, Switzerland

This is a reprint of articles from the Special Issue published online in the open access journal *Applied Sciences* (ISSN 2076-3417) (available at: https://www.mdpi.com/journal/applsci/special_issues/disasters_mapping_monitoring_assessing).

For citation purposes, cite each article independently as indicated on the article page online and as indicated below:

LastName, A.A.; LastName, B.B.; LastName, C.C. Article Title. *Journal Name* **Year**, *Volume Number*, Page Range.

ISBN 978-3-0365-6539-2 (Hbk)
ISBN 978-3-0365-6540-8 (PDF)

Contents

About the Editors

Spyridon Mavroulis

Dr. Spyridon Mavroulis is a Geologist with an MSc in the Prevention and Management of Natural Hazards, MSc in Environmental, Disasters and Crises Management Strategies and a PhD in Geology and Earth Sciences. His research interests focus mainly on the fields of neotectonics, tectonic geomorphology, hazard and risk assessment, disaster prevention and management of natural hazards and their effects on public health and on various elements of the built and the natural environment. He has participated in research projects on the integrated management of the geoenvironmental impact from natural hazards and related disasters. He has participated in over 20 scientific missions in Greece and worldwide in areas affected by earthquakes, landslides, tsunamis, volcano eruptions and megafires, to study the generation mechanism, assessing and mapping the impact on the geoenvironment and the built environment and evaluating actions and measures for disaster risk reduction. He has published articles on mapping, monitoring and assessing disasters in national and international scientific research journals and conferences.

Efthymios Lekkas

Dr. Efthymios Lekkas is a Professor of Dynamic, Tectonic, Applied Geology and Natural Disasters Management in the National and Kapodistrian University of Athens (NKUA) and the President of the Earthquake Planning and Protection Organization of Greece (EPPO). He is also Director of the Postgraduate Studies Program of "Environmental, Disasters and Crises Management Strategies" of the NKUA. He has participated as Principal Investigator and Lead Researcher in more than 300 applied research projects on earth sciences and disaster prevention and management. He has published books on natural and technological hazards and related disasters and scientific research publications in peer-reviewed journals and national and international conferences dealing with dynamic, tectonic, and applied geology, seismotectonics, earthquake engineering and the management of natural and technological disasters. He has participated in and/or coordinated over 70 scientific, operational and humanitarian missions for major disasters during the last 30 years in Greece and worldwide in order to assist local authorities and Civil Protection agencies and to offer scientific and technical assistance not only to the authorities of the affected countries but also to search and rescue teams and volunteers.

Preface to "Mapping, Monitoring and Assessing Disasters"

The detection, mapping and monitoring of disasters and assessing their impact play a key role in disaster management and disaster risk reduction. In the past, the acquisition of disaster data from the disaster-affected area, the rapid extraction of disaster information and the disaster situation reporting was a time consuming process, the results of which were available long after the completion of the immediate response phase.

In recent decades, the rapid scientific and technological developments and especially their synergy have contributed to the progression and evolution of the field of disaster prevention and management with significant results in disaster risk reduction. The geospatial technological advances have contributed significantly to the early detection of upcoming disasters, the automatic delineation of affected areas, the accurate mapping of disaster impact and the real-time or near-real-time monitoring of their evolution. Furthermore, they have enabled scientists to acquire related data in real time, to quickly analyze available information and timely disseminate critical information to the scientific community, the disasters and crises management agencies and the general public, aiming at the more effective mitigation of adverse effects on all sectors of human activity and on the majority of elements of the natural and built environment.

The purpose of this Special Issue "Mapping, Monitoring and Assessing Disasters" is to collect research advances in this scientific and operational field. It comprises 10 research articles addressing various issues of mapping and monitoring disasters and assessing their impact.

The phenomena studied by the authors comprise the seismic activity in the Ionian Islands (Western Greece) from 2014 to 2018, which included four earthquakes with magnitudes ranging from Mw=5.9 to 6.8, the seismic sequence of Thiva (Central Greece) from 2020 to 2021, the destructive 2021 Arkalochori (Crete, southern Greece) Mw=6.0 earthquake, landslides triggered by earthquakes in the Ionian Islands from historical times to the present day, the landslides caused by the Medicane Ianos that occurred in mid-September 2020 in western and central Greece, the fires in the wildland–urban interface (WUI) zones in the Attica Region (central Greece) during the 2021 spring and summer season (May-August), the megafire in the Maule region (Chile) during the 2021 summer season (January–February), which was one of the largest megafires in the history of the south central zone of the country, and the historic 1917 Samoa tsunami in the Pacific Ocean, which constituted the second most deadly recorded tsunami event in this area after the fatal 2009 event.

Joint pre-, co- and post-seismic ground deformation and seismological analysis was performed for the analysis of the seismicity in the Ionian Islands from 2014 to 2018. It was based on geodetic data from the commercial and institutional continuous Global Navigation Satellite System (GNSS) networks in the area, as well as seismological data from the Hellenic Unified Seismic Network (HUSN).

Double-difference relocation was utilized to assemble a high-resolution earthquake catalogue and to examine, in detail, the distribution of hypocenters and the spatiotemporal evolution of the 2020–2021 Thiva (central Greece) sequence. The local deformation was delineated by applying instrumental and imaging geodesy, and the long-term trends or anomalies that could have contributed to stress loading were identified.

Interdisciplinary research was conducted for the analysis of the 27 September 2021, Mw=6.0 Arkalochori (Crete, southern Greece) earthquake. It comprised synergistic geological mapping, tectonic analysis, fault photorealistic model creation by unmanned aerial system (UAS) data processing, as well as post-seismic surface deformation analysis by differential interferometry synthetic aperture radar (DInSAR) image interpretation coupled with accurately relocated epicenters recorded by locally established seismographs.

UAS-aided photogrammetry and terrestrial laser scanning (TLS) was applied to high-visit coastal areas in the western part of the Lefkada Island (Ionian Sea, western Greece), often affected by earthquake-triggered landslides (ETL). This application aimed to explore how the capabilities of these cutting-edge methodologies contribute to the improvement of our understanding and monitoring of the structural integrity of slopes. This approach allowed the initial identification of high-risk zones and the subsequent prioritization of measures and strategies for risk-mitigation-driven development.

Scientific publications and numerous contemporary sources were reevaluated for the compilation of the inventory of earthquake-triggered landslides (ETL) in the Cephalonia Island (Ionian Sea, western Greece). The related landslide susceptibility was examined by exploiting 10 landslide causal factors in the frame of a geographic information system (GIS)-based analytic hierarchy process (AHP). The comparison of the ETL inventory and the landslide susceptibility index (LSI) map highlight the high to critically high susceptible zones.

UAS photogrammetric survey guidance was developed for accurate landslide mapping and monitoring in steep terrains based on identical tests within landslide areas with different characteristics, with high-resolution orthophotos and digital surface models (DSMs) emerging from the UAS imagery processing through structure-from-motion (SfM) photogrammetry.

Landslides triggered by the September 2020 Ianos medicane were identified in Greece by using early remote sensing data and by conducting a series of post-event field surveys for verification. The rapid landslide recording was then compared with new methods of automated landslide mapping through the detection of changes in satellite imagery. All applied methods captured large events in mountainous areas and landslides with significant dimensions and/or long outflow distances.

The 1917 Samoa tsunamigenic earthquake was modeled from its origin to produce outputs of tsunami inundation extent and depth at a spatially flexible grid resolution, which were validated using available run up observations and tide gauge records. Then, the first detailed 1917 tsunami inundation model was combined with the digital distributions of buildings to produce exposure metrics to evaluate the likely impacts on present-day coastal assets and populations if a similar tsunami were to occur.

The effects of the 2021 wildfires in the Attica region (Greece) were examined based on Earth observation and GIS-based techniques for the development of a web app that included the derived knowledge. Sentinel-2 satellite imagery was used for extracting the burned area extent and severity using a normalized burn ratio (NBR)-based method. The erosion risk was modeled on a pre- and post-fire basis with the revised universal soil loss equation (RUSLE).

Several variables comprising human activity, geographic, topographic, and land coverage were used to develop a model of ignition risk in Maule (Chile), a large zone with a Mediterranean climate, which was affected by a megafire in 2017. The derived information should be integrated into decision-making processes.

The authors who contributed to this Special Issue come from universities, research centers, observatories and ministries in Greece, New Zealand and Chile. They are typical examples of countries that, from ancient times to the present day, have faced a multitude of disasters from geophysical hazards including earthquakes and their induced hazards (landslides, liquefaction, tsunami), to hydro-meteorological hazards (floods and fires among others), and have developed a culture of risk prevention.

This Special Issue is specifically addressed to scientists working in relevant scientific fields and to the staff of operational actors involved in each of the stages of the disaster management cycle, from prevention to recovery, and generally to those interested in effective environmental, disaster and crises management strategies.

Through its applications and examples, this Special Issue aims to provide the latest high-level knowledge on the above scientific fields and to update knowledge on relevant issues. Furthermore, it aims to inspire the further development of this scientific and operational field and the effective applications for disaster mapping and monitoring, and disaster impact assessment.

The Guest Editors of this Special Issue congratulate all the authors for their valuable contributions and all the reviewers for their valuable time and constructive comments that helped to improve the overall merit of this Special Issue. The Editorial Office of "*Applied Sciences*" is acknowledged for the collaboration and the Section Managing Editor for the excellent cooperation and continuous support throughout the preparation of this Special Issue.

Spyridon Mavroulis and Efthymios Lekkas
Editors

applied sciences

MDPI

Editorial

Special Issue on Mapping, Monitoring and Assessing Disasters

Spyridon Mavroulis * and Efthymios Lekkas

Section of Dynamic Tectonic Applied Geology, Faculty of Geology and Geoenvironment, School of Sciences, National and Kapodistrian University of Athens, Panepistimiopolis Zografou, 15784 Athens, Greece
* Correspondence: smavroulis@geol.uoa.gr

Mapping, monitoring, and assessing technologies and related studies and applications play a significant role in disaster management and disaster risk mitigation. In recent years, synergies of modern and innovative methodologies have augmented the efficiency of disaster mapping shortly after their advent and made it possible for involved scientists and researchers to acquire and analyze related data and to disseminate critical information to first responders during emergencies, authorities involved in disaster management and recovery processes, the affected population, and the general public.

This present Special Issue comprises 10 research papers addressing various issues of mapping and monitoring disasters and assessing their impact. They highlight recent advances in the field and are valuable for understanding the complexity of the generated phenomena. The phenomena and their effects included in this Special Issue can be divided into: (i) earthquakes and related ground deformation [1–3], (ii) earthquake-triggered landslides (ETL) [4–6], (iii) landslides triggered by hydrometeorological hazards comprising medicanes [6,7], (iv) tsunamis [8], and (v) wildfires [9,10].

Sakkas et al. [1] present the results of monitoring seismicity and ground deformation in the Ionian Islands (western Greece) during a period of intense seismic activity (2014–2018) with destructive earthquakes. Joint pre-, co-, and post-seismic ground deformation and seismological analysis is performed based on geodetic data from the commercial and institutional continuous Global Navigation Satellite System (GNSS) networks in the area, as well as seismological data from the Hellenic Unified Seismic Network (HUSN), respectively.

Kaviris et al. [2] monitored the 2020 to 2021 Thiva (Central Greece) earthquake sequence. They utilized double-difference relocation to assemble a high-resolution earthquake catalogue and examine, in detail, the distribution of hypocenters and the spatiotemporal evolution of the sequence. By applying instrumental and imaging geodesy, they delineated the local deformation and identified long-term trends that could have contributed to stress loading.

Vassilakis et al. [3] monitored the 27 September 2021, Mw = 6.0 Arkalochori (Crete, southern Greece) earthquake. They conducted interdisciplinary research comprising geological mapping, tectonic analysis, fault photorealistic model creation by unmanned aerial system (UAS) data processing, as well as post-seismic surface deformation analysis by differential interferometry synthetic aperture radar (DInSAR) image interpretation coupled with accurately relocated epicenters recorded by locally established seismographs.

Mavroulis et al. [4] applied UAS-aided photogrammetry and terrestrial laser scanning (TLS) to high-visit coastal areas in the western part of Lefkada Island (western Greece), often affected by ETL. This application aims to explore how the capabilities of these cutting-edge methodologies contribute to the improvement of our understanding on and monitoring of the structural integrity of slopes. This approach allows the initial identification of high-risk zones and the subsequent prioritization of measures and strategies for risk-mitigation-driven development.

Mavroulis et al. [5] studied ETL from historical times to present in Cephalonia Island (western Greece). Based on scientific publications and numerous contemporary sources, they compiled an inventory of sites affected by ETL, several of which caused human losses

Citation: Mavroulis, S.; Lekkas, E. Special Issue on Mapping, Monitoring and Assessing Disasters. *Appl. Sci.* **2023**, *13*, 963. https://doi.org/10.3390/app13020963

Received: 10 January 2023
Accepted: 10 January 2023
Published: 11 January 2023

and injuries. The study further examines the ETL susceptibility, exploiting 10 landslide causal factors in the frame of a geographic information system (GIS)-based analytic hierarchy process (AHP). The comparison of the ETL inventory and the landslide susceptibility index (LSI) map highlights the high to critically high susceptible zones and reveals that the majority of ETL was generated within the highlighted susceptible zones.

Nikolakopoulos et al. [6] focused on developing a UAS photogrammetric survey guidance for accurate landslide mapping and monitoring in steep terrains. They conducted four identical tests within landslide areas with different characteristics. High-resolution orthophotos and digital surface models (DSMs) emerge from the UAS imagery processing through structure-from-motion (SfM) photogrammetry. Accuracy assessment is carried out using quantitative and qualitative comparative approaches, and a strong relation is revealed between UAS acquisition geometry and landslide characteristics.

Valkaniotis et al. [7] identified the landslides triggered by the 2020 Ianos medicane by using early remote sensing data and conducting a series of post-event field surveys for verification. The rapid landslide recording is then compared with new methods of automated landslide mapping through the detection of changes in satellite imagery. All applied methods captured large events in mountainous areas and landslides with significant dimensions and/or long outflow distance. In terms of comparing the compiled inventory with past events, they concluded that the Ianos landslides were triggered along roughly the same locations of historical occurrences, revealing a relationship with long-term climatic and lithological/geomorphological conditions.

Sischka et al. [8] modeled the 1917 Samoa tsunamigenic earthquake from its origin to produce outputs of tsunami inundation extent and depth at spatially flexible grid resolution, which are validated using available run up observations and tide gauge records. Then, they combined the inundation model with digital distributions of buildings to produce exposure metrics for evaluating the likely impacts on present-day coastal assets and populations if a similar tsunami were to occur. They provided the first detailed 1917 tsunami inundation model, supporting an appreciation of the regional risk to local tsunamis.

Falaras et al. [9] examined the effects of the 2021 wildfires in the Attica region (Greece) based on Earth observation and GIS-based techniques for the development of a web app that includes the derived knowledge. The effects of wildfires are estimated with the use of Sentinel-2 satellite imagery concerning burned area extent and burn severity using a normalized burn ratio (NBR)-based method. In addition, the erosion risk is modeled on a pre-fire and post-fire basis with the revised universal soil loss equation (RUSLE). This study highlights the importance of assessing the effects of wildfires with a holistic approach to produce useful knowledge tools in post-fire impact assessment and restoration.

Azócar de la Cruz et al. [10] used several factors, such as human activity, geographic, topographic, and land cover variables to develop a model of ignition risk. The study area corresponds to Maule region (Chile), a large zone with a Mediterranean climate, affected by a megafire in 2017. Wildland fire management requires integrating this information into decision-making processes if we consider that the impact of climate change persists.

Funding: This research received no external funding.

Acknowledgments: We congratulate all authors for their valuable contributions to this Special Issue and to all reviewers for their valuable time and constructive comments that helped improve the overall merit of this Special Issue. We also thank the Editorial Team of "Applied Sciences" for the collaboration, and we express our appreciation to the Section Managing Editor for the excellent cooperation and continuous support throughout the preparation of the Special Issue. We hope that the papers will inspire further development and effective applications for disaster mapping and monitoring and impact assessment.

Conflicts of Interest: The author declares no conflict of interest.

References

1. Sakkas, V.; Kapetanidis, V.; Kaviris, G.; Spingos, I.; Mavroulis, S.; Diakakis, M.; Alexopoulos, J.D.; Kazantzidou-Firtinidou, D.; Kassaras, I.; Dilalos, S.; et al. Seismological and Ground Deformation Study of the Ionian Islands (W. Greece) during 2014–2018, a Period of Intense Seismic Activity. *Appl. Sci.* **2022**, 12, 2331. [CrossRef]
2. Kaviris, G.; Kapetanidis, V.; Spingos, I.; Sakellariou, N.; Karakonstantis, A.; Kouskouna, V.; Elias, P.; Karavias, A.; Sakkas, V.; Gatsios, T.; et al. Investigation of the Thiva 2020–2021 Earthquake Sequence Using Seismological Data and Space Techniques. *Appl. Sci.* **2022**, 12, 2630. [CrossRef]
3. Vassilakis, E.; Kaviris, G.; Kapetanidis, V.; Papageorgiou, E.; Foumelis, M.; Konsolaki, A.; Petrakis, S.; Evangelidis, C.P.; Alexopoulos, J.; Karastathis, V.; et al. The 27 September 2021 Earthquake in Central Crete (Greece)—Detailed Analysis of the Earthquake Sequence and Indications for Contemporary Arc-Parallel Extension to the Hellenic Arc. *Appl. Sci.* **2022**, 12, 2815. [CrossRef]
4. Mavroulis, S.; Vassilakis, E.; Diakakis, M.; Konsolaki, A.; Kaviris, G.; Kotsi, E.; Kapetanidis, V.; Sakkas, V.; Alexopoulos, J.D.; Lekkas, E.; et al. The Use of Innovative Techniques for Management of High-Risk Coastal Areas, Mitigation of Earthquake-Triggered Landslide Risk and Responsible Coastal Development. *Appl. Sci.* **2022**, 12, 2193. [CrossRef]
5. Mavroulis, S.; Diakakis, M.; Kranis, H.; Vassilakis, E.; Kapetanidis, V.; Spingos, I.; Kaviris, G.; Skourtsos, E.; Voulgaris, N.; Lekkas, E. Inventory of Historical and Recent Earthquake-Triggered Landslides and Assessment of Related Susceptibility by GIS-Based Analytic Hierarchy Process: The Case of Cephalonia (Ionian Islands, Western Greece). *Appl. Sci.* **2022**, 12, 2895. [CrossRef]
6. Nikolakopoulos, K.G.; Kyriou, A.; Koukouvelas, I.K. Developing a Guideline of Unmanned Aerial Vehicle's Acquisition Geometry for Landslide Mapping and Monitoring. *Appl. Sci.* **2022**, 12, 4598. [CrossRef]
7. Valkaniotis, S.; Papathanassiou, G.; Marinos, V.; Saroglou, C.; Zekkos, D.; Kallimogiannis, V.; Karantanellis, E.; Farmakis, I.; Zalachoris, G.; Manousakis, J.; et al. Landslides Triggered by Medicane Ianos in Greece, September 2020: Rapid Satellite Mapping and Field Survey. *Appl. Sci.* **2022**, 12, 12443. [CrossRef]
8. Sischka, L.; Bosserelle, C.; Williams, S.; Ting, J.C.; Paulik, R.; Whitworth, M.; Talia, L.; Viskovic, P. Reconstructing the 26 June 1917 Samoa Tsunami Disaster. *Appl. Sci.* **2022**, 12, 3389. [CrossRef]
9. Falaras, T.; Tselka, I.; Papadopoulos, I.; Nikolidaki, M.; Karavias, A.; Bafi, D.; Petani, A.; Krassakis, P.; Parcharidis, I. Operational Mapping and Post-Disaster Hazard Assessment by the Development of a Multiparametric Web App Using Geospatial Technologies and Data: Attica Region 2021 Wildfires (Greece). *Appl. Sci.* **2022**, 12, 7256. [CrossRef]
10. Azócar de la Cruz, G.; Alfaro, G.; Alonso, C.; Calvo, R.; Orellana, P. Modeling the Ignition Risk: Analysis before and after Megafire on Maule Region, Chile. *Appl. Sci.* **2022**, 12, 9353. [CrossRef]

applied sciences

MDPI

Article

The Use of Innovative Techniques for Management of High-Risk Coastal Areas, Mitigation of Earthquake-Triggered Landslide Risk and Responsible Coastal Development

Spyridon Mavroulis [1,*], Emmanuel Vassilakis [2], Michalis Diakakis [1], Aliki Konsolaki [2], George Kaviris [3], Evangelia Kotsi [1], Vasilis Kapetanidis [3], Vassilis Sakkas [3], John D. Alexopoulos [3], Efthymis Lekkas [1] and Nicholas Voulgaris [3]

1 Department of Dynamic Tectonic Applied Geology, Faculty of Geology and Geoenvironment, School of Sciences, National and Kapodistrian University of Athens, Panepistimiopolis Zografou, 15784 Athens, Greece; diakakism@geol.uoa.gr (M.D.); ekotsi@geol.uoa.gr (E.K.); elekkas@geol.uoa.gr (E.L.)
2 Department of Geography and Climatology, Faculty of Geology and Geoenvironment, School of Sciences, National and Kapodistrian University of Athens, Panepistimiopolis Zografou, 15784 Athens, Greece; evasilak@geol.uoa.gr (E.V.); alikikons@geol.uoa.gr (A.K.)
3 Department of Geophysics and Geothermy, Faculty of Geology and Geoenvironment, School of Sciences, National and Kapodistrian University of Athens, Panepistimiopolis Zografou, 15784 Athens, Greece; gkaviris@geol.uoa.gr (G.K.); vkapetan@geol.uoa.gr (V.K.); vsakkas@geol.uoa.gr (V.S.); jalexopoulos@geol.uoa.gr (J.D.A.); voulgaris@geol.uoa.gr (N.V.)
* Correspondence: smavroulis@geol.uoa.gr

Abstract: Coastal areas constitute a very dynamic environment, balancing between numerous natural and anthropogenic processes liable to sometimes hazardous geomorphic phenomena. Especially in tectonically active coastal regions and areas of high economic value, slope failures can have significant impacts and therefore need careful and detailed examination. This work uses Unmanned Aerial System (UAS)-aided photogrammetry and Terrestrial Laser Scanning (TLS) in tectonically active segments of the coastal zone of the Ionian Islands in Greece, to explore how their capabilities can help to improve our understanding of the structural integrity of the slopes. Results show that the two approaches are able to extract large numbers of discontinuity facets, in a more practical, rapid and safe way than conventional methods of rock slope stability analysis extending to unreachable yet important parts of the slope. Through this holistic record of the structural condition of the slope the two applications allow the identification of segments that are more prone to instability and failure. In this way, they improve our understanding of the prioritization of interventions aiming to enhance the prevention of slope failures, mitigating the associated risk and improving local development in these high-value locations.

Keywords: coastal areas; landslides; lidar; UAS; remote sensing; coastal development; TLS; Ionian

Citation: Mavroulis, S.; Vassilakis, E.; Diakakis, M.; Konsolaki, A.; Kaviris, G.; Kotsi, E.; Kapetanidis, V.; Sakkas, V.; Alexopoulos, J.D.; Lekkas, E.; et al. The Use of Innovative Techniques for Management of High-Risk Coastal Areas, Mitigation of Earthquake-Triggered Landslide Risk and Responsible Coastal Development. *Appl. Sci.* **2022**, *12*, 2193. https://doi.org/10.3390/app12042193

Academic Editor: Hyung-Sup Jung

Received: 18 January 2022
Accepted: 18 February 2022
Published: 20 February 2022

1. Introduction

Globally coastal areas balance in a highly dynamic regime formed by a variety of complex natural processes [1] and human intervention [2]. The evolution of these constantly changing environments is affected by numerous natural geomorphic mechanisms such as erosion, mass wasting, deposition, wave action, as well as tectonic and volcanic activity [3]. Despite these processes and the risks associated with them, coastal areas contain a large part of the world's population [4] and an important portion of socio-economic activities [5], including trade, tourism, transportation and others. This activity heavily influences [6] some segments of the coastline. Coastal zones are home to many interests including economic (ports, fishing, navigation), recreation, water quality and nature conservation. Assets developed in coastal areas can be very valuable [7] and the disruption of some of the activities can be very costly [8]. There are examples where the interests and exploitation schemes of these areas are conflicting with each other or lead to expansion into areas of high risk [9–11]. Thus, coastal zones are often characterized by both high exposure of valuable assets [8], as well as high hazard levels as a result of the various dynamic processes that take place.

Particularly, in seismically active areas, high inclination, deformation of geological formations and strong earthquakes lead to instability phenomena [12–14] that are often associated with significant damage to property and infrastructure [15,16] and loss of human lives [17,18]. On many occasions, these failures can become a very significant threat in coastal areas. The complexity of the processes together with the high risk caused by such mass-movement events create a demand for in-depth knowledge and understanding of the mechanisms (both natural and anthropogenic) that drive them [19]. For example, it is crucial to understand the role of discontinuities in geological formations that affect their integrity and stability, acting as catalysts in various geomorphic processes, including landslides. Monitoring this and other structural properties of rock formations is important to predict slope behavior in coastal areas.

Coastal slope failures are a significant source of hazard and a constraint for human activities in rocky coastal areas [20]. The most prevalent process responsible for the high risk of coastal cliffs is slope mass movements [21].

The extent and economic impact of this problem tend to increase over time, due to the general increasing trend in the use of coastal areas, especially steep and rocky coastlines, which correspond to one-third of the coasts worldwide [22,23]. The value of land in these locations is constantly increasing due to the high demand for exploitation, especially beaches of special beauty, which are often formed at the base of cliffs and are a tourism product of high value. There are also issues related to the presence of urban areas and archaeological and historical heritage sites close to the top of the cliffs [24,25].

In this context, coastal evolution and its processes have been examined through a variety of techniques, some of which belong to the field of remote sensing [26]. New techniques, such as photogrammetry and laser scanning, have shown interesting capabilities in studying changing terrain and particularly geomorphic processes, including slope stability and structural integrity [27,28].

For example, unmanned aerial systems (UASs) and UAS-aided photogrammetry have been used before in applications in coastal areas including surveying topographical changes [29], cliff erosion [30], coastline changes [31], and coastal floods [32–34]. In the field of mass movement, previous studies [35,36] have demonstrated that UAS, with the proper processing of data, can provide fairly accurate results [37,38].

Similarly, Terrestrial Laser Scanning (TLS) has demonstrated its capabilities in slope stability [39] including applications in coastal zones [40,41]. TLS has been used in numerous works to study landslides [42–44], cliff erosion [45–47] and rock formation discontinuities [48–50].

Despite the demonstrated capabilities of these technologies and their high-resolution imaging aptitude, there are very limited applications in highly visited areas or areas of high value that combine active tectonics and frequent instability phenomena. In addition, there are few applications that exploit outcomes of these techniques regarding slope stability to improve risk mitigation and enhance development.

Given that, in many cases, decisions on the exploitation of coastal zones are made with a poor understanding of the evolutionary processes and functioning of the coastal environments, in the case of high-value and highly active areas, it is considered crucial to gain an improved understanding of dangerous instability phenomena. Any sustainable and responsible effort of development on these types of zones requires a deep understanding of the complexity of geomorphic processes and the roles of influencing and instability-triggering factors [51–53], including tectonic and earthquake dynamics, mass wasting, erosion, deposition, land-use changes, human intervention, and others.

In this context, the aim of this work is to exploit innovative techniques such as UAS-aided photogrammetry and TLS to improve our understanding of the processes taking place, segments of the coastal zone of the Ionian Islands in Greece and, in particular, the island of Lefkada. The two study sites are situated in an earthquake-prone area, with frequent strong earthquakes leading to various mass-movement and slope failure phenomena, enhanced by active tectonics, but also subject to high-intensity storms (including Medicanes) that have also contributed to the high landslide risk. The present study aims to apply the two techniques at the Egremnoi and Porto Katsiki beaches, two highly visited tourist attractions both characterized by noteworthy economic and ecological value. The study aims to determine the zones of higher risk through the two applications and discuss the outcomes in relation to local development, considering that Lefkada is a typical example of a Mediterranean busy coastal zone, home to hundreds of thousands of visitors each year.

The study is organized as follows. First, we present the broader area and the study site, in terms of its geodynamic characteristics, then we describe the methodology for the UAS and the TLS applications, followed by the results. Then we discuss the practical implications of the findings and describe the most important conclusions in the last section.

2. Study Area

2.1. Geological and Seismotectonic Setting

Lefkada Island is located in the southern Ionian Islands (Western Greece), in the front of the Hellenides orogenic belt, in a region dominated by the presence of the Cephalonia Transform Fault Zone (CTFZ) [54,55]; (Figure 1). The latter is a 140-km long right-lateral fault zone, as evidenced by a large number of focal mechanisms [56–59]. The study area lies in the transition zone between the subduction of Tethys beneath the Aegean microplate to the south and the continental collision to the north (Figure 1). The CTFZ consists of two major segments, i.e., the Cephalonia Segment (CS) to the south and the Lefkada Segment (LS) to the north.

The proximity of Lefkada to the CTFZ is the main reason the island is among the most seismogenic areas in Europe. Regarding the historical era, large and destructive earthquakes have occurred during the last six centuries [60,61] (Figure 2a). Most of these events mainly affected the western coast of the island, such as the 1704 (M = 6.3), 1769 (M = 6.7), 1783 (M = 6.7) and 1869 (M = 6.4) earthquakes [62–64], (Figure 2a). Thirteen disastrous events that took place at the western onshore area of Lefkada occurred between 1612 and 1869 [64,65]. On the contrary, the 1723 (M = 6.7) earthquake was the sole historical event that occurred at the southwestern part of Lefkada Island [58].

Intense seismic activity has been recorded in the Lefkada region during the 20th century, characterized by the frequent occurrence of destructive events [66]. More specifically, 87 events with Mw $\geq$ 4.1 occurred between 1911 and 1998, 36 of which with Mw $\geq$ 5.0 and 5 with Mw $\geq$ 6.0 (Figure 2a). Both of the largest events of the 20th century, with Mw = 6.5, occurred in 1948. The first event of 22 April destroyed the southwestern part of the island, whereas the second on 30 June caused damage to the northwestern part [66].

Two large events occurred in the vicinity of Lefkada Island during the 21st century. The first was the Mw = 6.3 on 14 August 2003, located northwest of the island at a depth of 9 km. Its focal mechanism indicated dextral strike-slip faulting (e.g., [58,67,68]).

The most recent destructive event occurred on 17 November 2015 (Mw = 6.4) close to Athani, at the southwestern coast of Lefkada, on an almost vertical dextral strike-slip fault [55,69]. Most aftershocks were aligned parallel to the western coast of Lefkada Island, along an NNE-SSW direction, with another group of epicenters located close to the northern part of Cephalonia Island [63]. The seismogenic layer ranged from 3 to 16 km in depth, whereas more than one fault plane with different strike and dip values were activated [64]. Despite its larger magnitude, this event caused less damage than the 2003 earthquake, proving the structural efficiency of the local buildings [69].

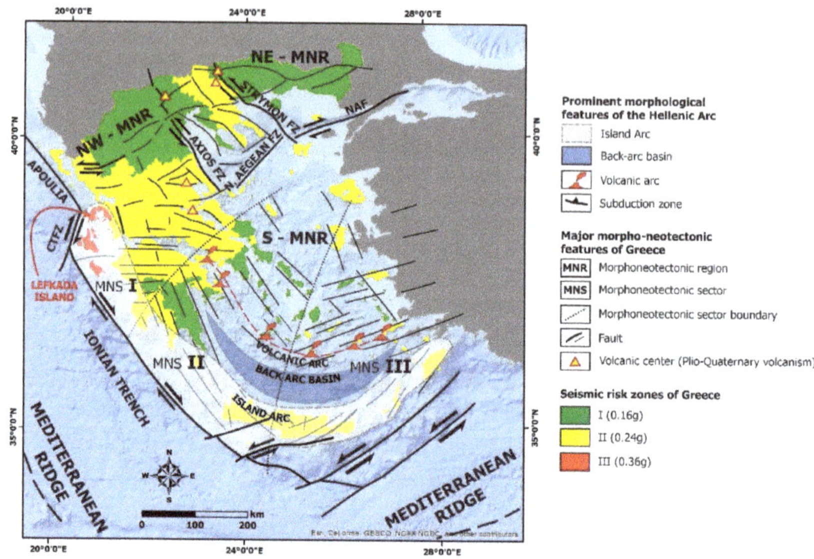

Figure 1. Map of the Hellenic Arc showing the location of Lefkada Island at the northwesternmost part of the Hellenic Arc along with the prominent morphological features of the Hellenic Arc and the major morphoneotectonic features based on Mariolakos and Papanikolaou [70,71] and the seismic risk zones of Greece. Lefkada belongs to zone III of the current Greek Building Code with a Peak Ground Acceleration (PGA) value of 0.36 g for a return period of 475 years.

The geological structure of the southern Ionian Islands is characterized by the occurrence of alpine formations of the Ionian and Paxoi geotectonic units and post-alpine deposits lying unconformably on the alpine basement [72–77]. Similarly, Lefkada is composed of (a) alpine formations that belong to Ionian and Paxoi geotectonic units, (b) molassic formations and (c) recent deposits that lie unconformably on the previous formations [74,78] (Figure 2a). Ionian formations are observed in the largest part of the island, while Paxoi formations are only observed in the southwestern part of the island and more specifically in the Lefkata peninsula (Figure 2a). As regards the tectonic structure, the faults dissecting the island are mainly normal or strike-slip with a sinistral or dextral sense of shear [77,79,80]. The active structures are major faults detected mainly along the margins of fault blocks [72,77].

In particular, Lefkada is composed of the following 8 fault blocks [78] (Figure 2a):

1. The fault block of Lefkada town, located in the northeastern part of the island and bound to the south by the Frini-Apolpaena fault zone (FAFZ).
2. The Tsoukalades-Katouna fault block, located to the south of the previous fault block and bounded to the north by the FAFZ, to the south by the Pigadisanoi-Fraxi fault zone (PFFZ) and to the west by the Tsoukalades-Agios Nikitas fault zone (TANFZ).
3. The Agios Nikitas fault block, located in the northwestern part of the island and bounded to the east by the Agios TANFZ and to the west by an almost N–S striking fault zone parallel to the coast.
4. The Drymonas fault block, located east of the Agios Nikitas fault block and bounded to the west by the Agios Nikitas fault zone (ANFZ), to the east by the N–S striking Drymonas fault zone (DFZ) and to the south by the NW–SE striking Kalamitsi-Exantheia fault zone (KEFZ).
5. The Mega Oros-Skaroi fault block, located east of the Drymonas fault block and bounded to the north by the NW–SE striking PFFZ and to the south by the NE–SW striking Sivros-Nidri fault zone (SNFZ).

6. The Vlicho-Poros fault block located in the southeastern part of the island and bounded to the northwest by the NE–SW striking SNFZ and Vassiliki fault zone (VFZ) and to the southeast by the NW–SE striking Syvota-Sivros fault zone (SSFZ).
7. The Sykeros-Achrada fault block, located in the southern-central part of the island and bounded to the northwest by the NE–SW striking VFZ and to the northeast by the NW–SE striking SSFZ.
8. The Lefkata fault block located, in the southwestern part of the island and bounded to the north by the NW–SE striking KEFZ and the Ionian thrust on to Paxoi unit, to the east by the NE–SW striking VFZ and to the west by the dextral strike-slip Athani fault zone (AFZ).

Some of the aforementioned main fault blocks are composed of smaller fault blocks. The most characteristic case is the Dragano-Athani graben within the fault block of the Lefkata peninsula. This graben is bounded to the west and east by NNE-SSW striking faults. Its western margin is defined by the AFZ with a total horizontal displacement of approximately 860 m corresponding to a Quaternary slip rate of about 8 mm/year [81] and similar geometric and kinematic characteristics with the Lefkada segment of the Cephalonia Transform Fault Zone [79].

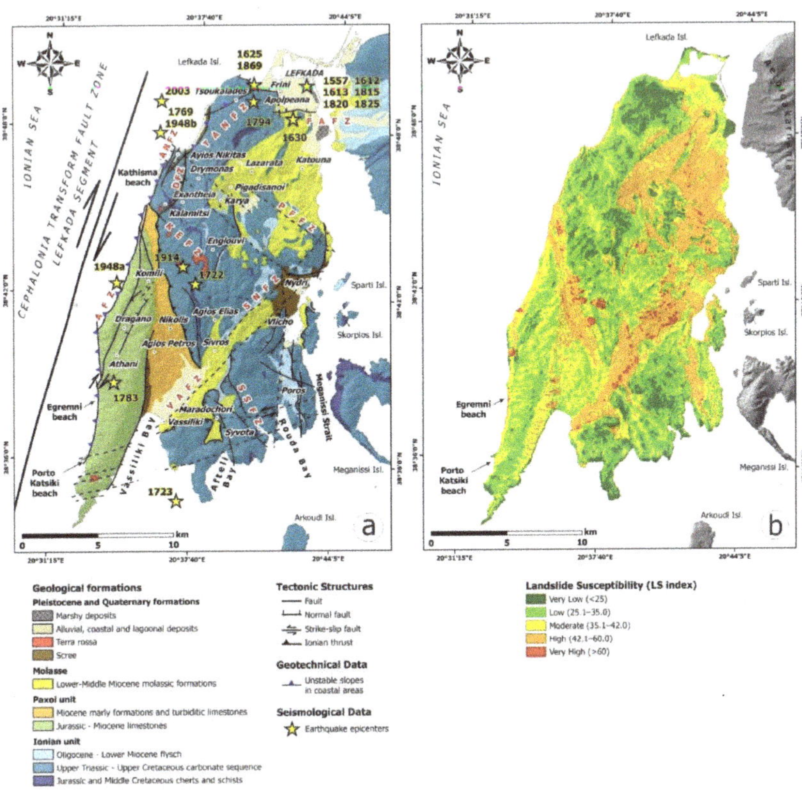

Figure 2. (**a**) The neotectonic map of Lefkada Island from Lekkas et al. [77]. The Egremnoi and Porto Katsiki coastal areas are located in the southwestern part of the fault block of Lefkata peninsula and are composed of Jurassic–Miocene limestones of the Paxoi geotectonic unit. (**b**) Based on the landslide susceptibility map of the Ionian Islands compiled by Mavroulis et al. [82], these slopes are formed in coastal areas highly susceptible to failure.

2.2. Study Sites

In the frame of susceptibility and hazard assessment in the Ionian Islands for highlighting sites of earthquake-related hazards, Mavroulis et al. [82] studied the landslide susceptibility of the Ionian Islands by applying the Analytical Hierarchical Process (AHP) used along with the Weighted Linear Combination (WLC) method in the context of multicriteria decision analysis for the calculation of the Landslide Susceptibility Index (LSI). Based on the applied methodology and the respective results, it is concluded that the abrupt coastal slopes and scarps in the western part of Lefkada are characterized by high and very high susceptibility to earthquake-triggered landslides (Figure 2b).

Indeed, the western coastal part of the island has been heavily affected by landslides triggered by historical and recent earthquakes [64,79,83]. In particular, landslides comprising mainly rockfalls have been triggered by earthquakes on:

- 22 November 1704 (M = 6.3, I = IX).
- 22 and 23 March 1783 (M = 6.7, I = X for the event on 23 March).
- 14 December 1885 (Ms = 5.7, Mw = 5.1).
- 27 November 1914 (M = 6.3, I = IX).
- 22 April 1948 (Mw = 6.5, I = IX).
- 30 June 1948 (Mw = 6.4).
- 14 August 2003 (Mw = 6.2, Ms = 6.4, I = VIII).
- 17 November 2015 (Mw = 6.4, I = VIII).

Among the areas affected by earthquake-triggered landslides are the coastal areas of Egremnoi and Porto Katsiki (Figure 3), located in southwestern coastal Lefkada [64,84,85]. They constitute typical cases of extensive coastal slopes, where the change in morphological slope is so abrupt that almost vertical planes are formed. In addition to faults and joints, the continuous marine and terrestrial erosion processes play an important role in the evolution of these coastal slopes.

The geological structure of Egremnoi and Porto Katsiki is similar, as both are composed of Jurassic-Miocene limestone of the Paxoi unit. They are intensively faulted resulting in their disintegration and brecciation, while locally completely pulverized and extremely unstable. They are extremely susceptible to landslides, a fact that has been revealed not only by the landslide susceptibility assessment of Mavroulis et al. [82], but also by the triggering of landslides not only by nearby earthquakes but also by distant earthquakes (Figure 3).

Typical examples of landslides induced by nearby earthquakes are the rockfalls and slides caused by the earthquake on 17 November 2015 (Figure 3). In the Egremnoi area, a large mass comprising loose limestone blocks, semi-cohesive scree with limestone breccia and red clay was mobilized during the 2015 earthquake and rolled down towards the adjacent beach, which was almost entirely covered by landslide material [78] (Figure 3a–c). Furthermore, the part of the road network leading to the beach was completely destroyed, and some buildings found on the edge of the slope were at risk of total collapse [86,87]. As regards the presence of residents and tourists in this area, it was fortunate that the earthquake was generated during the winter season and the beach was empty.

In Porto Katsiki, loose brecciated limestone blocks were detached from the abrupt limestone slope and fell on the narrow beach [78] (Figure 3f). It is significant to note that both beaches are amongst the most beautiful and visited in the Mediterranean with thousands of locals and tourists attracted during the summer season staying close to the steep limestone slopes or inside coastal caves. Fortunately, the 2015 earthquake was generated during the winter season with very limited tourist flow in Lefkada and both beaches were totally empty.

A typical example of landslides induced in these areas by distant earthquakes comprises the rockfalls triggered in the large coastal slope of Porto Katsiki by the 8 June 2008, Mw = 6.8, Andravida (NW Peloponnese, Western Greece) strike-slip earthquake [88] (Figure 3d). The epicenter was reported at a distance of 110 km southeast of the affected slope. Due to this large distance from the epicenter and due to the absence of similar phenomena in the intermediate areas with similar geological and morphological properties, these rockfalls were considered to be far-field earthquake environmental effects, attributed to the combined effect of the rugged morphology, the intensively faulted and eroded limestones and the pre-existing instability conditions along the abrupt coastal slopes in the western part of the Lefkata peninsula [88].

Figure 3. Views of the Egremnoi (**a–c**) and Porto Katsiki (**d–f**) slopes in the southwestern part of Lefkada Island. (**a**) The Egremnoi slope and the adjacent impressive beach were heavily affected by the 2015 Mw = 6.4 Lefkada earthquake. Extended landslides were induced (**b**) and covered almost the entire longitudinal narrow beach (**c**) leaving only small parts intact and inaccessible after the earthquake. Due to the fact that the shock occurred in November, the beach was empty, and no effects were reported to people. However, the road leading from the upper parts of the slope to the beach was totally destroyed and houses founded on the slope were found on the edge of the cliff and on the verge of collapse. (**d**) The Porto Katsiki slope has been affected not only by nearby earthquakes but also by distant earthquakes. The 8 June 2008, Mw = 6.8 earthquake with the epicenter located 110 km southeast of Porto Katsiki triggered rockfalls in the southern segment of the slope, while the mobilized material ended up on the adjacent part of the beach fortunately without fatalities and injuries. (**e**) Boulders and smaller fragments are concentrated mainly in the southern part of the beach. (**f**) Rockfalls triggered by the 17 November 2015, Mw = 6.4 Lefkada earthquake. They were mainly generated in the southern segment of the slope, while they were limited or absent in the central and the northern segment respectively. Since 2015, protection marking on the beach with poles and ropes (red dashed line) and rockfall warning signs have been placed for the safety of visitors.

3. Methodology

In this context, the coastal morphology evolution, including short-term onshore changes and processes, has been examined through several remote sensing techniques based on data acquisition with the use of state-of-the-art equipment. The swift development in sensor technology and geomatics during the last decade has improved the acquisition and processing of remotely sensed information, especially regarding the monitoring of infrastructure as well as medium-sized territorial zones due to geological risks [89]. Rather prominent techniques include the generation of point clouds of the areas of interest, which can be composed either directly by using surveying equipment such as Terrestrial Laser Scanners (TLSs) or indirectly by processing numerous photographs taken by Unmanned Aerial Systems (UASs). Both Light Detection and Range (LiDAR) and Structure from Motion (SfM) processing techniques can provide large amounts of digital data of very high resolution from which point clouds of similar quality can be generated. High-resolution 3D point clouds and 3D meshes are the fundamental types of data for any quantitative structural analysis afterwards.

Laser scanning is an optimized technique for obtaining 3D surveying data. Usually, TLS has a spatial resolution ranging from millimeters to centimeters. Regarding this, the widespread use of UASs, together with SfM and Multi-View Stereo (MVS) technologies, has improved the flexibility of topographic surveys and structural analysis. In fact, nowadays, due to swift technological development, it can be used rather easily under adverse conditions, as nadiral, oblique, or frontal images can be acquired with minimal effort. Therefore, cliffs, steep slopes, hanging rocks, and morphological discontinuities in general, are either being scanned or photographed for the generation of a basement dataset on which further processing will be applied. The mapping of structural failures and rock discontinuity density at coastal cliffs proves to be very critical concerning the safety regime in the beaches beneath them, especially the popular ones [90].

During the last few years, there has been an ongoing debate as to whether the direct TLS point clouds are more accurate and precise than the indirect data constructed after photogrammetric processing [91–93]. In this context, more attention has been given to the latter, which can also generate high-resolution 3D models of slopes, focusing on the advantage of low costs, flexible oblique view sensing and photo-realistic vision information when compared with terrestrial LiDAR point clouds [94], which are produced by rather expensive equipment. On the other hand, the accuracy of these is much more adequate, not to mention the data density, which is also difficult to be obtained with UAS flights at a reasonable elevation (e.g., 75 m).

Furthermore, the use of UASs is a practical way of mapping areas larger than TLS can provide, and it has been proved to be optimal for landslides and rockfalls that often cover areas that range from less than one square kilometer up to a few square kilometers. Additionally, the limitations of beach width where in situ observations or the use of other techniques such as TLS are unattainable can be overcome via UAS and restrictions associated with LiDAR instrumentation (e.g., high weight and occlusion areas), since the final product is very similar, as mentioned above.

In this context, we chose to apply the two different approaches at the two case studies of Egremnoi and Porto Katsiki, taking into consideration the most appropriate technique for each case. A point cloud derived from SfM processing was utilized for creating the basement datasets (Digital Surface Model and ortho-photograph) for the Egremnoi coastal zone, whilst LiDAR scanning was used for direct point cloud acquisition at the popular Porto Katsiki beach. The flowchart below illustrates the main steps of the methodology along with the software solutions used for each one of them (Figure 4).

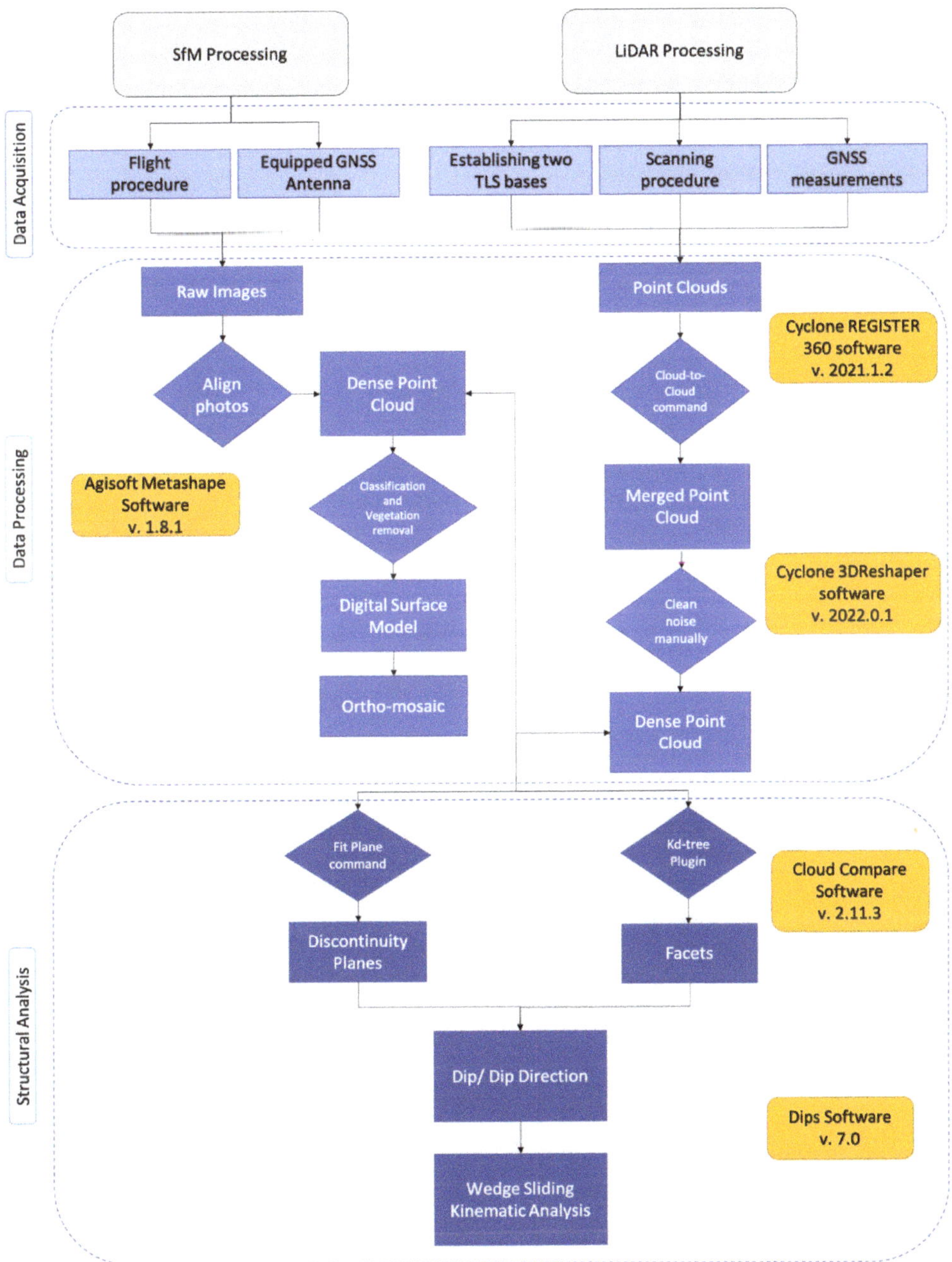

Figure 4. Basic steps of the applied methodology.

3.1. *UAS-Aided Survey at Egremnoi*

The raw image data acquisition was conducted with a rotor-wing UAS (DJI Phantom 4 RTK) equipped with a stabilized, built-in 20M camera (of 8.8 mm focal length) bundled on a two-axis gimbal. This unmanned platform has been chosen due to its relatively reasonable cost and easy on-site operation in combination with the equipped miniaturized

Geodetic Navigation Satellite System (GNSS) antenna, which provides sufficient precision, particularly for the horizontal positioning required to facilitate proper alignment of the images captured during the survey [95]. Although ground control points (GCPs) are a vital tool for ensuring geolocation accuracy, in this case study, in which the morphology of the area consists of very steep slopes with limited accessibility, the use of the RTK antenna with which the drone is equipped is efficient enough for highly accurate results. Nevertheless, we used the GNSS receiver in the Network Real-Time Kinematic (NRTK) mode, connected to the SmartNET provider accuracy service [96] for obtaining subjective geolocation, which is also necessary for the comparison between independent surveys [49].

The UAV flight survey was designed to cover an area of 0.3 km^2, of a 160 m steep slope consisting of fragmented limestones that fall on the coastal zone, especially after the occurrence of earthquakes [78]. The take-off point was placed at an elevation of 190 m, and the UAV was programmed to fly along four transects almost parallel to the coastline, at the absolute height of 230 m. The camera was set pointing 30 degrees off nadir, with an image overlap value of 75%. This resulted in the acquisition of 592 images that were processed within Agisoft Metashape photogrammetry software, to produce (a) DSM (0.12 m resolution), (b) a vertical ortho-mosaic image of the coastal zone (0.06 m resolution) and (c) an ortho-mosaic image normal to the steep morphological discontinuity (0.06 m resolution) (Figure 5), based on a dense point cloud that consisted of 31 million points. Regarding accuracy, the processing resulted in a root mean square error (RMSE) of 0.08 m on the xy-axes and 0.18 m on the z-axis, which are rather acceptable values.

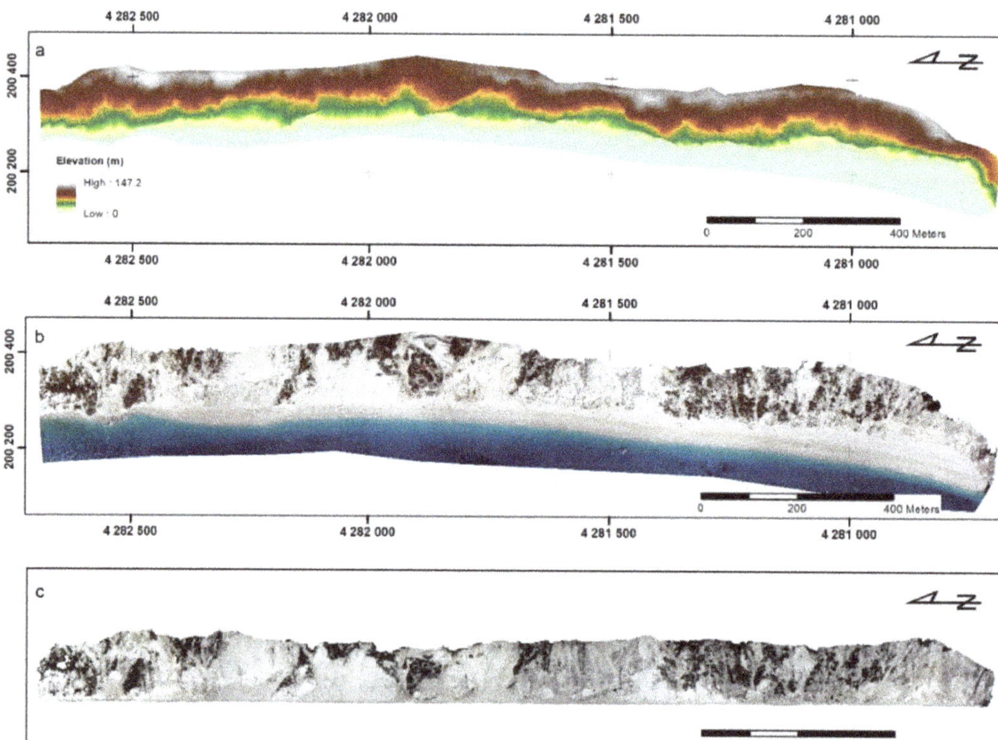

Figure 5. The SfM technique led to the construction of several high-resolution products such as (**a**) DSM, (**b**) ortho-image and (**c**) vertical photo-mosaic of the Egremnoi steep slope.

3.2. TLS-Aided Survey at Porto Katsiki

A series of reasons led us to use a TLS instead of a UAV for the point cloud generation at the Porto Katsiki white limestone steep slope, which bounds the beach zone and increases the rockfall risk, especially during the tourist season. The crowd presence during the data acquisition and the unstable NRTK signal due to the bad reception of the cellular network prevented the surveying team from using a UAV at this site. Thankfully, the morphology of the surrounding area provided us with a rather ideal place for establishing the LiDAR equipment and using it as a base for the tripod on which the TLS was placed (Figure 6).

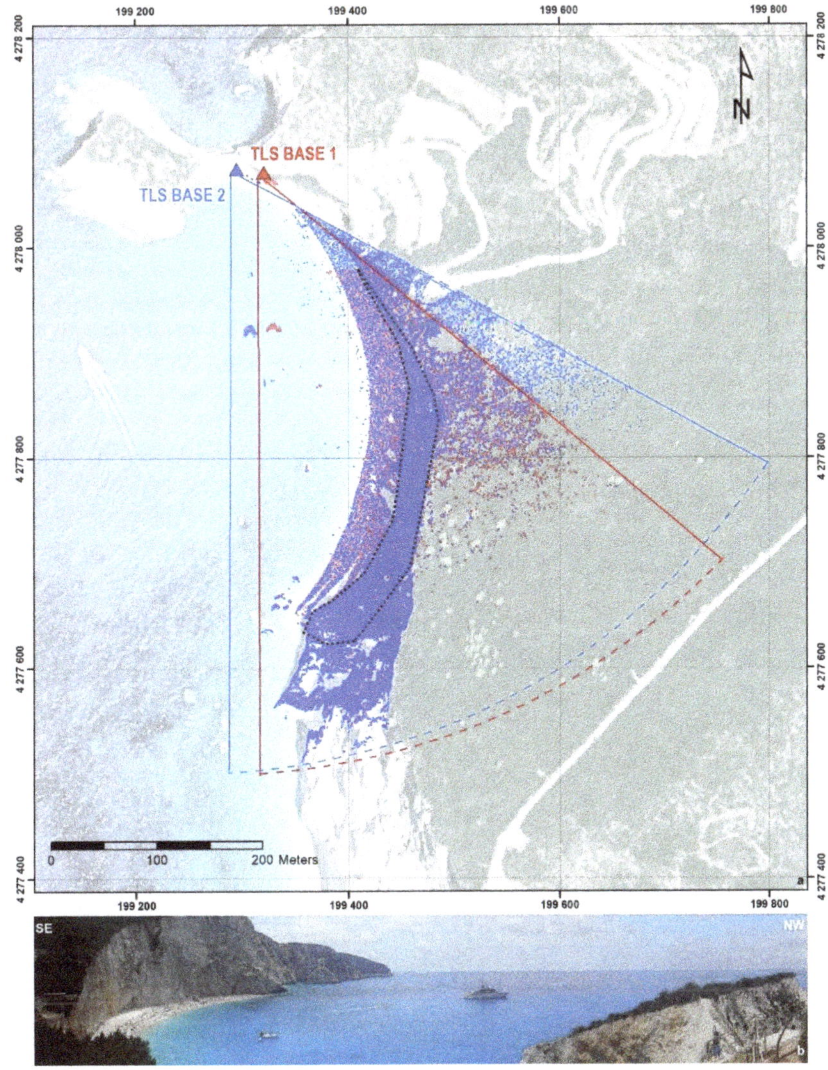

Figure 6. (**a**) Spatial location of the TLS point clouds acquired from two base stations (in red and blue colors). The black dots delineate the common scanned area of the steep slope with the minimal shadowing effect. (**b**) Panoramic view of Porto Katsiki beach during the fieldwork. Note the scanning equipment established at the bottom right (TLS base 2).

The state-of-the-art, long-range Terrestrial Laser Scanner, Leica P50, was used, with the ability to acquire high-precision point data at distances up to 1 km. According to its specifications, the point accuracy may reach the order of 3 mm over the range that it was used at Porto Katsiki, since the TLS base was established at a distance between 120 and 430 m opposite the steep slope. This resulted in a rather dense point cloud as the laser beam adds a point almost every 0.9 cm at a distance of 120 m or every 3.6 cm at the furthest slope areas, producing a rather reliable and valuable dataset for further processing. For the latter, in addition to the xyz coordinates, RGB values (using the internal 4MP camera) are included along with intensity laser signal values as well. We acquired two point clouds from two different TLS bases, with known coordinates, 25 m away from each other (Figure 6) in order to change the acquisition angle towards the slope as much as possible and therefore reduce the shadowing during the scanning. Both point clouds were processed together and merged, after co-registration and geolocation.

The final dataset consisted of 90 million points, and after the cleaning procedure, a point cloud of 65.5 million was generated for further interpretation (Figure 7). The latter includes several types of point classification and meshing, leading to detailed morphotectonic analysis of the rock slope and calculation of structural discontinuities' orientation (Figure 7).

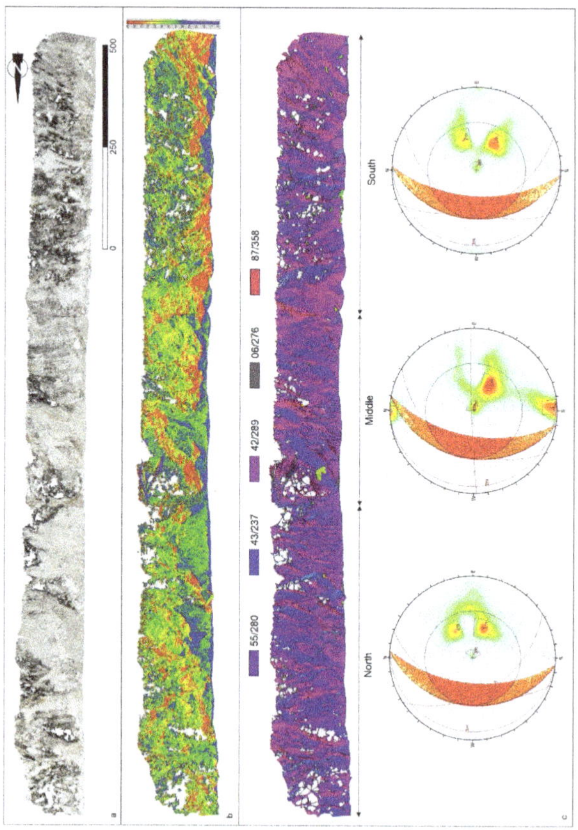

Figure 7. (**a**) The point cloud of the cliff uphill from the Egremnoi beach, cleaned from the noise caused by the vegetation, which might alter its steep topography. Classification of the extracted facets according to their calculated dip in degrees (**b**) and each one's dip direction that is color-coded regarding the statistical analysis (**c**). The cliff was segmented into three segments (North, Middle and South) and below each one, the wedge-sliding kinematic analysis is presented.

4. Results

The point clouds that were generated at both areas were used for further analysis, which was rather common for both datasets, with no restrictions concerning their origin (UAS and TLS, respectively).

The structural analysis was carried out using the facets/fracture detection and more specifically following the Kd-tree method [97] of the freeware CloudCompare [98]. The Kd-tree method is a kind of high-dimension binary tree, and by using it, one can search the nearest neighbor points in a relatively quick way [99] based on splitting the space of the mesh into two parts. The top node splits the space in one dimension, and the next nodes spilt the space in another dimension. The splitting causes approximately half of the points to be stored in the left subtree, and the other half is stored in the right subtree. It stops splitting when the points in one node achieve the given maximum count [99]. In this case study, the Kd-tree method has been used for the extraction of the orientation at each discontinuity located on the slopes under investigation [100,101]. Since the original data are points, the Kd-tree was used to organize them in k-dimensional space, searching for the neighboring points. Afterward, this point organization is used for dividing the cloud into small planar patches, which are then grouped into facets.

The Egremnoi Beach dataset consisted of the final point-cloud, which was generated after the application of the SfM method, followed by the construction of a 3D model using Agisoft Metashape software in mesh format (i.e., Triangular irregular network—TIN) with almost 25 million facets, after excluding large parts of the slope that were covered by vegetation, which altered the slope morphology (Figure 7a). The vegetation removal succeeded after classifying the points of the cloud, due to relatively high values of green in the Red–Green–Blue color model. The resulting dataset was used at the next stage for the structural analysis of the steep slopes along the coastline, considering that a large accumulation of debris occurred at certain locations next to the slope's prone base. The dip of every facet was calculated by using CloudCompare [98] algorithms (Figure 7b), and the patterns that were identified at the dip profile map proved to be more than helpful for the prone segmentation, as it was quite large for further processing. Therefore, it was divided into three segments and the discontinuity planes were statistically interpreted individually (Figure 7c).

The average geometric characteristics of the topography of the northern segment are 55/275, and the orientation of 4141 discontinuity planes that were extracted from the point-cloud processing was concentrated in two main groups at 24/289 and 43/237. The kinematic analysis of the discontinuity poles using Dips software v.7.0 [102], in the context of the wedge-sliding risk, indicated that 29% of the facet pairings are under serious rockfall potential, considering that the friction angle is set at 30° [87,103]. The 3259 facets identified at the southern segment are concentrated more or less in the same orientations (42/302 and 43/246) and the morphology is approximately identical (55/273). The kinematic analysis for wedge sliding showed that 25% of the discontinuity pairings could lead to failure, especially after an earthquake occurrence. The respective percentage of potential rockfall at the median segment of the Egremnoi cliff is lower (23%), even though the statistical significance of one of the main facet groups was dramatically decreased and substituted from a new, almost vertical one, with E-W trending (87/358). The latter is the result of the kinematic analysis of 4058 discontinuity planes and the second pole concentration is at 39/301, which is very similar to those concentrations identified at the other two segments' interpretations.

A similar processing methodology was followed with the Porto Katsiki cliff TLS data, even though it was based on more objective management of the mesh product, due to higher point density and a lack of any kind of vegetation altering the slope morphology, which led to impressive and much more clear results (Figure 8a). A large number of facets (23,380 records) were identified, each one characterized by dip and dip-direction measurements, among other location and precision information, which was extracted following the above-mentioned methodology. Several groups of oriented facets were recognized and classified according to their dip (shown with color coding in Figure 8b). The facets were grouped based on their dip direction in six categories shown with different colors in Figure 8c. Each classification contained classes that were assigned different colors, providing images where sectors with similar rock mass characteristics regarding the discontinuity orientation were defined (Figure 8d). At the next stage, the prone area was segmented into each one of those sectors, which seem to behave homogenously in the context of safety due to rockfalls. The average slope orientation characteristics were calculated using tools provided by CloudCompare freeware [98], and the facet list was filtered by excluding the morphology facets and keeping the facets that coincide with rock fracture discontinuities (6175 records). The latter was used at the next step of the kinematic analysis related to the wedge-sliding risk along the prone.

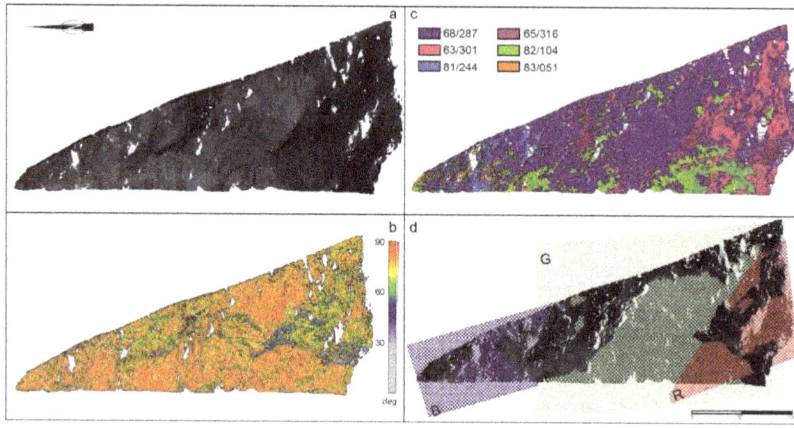

Figure 8. The geometrically corrected point cloud of the Porto Katsiki cliff, above the beach, in real color (**a**). The generated mesh file was classified according to each facet's dip in degrees (**b**) and each one's dip direction (**c**). The latter led to (**d**) a cliff segmentation into three generic morphological planes; Blue (81/244), Green (68/287) and Red (63/301).

Specifically, at the northern morphological segment (Blue plane at Figure 8d) where the plane of the average slope is calculated at 81/244, no more than 21.5% of the discontinuity plane combinations would lead to wedge sliding (Figure 9), which is the most common in this prone area, considering that the limestone friction angle is set at 30°. The percentage rises to 27% for the median segment (Green plane at Figure 8d), where the average slope is oriented at 68/287 and almost reaches 32% for the southern segment (Red plane at Figure 8d). Therefore, it is apparent that the factor of safety in the context of visiting this beach is rather critical, but it is significantly increased while moving to the southernmost segments of the beachfront. Additionally, a back analysis indicates that the rock mass sensitivity to failure increases at the intersection of the three segments since several fallen blocks are found adjacent to these locations.

Taking into account the results from the UAV and TLS surveys in Egremnoi and Porto Katsiki coastal areas, respectively, as well as the data obtained from the statistical analysis of the slope discontinuities, we can draw important conclusions about the condition of the slopes and their response to a possible strong earthquake in the future.

As emerged from the statistical analysis of the discontinuities and the kinematic analysis of the wedge-sliding risk for the Egremnoi slope, the percentages of facet pairings, which are under serious rockfall potential in the case of an earthquake occurrence, vary from 23 to 29%. Based on this small difference from section to section of the slope, as well as its overall response during the Lefkada earthquake in November 2015, as shown in Figure 3b,c, it is concluded that the slope is characterized by a uniformity in the discontinuities' orientation and its response to strong shocks.

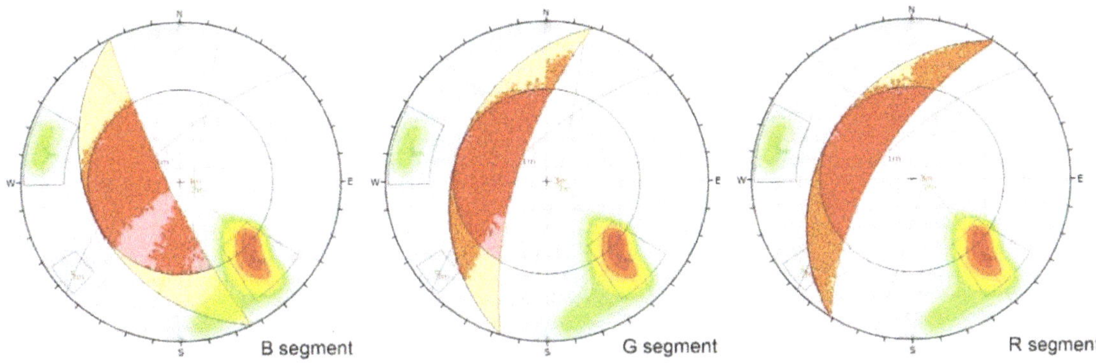

Figure 9. Wedge-sliding risk analysis for the three individual segments of Porto Katsiki beach (B for blue morphological plane, G for green morphological plane, R for red morphological plane of Figure 8d).

However, nothing similar emerges for the coastal slope in Porto Katsiki. Although particularly susceptible to earthquake-triggered rockfalls, effects that have already occurred during both nearby and distant earthquakes, the Porto Katsiki slope has not been shown in the past and is not expected to show a uniform response to a future earthquake, regardless of the distance from the epicenter. From the statistical analysis of the discontinuities and the kinematic analysis of the wedge-sliding risk for the slope in Porto Katsiki, it is found that the parts more susceptible to falls are located along the intersection of the three generic morphological planes (blue plane—northern segment, green plane—middle segment and red plane—southern segment in Figure 10). This fact, in combination with the generation of failures and the increased presence of boulders and smaller fragments from previous earthquakes (2003 and 2015 Lefkada and 2008 Andravida events) in the southern part of the slope and the beach, respectively, indicates that this part presents the highest susceptibility to slope failures. On the contrary, this does not apply to the middle and northern parts of the slope. In the middle part, large rock fragments resulting from failures are very limited, while in the northern part, they are entirely absent. This indicates the best response of these segments during the applied earthquake loads. Therefore, it is concluded that the rockfall risk during earthquakes in the southern part of Porto Katsiki is high.

The effects on people staying in the adjacent narrow beach are inevitable when rockfalls occur unless special care and preventive measures are taken for their mitigation by the bodies and municipal authorities involved in the prevention and management of disasters caused by earthquakes and earthquake-related hazards. Even before the 2015 Lefkada earthquake, the beach had been divided into two zones of safe and unsafe, the distinction of which was made by placing protection poles and ropes, while there are also visible rockfall warning signs for visitors. However, the monitoring and updating of measures must be continuous, always taking into account all the factors that can affect the slope and increase the potential for slope failures and impact on the adjacent beach.

Figure 10. The three morphological planes detected from the TLS survey in the Porto Katsiki coastal area. The blue plane corresponds to the northern segment, the green to the middle and the red to the southern. The discontinuities detected along the planes' intersection create conditions favorable to failure. Taking into account the presence of boulders and other smaller fragments along the beach, it is concluded that the southern segment of the slope (red plane) is characterized by the highest susceptibility to large rockfalls with impact on the adjacent beach. Please note the colors used correspond to the three generic morphological planes detected through the TLS application and are shown in Figure 8d.

5. Discussion

This work used UAS-aided photogrammetry and TLS in tectonically active segments of the coastal zone as a tool to improve our understanding of the mass-movement processes and the spatial dimension of landslide risk through a highly localized lens, as an approach that improves our capabilities in coastal management in slope-failure-prone areas.

The outcome of the two applications at Egremnoi and Porto Katsiki beaches is a strong indication these cutting-edge technologies have the capabilities to produce valuable results in the field of slope failures, allowing the identification of higher-risk zones and relative prioritization of any interventions.

Given the significance of slope failures as a hazard in coastal cliffs and the extent of its impact acknowledged in the introduction of the present study, rock slope stability analysis is considered very important for reducing the effects of instabilities and increasing the safety of adjacent or neighboring areas, while preserving the value of the sites in question. However, the classical survey methods comprising conventional data collection, which have been proposed and applied in various environments and settings (e.g., [104,105], present significant problems. These problems have to do with the time and cost needed for implementing fieldwork, the safety of researchers during data acquisition and the coverage of the data collection [43,106–108].

The visual inspections close to steep slopes, scarps and cliffs and direct measurement conducted by the researchers, who act as rock climbers, affect their safety during the field survey by exposing them to potential failure impacts. Furthermore, some discontinuities cannot be directly observed during visual inspections and are partially recorded and mapped, unless suitable aerial or satellite imagery is used. Consequently, a limited number of manual compass data measurements of dip and dip direction can be collected during the implementation of the classical survey methods in steep slopes, affecting the results as discontinuities measured in small areas are not representative of the whole rock masses. If an effort is made to fully cover the slopes, then field data acquisition may be a time-consuming and expensive procedure.

In the case of rock slope stability analysis on the coastal slopes of Egremnoi and Porto Katsiki, the classical methods, which include surveys with a geological compass measuring dip and dip direction directly on the discontinuity, face many problems, which are strongly related and attributed to the geological and geomorphological structure.

These issues are strongly related to the current geological and geomorphological setting of western Lefkada. Its tectonic structure is characterized by active and seismic faults, which, in combination with other morphological discontinuities, form a geotechnically unstable area with high and steep slopes composed of highly fractured, brecciated, unsupported and almost powdered geological formations, mainly limestones of the Paxoi geotectonic unit with secondary welding and extensive semi-cohesive scree with limestone breccia and red-clay-filled fractures as well as similar geometries of beds and discontinuities. These intense tectonic processes have a great impact on the geomorphology of the study area. They have increased the height and inclination of slopes resulting in almost vertical slopes with maximum elevation larger than 100m in several sites. In addition, the combined action of (i) endogenous processes, including deformation, which has caused lithological heterogeneity and mechanical anisotropy of the alpine formations and post-alpine deposits, and (ii) surficial processes including repeated cycles of mechanical, chemical and organic weathering and marine and aeolian erosion, have contributed to decreased cohesion and formation loosening along the steep coastal slopes.

All the aforementioned factors make it excessively difficult or even impossible to fully apply conventional methods for rock slope stability analysis in sites such as Egremnoi and Porto Katsiki. They would be time-consuming and expensive due to the equipment that must be used for the researchers' safety during field survey, while the results will only refer to accessible parts of the slopes as most of them will be excluded due to inaccessibility. On the contrary, the application of UAS-aided photogrammetry and TLS surveying offer examination of the Egremnoi and Porto Katsiki slopes, through a highly localized lens across all parts of the surfaces under study.

Furthermore, the large number of planes identified in the studied slopes (14,849 discontinuity planes from UAS in Egremnoi and 23,380 facets from TLS in Porto Katsiki) is at first essential for the thorough study of the deformation and then for the determination of the potential for failures along the slopes, either by one structure or by the synergy of two or more structures, which in turn reveals individual parts of the slopes susceptible to failures in the case of important external loads applied during an earthquake. The large number of data recorded during the application of these methods as well as the speed with which these data are processed is a significant advantage over conventional methods [109].

In addition, it should be noted that the UAS- and TLS-derived datasets are available for capturing on-demand data in the sense that the surveyor chooses the exact desired time of recording the study area and all of its features. This is an advantage compared to other image-exploitation approaches (e.g., satellite imagery analysis), especially when it comes to disaster management, in which capturing the natural processes is time-sensitive. In addition, UAS and TLS data capturing do not have the limitations of cloud cover. Furthermore, the capabilities of the UAS and TLS surveys fit the opportunistic nature of surveying a geomorphic phenomenon in the sense that the required equipment can be rapidly deployed in the field to collect information (e.g., after morphological changes), within the short time window that it is available (i.e., slopes affected by cleanup works).

With respect to the field data gathering process, the suggested approaches have important benefits. Firstly, the segment of the coast under examination can be revisited virtually (back at the lab) at a later time, as the geometrical data captured are stored and can be reused, which is also essential for multi-temporal processing. This is particularly useful in the case of remote sites or unsafe areas. This availability of terrain data for a part of the coastal cliffs/zone allows the processing of multiple parts, permitting a more holistic study of the conditions. In this way, the captured scarp information in the form of a DSM offers increased flexibility and provides the opportunity to perform different trials on the study site, depending on the prevalent risk (e.g., a housing complex on top of the hill or an area filled with visitors).

The very high resolution provides a description of the slope in very high detail, with nodes every few centimeters. This is a strong advantage of this approach, as it increases practicality, reduces time and personnel needs and minimizes safety concerns, especially considering steep slopes and inaccessible areas that would otherwise be surveyed using handheld instruments, putting surveyors at risk.

In highly touristic areas of special natural beauty, such as the coastal areas in the western part of Lefkada, the interpretation of data originating from either UAS-aided photogrammetry or LiDAR scanning is not only an important scientific innovation for assessing the landslide susceptibility of slopes, but also a significant tool in the direction of responsible and effective management of coastal areas through the cooperation of authorities involved in the prevention and management of disasters from natural hazards. This effectiveness is achieved through the adoption of invasive measures to reduce and eliminate the adverse effects of the occurrence of such phenomena on infrastructure and visitors. These measures may require prohibiting access to the adjacent beaches until their completion, thus imposing enormous constraints on the socio-economic development of the highly touristic areas. However, landslide risk mitigation measures could be of low cost and without major interventions in the landscape. They do not presuppose a total ban on access to susceptible areas and therefore they do not cause any disruption to tourism and socio-economic activities of the areas where they are applied, in addition to the enhanced safety that they provide.

The most suitable approach for the beaches adjacent to the Egremnoi and Porto Katsiki slopes is their division into access zones. This distinction mainly takes into account the results of the UAS and TLS survey and especially the landslide-susceptible parts of the slopes, which are adjacent to the beaches. Additional elements that are evaluated are the presence of unstable fragments on slopes with a high potential to detach and fall on or slide to the beaches in the case of a strong earthquake, as well as the location of the fragments, which have ended up on the beach from previous earthquakes or other episodes that also triggered landslides, such as heavy and rapid rainfall. After evaluating these data, the beaches could be divided into three zones: (1) The free-access zone of low landslide risk, (2) the restricted-access zone of moderate risk and (3) the prohibited-access zone of high risk. The boundaries of the zones can be modified at any time depending on whether the conditions that favor the occurrence of landslides are aggravated or not. Thus, prohibited- and restricted-access zones can be merged into a single access zone of high risk after the occurrence of a major earthquake, during a prolonged aftershock sequence, as well as during or after a prolonged period of heavy rainfall.

This high-spatial-resolution mapping of slope deformations can be used as a landslide precursor, assisting prevention measures. This zonation enables stakeholders to reduce the adverse effects of potential landslides without, however, barring access to beaches for long periods of time, which are required by the implementation of large-scale projects and interventions.

Additionally, in the context of the multi-temporal approach, repeated data acquisitions could be scheduled at regular time periods or after triggering episodes, which might potentially disrupt the slope stability. A comparative process could quite easily be merged into an automated procedure, and an alert might be generated for the re-establishment of the safety zones.

Even though the point-cloud differences are not crucial, regardless of their origin, a matter of point density arises. TLS scanning provides denser datasets, with higher accuracies (depending on the specifications of the LiDAR) provided that there are suitable steady spots for establishing the equipment, at small distances, opposite the monitored cliff behind the beachfront. The latter adds credit to continuous multi-temporal observation and the quantification of any displacement, which might spur strong objections when it comes to the exact positioning of UAS imagery and therefore the overall model accuracy. Moreover, restrictions for UAS flying admittance due to safety may apply in no-fly zones or in areas with a human presence.

A strong argument for preferring UAS photogrammetry products instead of directly acquired data from TLS scanning would be the presence of vegetation cover on the cliff slopes. Even though its existence is an objective parameter that might affect both point-cloud datasets by—sometimes—hiding important fractures, its removal during the photogrammetric processing seems to be rather more harmless. That is due to multiple angles of imagery acquisition, which allows for information of point generation to be available even after the removal of points that are classified as vegetation. On the contrary, the canopy at the vegetated areas that are scanned with TLS does not permit laser penetration, and rather large areas of "no-data" appear, with no participation in further processing.

6. Conclusions

This study applies two cutting-edge methodologies that enhance our capabilities to understand slope stability and monitoring in tectonically active coastal areas. The two sites studied in the present work are a very common setting in the scenic coastline of the highly touristic Ionian Islands and other parts of the Mediterranean, where slope failures and instabilities threaten high-value locations and their visitors.

The application demonstrates that the UAS- and TLS-aided surveys are able to provide a very detailed description of the structural condition of the slopes, in terms of discontinuities, using large numbers of measurements in a more practical, reliable and safe way in comparison with conventional surveys.

Evaluation of the outcome of the surveys allows the identification of specific segments more prone to slope failures, improving the understanding of risk in the study sites and creating circumstances for the prioritization of interventions. This improved understanding is feasible because of the applied technologies. Regarding future research in the field, it will be interesting to further explore the combined use of UAS and TLS applications and the evaluation of the respective outcomes, as these first results show that it has the potential to maximize the information content provided to stakeholders and decision-makers. In this way, the synergy of the two applications has the potential to provide related knowledge in order to adopt site-specific mitigation measures and strategies and can be an effective and responsible method for risk-mitigation-driven development of high-value coastal areas. Further research could also explore how these applications would compare in terms of performance against conventional rock slope stability analysis and ground-based slope-failure monitoring techniques.

The approaches presented can be a useful, efficient and transferable tool that would benefit risk-mitigation efforts and the design of protection measures by expanding applications to other sites with similar geological and geomorphological properties worldwide.

Author Contributions: Conceptualization, E.V. and S.M.; methodology E.V., A.K., E.K., S.M. and M.D.; software, E.V., A.K., S.M. and M.D.; validation, E.V.; formal analysis, E.V., S.M., M.D. and G.K.; investigation, E.V., S.M., M.D. and G.K.; resources, N.V. and E.L.; data curation, E.V., A.K., E.K., S.M. and M.D.; writing—original draft preparation, E.V., S.M., M.D. and G.K.; writing—review and editing, E.V., S.M., M.D., G.K., V.K., V.S., J.D.A., E.L. and N.V.; visualization, E.V. and S.M.; supervision, S.M. and M.D.; project administration, N.V. All authors have read and agreed to the published version of the manuscript.

Funding: This research was funded by the "Telemachus–Innovative Operational Seismic Risk Management System of the Ionian Islands" project, included in the Priority Axis "Environmental Protection and Sustainable Development" of the Regional Operational Programme "Ionian Islands 2014–2020", grant number MIS 5007986, which is funded by the European Regional Development Fund (ERDF) and National Resources under the National Strategic Reference Framework NSRF 2014–2020.

Institutional Review Board Statement: Not applicable.

Informed Consent Statement: Not applicable.

Data Availability Statement: Data available on request from the corresponding author.

Acknowledgments: Special thanks to Metrica S.A. for providing support during the field data acquisition and processing in order to achieve maximum precision and accuracy of the TLS data.

Conflicts of Interest: The authors declare no conflict of interest.

References

1. Cooper, M.J.P.; Beevers, M.D.; Oppenheimer, M. The potential impacts of sea level rise on the coastal region of New Jersey, USA. *Clim. Chang.* **2008**, *90*, 475. [CrossRef]
2. Zhang, S.; Zhang, L.; Chen, H.X. Relationships among three repeated large-scale debris flows at Pubugou Ravine in the Wenchuan earthquake zone. *Can. Geotech. J.* **2014**, *51*, 951–965. [CrossRef]
3. Quartau, R.; Tempera, F.; Mitchell, N.C.; Pinheiro, L.M.; Duarte, H.; Brito, P.O.; Bates, C.R.; Monteiro, J.H. Morphology of the Faial Island shelf (Azores): The interplay between volcanic, erosional, depositional, tectonic and mass-wasting processes. Geochemistry. *Geochem. Geophys. Geosyst.* **2012**, *13*, 1–30. [CrossRef]
4. Neumann, B.; Vafeidis, A.T.; Zimmermann, J.; Nicholls, R.J. Future coastal population growth and exposure to sea-level rise and coastal flooding–A global assessment. *PLoS ONE* **2015**, *10*, e0118571. [CrossRef]
5. Taussik, J. Development Plans and the Coastal Zone. *Town Plan. Rev.* **1996**, *67*, 397–420. [CrossRef]
6. Gormsen, E. The impact of tourism on coastal areas. *GeoJournal* **1997**, *42*, 39–54. [CrossRef]
7. Klein, R.J.T.; Nicholls, R.J.; Thomalla, F. Resilience to natural hazards: How useful is this concept? *Glob. Environ. Chang. B Environ. Hazards* **2003**, *5*, 35–45. [CrossRef]
8. Small, C.; Nicholls, R.J. A global analysis of human settlement in coastal zones. *J. Coast. Res.* **2003**, *19*, 584–599.
9. Lekkas, E. Landslide hazard and risk in geologically active areas. The case of the caldera of the Santorini (Thera) volcano island complex (Greece). In Proceedings of the 7th Asian Regional Conference for IAEG, Chengdu, China, 9–11 September 2009; pp. 417–423.
10. Antoniou, A.A.; Lekkas, E. Rockfall susceptibility map for Athinios port, Santorini Island, Greece. *Geomorphology* **2010**, *118*, 152–166. [CrossRef]
11. Paton, D.; Johnston, D. Disasters and communities: Vulnerability, resilience and preparedness. *Disaster Prev. Manag. Int. J.* **2001**, *10*, 270–277. [CrossRef]

12. Robinson, T.R.; Rosser, N.J.; Densmore, A.L.; Williams, J.G.; Kincey, M.E.; Benjamin, J.; Bell, H.J.A. Rapid post-earthquake modelling of coseismic landslide intensity and distribution for emergency response decision support. *Nat. Hazards Earth Syst. Sci.* **2017**, *17*, 1521–1540. [CrossRef]
13. Tapas, M.R.; Priyom, R.; Ritwik, M.; Govindharaj, K.B.; Kumar, K.V. Spatial characteristics of landslides triggered by the 2015 Mw 7.8 (Gorkha) and Mw 7.3 (Dolakha) earthquakes in Nepal. *Landslides* **2017**, *14*, 697–704. [CrossRef]
14. Meunier, P.; Uchida, T.; Hovius, N. Landslide patterns reveal the sources of large earthquakes. *Earth Planet. Sci. Lett.* **2013**, *363*, 27–33. [CrossRef]
15. Rahman, A.-U.; Khan, A.N.; Collins, A.E. Analysis of landslide causes and associated damages in the Kashmir Himalayas of Pakistan. *Nat. Hazards* **2014**, *71*, 803–821. [CrossRef]
16. Widiyanto, W.; Santoso, P.; Hsiao, S.-C.; Imananta, R. Post-event Field Survey of 28 September 2018 Sulawesi Earthquake and Tsunami. *Nat. Hazards Earth Syst. Sci.* **2019**, *19*, 2781–2794. [CrossRef]
17. Budimir, M.E.A.; Atkinson, P.M.; Lewis, H.G. Earthquake-and-landslide events are associated with more fatalities than earthquakes alone. *Nat. Hazards* **2014**, *72*, 895–914. [CrossRef]
18. Allen, T.I.; Marano, K.D.; Earle, P.S.; Wald, D.J. PAGER-CAT: A composite earthquake catalog for calibrating global fatality models. *Seismol. Res. Lett.* **2009**, *80*, 57–82. [CrossRef]
19. Xie, S.P.; Deser, C.; Vecchi, G.A.; Ma, J.; Teng, H.; Wittenberg, A.T. Global warming pattern formation: Sea surface temperature and rainfall. *J. Clim.* **2010**, *23*, 966–986. [CrossRef]
20. Moore, L.J.; Griggs, G.B. Long-term cliff retreat and erosion hotspots along the central shores of the Monterey Bay National Marine Sanctuary. *Mar. Geol.* **2002**, *181*, 265–283. [CrossRef]
21. Sunamura, T. *Geomorphology of Rocky Coasts*; John Wiley and Sons: Chichester, UK, 1992.
22. Emery, K.O.; Kuhn, G.G. Sea cliffs: Their processes, profiles, and classification. *Geol. Soc. Am. Bull.* **1982**, *93*, 644–654. [CrossRef]
23. Bird, E.C.F. *Coastal Geomorphology: An Introduction*; John Wiley and Sons: Chichester, UK, 2000.
24. Bromhead, E.N.; Ibsen, M.L. A review of landsliding and coastal erosion damage to historic fortifications in South East England. *Landslides* **2006**, *3*, 341–347. [CrossRef]
25. Carrasco, A.R.; Ferreira, Ó.; Matias, A.; Dias, J.A. Historic monuments threatened by coastal hazards at Boca do Rio, Algarve, Portugal. *Coast. Manag.* **2007**, *35*, 163–179. [CrossRef]
26. Casagli, N.; Frodella, W.; Morelli, S.; Tofani, V.; Ciampalini, A.; Intrieri, E.; Raspini, F.; Rossi, G.; Tanteri, L.; Lu, P. Spaceborne, UAV and ground-based remote sensing techniques for landslide mapping, monitoring and early warning. *Geoenviron. Dis.* **2017**, *4*, 9. [CrossRef]
27. Riquelme, A.J.; Tomás, R.; Abellán, A. Characterization of rock slopes through slope mass rating using 3D point clouds. *Int. J. Rock Mech. Min. Sci.* **2016**, *84*, 165–176. [CrossRef]
28. Riquelme, A.; Cano, M.; Tomás, R.; Abellán, A. Identification of Rock Slope Discontinuity Sets from Laser Scanner and Photogrammetric Point Clouds: A Comparative Analysis. *Procedia Eng.* **2017**, *191*, 838–845. [CrossRef]
29. Gonçalves, J.A.; Henriques, R. UAV photogrammetry for topographic monitoring of coastal areas. *ISPRS J. Photogramm. Remote Sens.* **2015**, *104*, 101–111. [CrossRef]
30. Barlow, J.; Gilham, J.; Ibarra Cofrã, I. Kinematic analysis of sea cliff stability using UAV photogrammetry. *Int. J. Remote Sens.* **2017**, *38*, 2464–2479. [CrossRef]
31. Papakonstantinou, A.; Topouzelis, K.; Pavlogeorgatos, G. Coastline zones identification and 3D coastal mapping using UAV spatial data. *ISPRS Int. J. Geo-Inf.* **2016**, *5*, 75. [CrossRef]
32. Klemas, V. Remote sensing of floods and flood-prone areas: An overview. *J. Coast. Res.* **2015**, *31*, 1005–1013. [CrossRef]
33. Drummond, C.D.; Harley, M.D.; Turner, I.L.; Matheen, A.N.A.; Glamore, W.C. UAV applications to coastal engineering. In Proceedings of the Australasian Coasts and Ports Conference 2015, Auckland, New Zealand, 15–18 September 2015.
34. Nex, F.; Remondino, F. UAV for 3D mapping applications: A review. *Appl. Geomat.* **2014**, *6*, 1–15. [CrossRef]
35. Lucieer, A.; de Jong, S.M.; Turner, D. Mapping landslide displacements using Structure from Motion (SfM) and image correlation of multi-temporal UAV photography. *Prog. Phys. Geogr. Earth Environ.* **2014**, *38*, 97–116. [CrossRef]
36. Vassilakis, E.; Foumelis, M.; Erkeki, A.; Kotsi, E.; Lekkas, E. Post-event surface deformation of Amyntaio slide (Greece) by complementary analysis of Remotely Piloted Airborne System imagery and SAR interferometry. *Appl. Geomat.* **2020**, *13*, 65–75. [CrossRef]
37. Giordan, D.; Hayakawa, Y.S.; Nex, F.; Tarolli, P. The use of remotely piloted aircraft systems (RPAS) in monitoring applications and management of natural hazards. *Nat. Hazards Earth Syst. Sci.* **2018**, *18*, 3085–3087. [CrossRef]
38. Karantanellis, E.; Marinos, V.; Vassilakis, E. 3D hazard analysis and object-based characterization of landslide motion mechanism using uav imagery. *Int. Arch. Photogramm. Remote Sens. Spat. Inf. Sci.* **2019**, *42*, 425–430. [CrossRef]
39. Slob, S.; Van Knapen, B.; Hack, R.; Turner, K.; Kemeny, J. Method for automated discontinuity analysis of rock slopes with three-dimensional laser scanning. *Transp. Res. Rec. Transp. Res. Board* **2005**, *1913*, 187–194. [CrossRef]
40. O'Dea, A.; Brodie, K.L.; Hartzell, P. Continuous Coastal Monitoring with an Automated Terrestrial Lidar Scanner. *J. Mar. Sci. Eng.* **2019**, *7*, 37. [CrossRef]
41. Fabbri, S.; Giambastiani, B.M.S.; Sistilli, F.; Scarelli, F.; Gabbianelli, G. Geomorphological analysis and classification of foredune ridges based on Terrestrial Laser Scanning (TLS) technology. *Geomorphology* **2017**, *295*, 436–451. [CrossRef]

42. Derron, M.H.; Jaboyedoff, M. LIDAR and DEM techniques for landslides monitoring and characterization. *Nat. Hazards Earth Syst. Sci.* **2010**, *10*, 1877–1879. [CrossRef]
43. Sturzenegger, M.; Stead, D. Quantifying discontinuity orientation and persistence on high mountain rock slopes and large landslides using terrestrial remote sensing techniques. *Nat. Hazards Earth Syst. Sci.* **2009**, *9*, 267–287. [CrossRef]
44. Fiorucci, F.; Giordan, D.; Santangelo, M.; Dutto, F.; Rossi, M.; Guzzetti, F. Criteria for the optimal selection of remote sensing optical images to map event landslides. *Nat. Hazards Earth Syst. Sci.* **2018**, *18*, 405–417. [CrossRef]
45. Rosser, N.J.; Petley, D.N.; Lim, M.; Dunning, S.A.; Allison, R.J. Terrestrial laser scanning for monitoring the process of hard rock coastal cliff erosion. *Q. J. Eng. Geol. Hydrogeol.* **2005**, *38*, 363–375. [CrossRef]
46. Westoby, M.J.; Lim, M.; Hogg, M.; Pound, M.J.; Dunlop, L.; Woodward, J. Cost-effective erosion monitoring of coastal cliffs. *Coast. Eng.* **2018**, *138*, 152–164. [CrossRef]
47. De Rose, R.C.; Basher, L.R. Measurement of river bank and cliff erosion from sequential LIDAR and historical aerial photography. *Geomorphology* **2011**, *126*, 132–147. [CrossRef]
48. Westoby, M.J.; Brasington, J.; Glasser, N.F.; Hambrey, M.J.; Reynolds, J.M. "Structure-from-Motion" photogrammetry: A low-cost, effective tool for geoscience applications. *Geomorphology* **2012**, *179*, 300–314. [CrossRef]
49. Fugazza, D.; Scaioni, M.; Corti, M.; D'Agata, C.; Azzoni, R.S.; Cernuschi, M.; Smiraglia, C.; Diolaiuti, G.A. Combination of UAV and terrestrial photogrammetry to assess rapid glacier evolution and map glacier hazards. *Nat. Hazards Earth Syst. Sci.* **2018**, *18*, 1055–1071. [CrossRef]
50. Viero, A.; Teza, G.; Massironi, M.; Jaboyedoff, M.; Galgaro, A. Laser scanning-based recognition of rotational movements on a deep seated gravitational instability: The Cinque Torri case (North-Eastern Italian Alps). *Geomorphology* **2010**, *122*, 191–204. [CrossRef]
51. Pye, K.; Allen, J.R.L. Past, present and future interactions, management challenges and research needs in coastal and estuarine environments. *Geol. Soc. Spec. Publ.* **2000**, *175*, 1–4. [CrossRef]
52. Baily, B.; Nowell, D. Techniques for monitoring coastal change: A review and case study. *Ocean Coast. Manag.* **1996**, *32*, 85–95. [CrossRef]
53. Barrett, M.G. *Coastal Zone Planning and Management*; Thomas Telford: London, UK, 1992.
54. Scordilis, E.M.; Karakaisis, G.F.; Karacostas, B.G.; Panagiotopoulos, D.G.; Comninakis, P.E.; Papazachos, B.C. Evidence for transform faulting in the Ionian sea: The Cephalonia island earthquake sequence of 1983. *Pure Appl. Geophys.* **1985**, *123*, 388–397. [CrossRef]
55. Ganas, A.; Elias, P.; Bozionelos, G.; Papathanassiou, G.; Avallone, A.; Papastergios, A.; Valkaniotis, S.; Parcharidis, I.; Briole, P. Coseismic deformation, field observations and seismic fault of the 17 November 2015 M = 6.5, Lefkada Island, Greece earthquake. *Tectonophysics* **2016**, *687*, 210–222. [CrossRef]
56. Kiratzi, A.A.; Wagner, G.S.; Langston, C.A. Source parameters of some large earthquakes in Northern Aegean determined by body waveform inversion. *Pure Appl. Geophys.* **1991**, *135*, 515–527. [CrossRef]
57. Louvari, E.; Kiratzi, A.A.; Papazachos, B.C. The Cephalonia Transform Fault and its extension to western Lefkada Island (Greece). *Tectonophysics* **1999**, *308*, 223–236. [CrossRef]
58. Papadimitriou, P.; Kaviris, G.; Makropoulos, K. The MW = 6.3 2003 Lefkada earthquake (Greece) and induced stress transfer changes. *Tectonophysics* **2006**, *423*, 73–82. [CrossRef]
59. Kapetanidis, V.; Kassaras, I. Contemporary crustal stress of the Greek region deduced from earthquake focal mechanisms. *J. Geodyn.* **2019**, *123*, 55–82. [CrossRef]
60. Papazachos, V.; Papazachou, C. *The Earthquakes of Greece*; Editions Ziti: Thessaloniki, Greece, 1997.
61. Stucchi, M.; Rovida, A.; Gomez Capera, A.A.; Alexandre, P.; Camelbeeck, T.; Demircioglu, M.B.; Gasperini, P.; Kouskouna, V.; Musson, R.M.W.; Radulian, M.; et al. The SHARE European Earthquake Catalogue (SHEEC) 1000–1899. *J. Seismol.* **2013**, *17*, 523–544. [CrossRef]
62. Kouskouna, V.; Makropoulos, K.; Tsiknakis, K. Contribution of historical information to a realistic seismicity and hazard assessment of an area. The Ionian Islands earthquakes of 1767 and 1769: Historical investigation. In *Review of Historical Seismicity in Europe*; University of Athens: Athens, Greece, 1993; pp. 195–206.
63. Papadimitriou, P.; Karakonstantis, A.; Bozionelos, G.; Kapetanidis, V.; Kaviris, G.; Spingos, I.; Millas, C.; Kassaras, I.; Voulgaris, N. Preliminary Report on the Lefkada 17 November 2015 Mw = 6.4 Earthquake; Athens. Available online: https://www.emsc-csem.org/Doc/Additional_Earthquake_Report/470390/20151117_lefkada_report_nkua.pdf (accessed on 20 December 2021).
64. Mavroulis, S. Environmental Effects and Evaluation of Environmental Intensities of Historic and Recent Earthquakes in Western Greece (Western Peloponnese and Central Ionian Islands) and Correlation with Active Tectonics and Seismological Parameters. Ph.D. Thesis, National and Kapodistrian University of Athens, Athens, Greece, 2020.
65. Papadimitriou, E.; Karakostas, V.; Mesimeri, M.; Chouliaras, G.; Kourouklas, C. The Mw6.5 17 November 2015 Lefkada (Greece) Earthquake: Structural Interpretation by Means of the Aftershock Analysis. *Pure Appl. Geophys.* **2017**, *174*, 3869–3888. [CrossRef]
66. Makropoulos, K.; Kaviris, G.; Kouskouna, V. An updated and extended earthquake catalogue for Greece and adjacent areas since 1900. *Nat. Hazards Earth Syst. Sci.* **2012**, *12*, 1425–1430. [CrossRef]
67. Karakostas, V.G. Properties of the 2003 Lefkada, Ionian Islands, Greece, earthquake seismic sequence and seismicity triggering. *Bull. Seism. Soc. Am.* **2004**, *94*, 1976–1981. [CrossRef]

68. Karakostas, V.G.; Papadimitriou, E.E. Fault complexity associated with the 14 August 2003 Mw6.2 Lefkada, Greece, aftershock sequence. *Acta Geophys.* **2010**, *58*, 838–854. [CrossRef]
69. Sokos, E.; Zahradnik, J.; Gallovic, F.; Serpetsidaki, A.; Plicka, V.; Kiratzi, A. Asperity break after 12years: The Mw6.4 2015 Lefkada (Greece) earthquake. *Geophys. Res. Lett.* **2016**, *43*, 6137–6145. [CrossRef]
70. Mariolakos, I.; Papanikolaou, D. The Neogene basins of the Aegean Arc from the Paleogeographic and the Geodynamic point of view. In Proceedings of the International Symposium on the Hellenic Arc and Trench (HEAT), Athens, Greece, 8–10 April 1981; Volume 1, pp. 383–399.
71. Mariolakos, I.; Papanikolaou, D. Deformation pattern and relation between deformation and seismicity in the Hellenic arc. *Bull. Geol. Soc. Greece* **1987**, *19*, 59–76.
72. Kassaras, I.; Kazantzidou Firtinidou, D., Ganas, A.; Tonna, S.; Pomonis, A.; Karakostas, C.; Papadatou-Giannopoulou, C.; Psarris, D.; Lekkas, E.; Makropoulos, K. On the Lefkas (Ionian Sea) 17 November 2015 Mw = 6.5 Earthquake Macroseismic Effects. *J. Earthq. Eng.* **2020**, *24*, 1913–1943. [CrossRef]
73. Lekkas, E.L.; Mavroulis, S.D. Earthquake environmental effects and ESI 2007 seismic intensities of the early 2014 Cephalonia (Ionian Sea, western Greece) earthquakes (January 26 and February 3, Mw 6.0). *Nat. Hazards* **2015**, *78*, 1517–1544. [CrossRef]
74. Lekkas, E.L.; Mavroulis, S.D. Fault zones ruptured during the early 2014 Cephalonia Island (Ionian Sea, Western Greece) earthquakes (January 26 and February 3, Mw 6.0) based on the associated co-seismic surface ruptures. *J. Seismol.* **2016**, *20*, 63–78. [CrossRef]
75. Mavroulis, S.; Stanota, E.S.; Lekkas, E. Evaluation of environmental seismic intensities of all known historical and recent earthquakes felt in Zakynthos Island, Greece using the Environmental Seismic Intensity (ESI 2007) scale. *Quat. Int.* **2019**, *532*, 1–22. [CrossRef]
76. Mavroulis, S.; Lekkas, E. Revisiting the most destructive earthquake sequence in the recent history of Greece: Environmental effects induced by the 9, 11 and 12 August 1953 Ionian Sea earthquakes. *Appl. Sci.* **2021**, *11*, 8429. [CrossRef]
77. Lekkas, E.; Danamos, G.D.; Lozios, S. Neotectonic structure and evolution of Lefkada island. *Bull. Geol. Soc. Greece* **2001**, *34*, 157–163. [CrossRef]
78. Lekkas, E.; Mavroulis, S.; Carydis, P.; Alexoudi, V. The 17 November 2015 Mw 6.4 Lefkas (Ionian Sea, Western Greece) Earthquake: Impact on Environment and Buildings. *Geotech. Geol. Eng.* **2018**, *36*, 2109–2142. [CrossRef]
79. Rondoyanni, T.; Sakellariou, M.; Baskoutas, J.; Christodoulou, N. Evaluation of active faulting and earthquake secondary effects in Lefkada Island, Ionian Sea, Greece: An overview. *Nat. Hazards* **2012**, *61*, 843–860. [CrossRef]
80. Rondoyanni, T. Les seismes et l'environnement geólogique de l'île de Lefkade, Grèce: Passe et Futur. In *Engineering Geology and the Environment, Proceedings of the International Symposium on Engineering Geology and the Environment, Athens, Greece, 23–27 June 1997*; Marinos, P., Koukis, G., Tsiambaos, G., Stournaras, G., Eds.; A.A. Balkema: Rotterdam, The Netherlands, 1997; pp. 1469–1474.
81. Cushing, M. Evolution Structurale de la Marge Nord-Ouest Hellénique dans l'île de Lefkas et ses Environs (Grèce Nordoccidentale). Ph.D. Thesis, Paris-Sud University, Bures-sur-Yvette, France, 1985.
82. Mavroulis, S.; Diakakis, M.; Kotsi, E.; Vassilakis, E.; Lekkas, E. Susceptibility and hazard assessment in the Ionian Islands for highlighting sites of significant earthquake-related hazards. In Proceedings of the Safe Corfu 2019–6th International Conference on Civil Protection & New Technologies, Corfu, Greece, 6–9 November 2019; Volume 1, pp. 13–16.
83. Papathanassiou, G.; Pavlides, S. Using the INQUA scale for the assessment of intensity: Case study of the 2003 Lefkada (Ionian Islands), Greece earthquake. *Quat. Int.* **2007**, *173–174*, 4–14. [CrossRef]
84. Valkaniotis, S.; Papathanassiou, G.; Ganas, A. Mapping an earthquake-induced landslide based on UAV imagery; case study of the 2015 Okeanos landslide, Lefkada, Greece. *Eng. Geol.* **2018**, *245*, 141–152. [CrossRef]
85. Roufi, A.; Vassilakis, E.; Poulos, S. Western Lefkada Shoreline Displacement Rates Based on Photogrammetric Processing of Remote Sensing Datasets from Various Sources. In Proceedings of the 15th International Congress of the Geological Society of Greece, Athens, Greece, 22–24 May 2019; Volume 1, pp. 512–513.
86. Saroglou, C. GIS-based rockfall susceptibility zoning in Greece. *Geosciences* **2019**, *9*, 163. [CrossRef]
87. Gong, W.; Zekkos, D.; Clark, M.; Manousakis, J.; Kirshbaum, D. Regional 3D Stability Analyses of the Egkremnoi Coastline and Comparison with Landslides Caused by the 2015 Lefkada Earthquake. In Proceedings of the International Foundations Congress and Equipment Expo 2021, Dallas, TX, USA, 10–14 May 2021; Volume 1, pp. 130–138.
88. Mavroulis, S.D.; Fountoulis, I.G.; Skourtsos, E.N.; Lekkas, E.L.; Papanikolaou, I.D. Seismic intensity assignments for the 2008 Andravida (NW Peloponnese, Greece) strike-slip event (8 June, MW = 6.4) based on the application of the environmental seismic intensity scale (ESI 2007) and the European Macroseismic scale (EMS-98). *Ann. Geophys.* **2013**, *56*. [CrossRef]
89. Vanneschi, C.; Di Camillo, M.; Aiello, E.; Bonciani, F.; Salvini, R. SFM-MVS photogrammetry for rockfall analysis and hazard assessment along the ancient roman via Flaminia road at the Furlo gorge (Italy). *ISPRS Int. J. Geo Inf.* **2019**, *8*, 325. [CrossRef]
90. Loiotine, L.; Andriani, G.F.; Jaboyedoff, M.; Parise, M.; Derron, M.-H. Comparison of Remote Sensing Techniques for Geostructural Analysis and Cliff Monitoring in Coastal Areas of High Tourist Attraction: The Case Study of Polignano a Mare (Southern Italy). *Remote Sens.* **2021**, *13*, 5045. [CrossRef]
91. Del Río, L.; Posanski, D.; Gracia, F.J.; Pérez-Romero, A.M. A comparative approach of monitoring techniques to assess erosion processes on soft cliffs. *Bull. Eng. Geol. Environ.* **2020**, *79*, 1797–1814. [CrossRef]
92. Mazzanti, P.; Caporossi, P.; Brunetti, A.; Mohammadi, F.I.; Bozzano, F. Short-term geomorphological evolution of the Poggio Baldi landslide upper scarp via 3D change detection. *Landslides* **2021**, *18*, 2367–2381. [CrossRef]

93. Forte, G.; Verrucci, L.; Di Giulio, A.; De Falco, M.; Tommasi, P.; Lanzo, G.; Franke, K.W.; Santo, A. Analysis of major rock slides that occurred during the 2016–2017 Central Italy seismic sequence. *Eng. Geol.* **2021**, *290*, 106194. [CrossRef]
94. Zhang, Y.; Yue, P.; Zhang, G.; Guan, T.; Lv, M.; Zhong, D. Augmented reality mapping of rock mass discontinuities and rockfall susceptibility based on unmanned aerial vehicle photogrammetry. *Remote Sens.* **2019**, *11*, 1311. [CrossRef]
95. Fonstad, M.A.; Dietrich, J.T.; Courville, B.C.; Jensen, J.L.; Carbonneau, P.E. Topographic structure from motion: A new development in photogrammetric measurement. Earth Surf. Process. *Landforms* **2013**, *38*, 421–430. [CrossRef]
96. Panagiotopoulou, S.; Erkeki, A.; Antonakakis, A.; Grigorakakis, P.; Protopapa, V.; Tsiostas, G.; Vlachou, K.; Vassilakis, E. Evaluation of Network Real Time Kinematics contribution to the accuracy/productivity ratio for UAS-SfM Photogrammetry. In Proceedings of the 2020 European Navigation Conference (ENC), Dresden, Germany, 23–24 November 2020.
97. Friedman, J.H.; Bentley, J.L.; Finker, R.A. An algorithm for finding best matches in logarithmic expected time. *ACM Trans. Math. Softw.* **1977**, *3*, 209–226. [CrossRef]
98. Girardeau-Montaut, D. Cloud Compare. 2011. Available online: https://www.danielgm.net/cc/ (accessed on 15 October 2021).
99. Xiao, Z.; Huang, W. Kd-tree Based Nonuniform Simplification of 3D Point Cloud. In Proceedings of the 2009 Third International Conference on Genetic and Evolutionary Computing, Guilin, China, 14–17 October 2009; pp. 339–342.
100. Török, Á.; Barsi, Á.; Bögöly, G.; Lovas, T.; Somogyi, Á.; Görög, P. Slope stability and rockfall assessment of volcanic tuffs using RPAS with 2-D FEM slope modelling. *Nat. Hazards Earth Syst. Sci.* **2018**, *18*, 583–597. [CrossRef]
101. Mineo, S.; Pappalardo, G.; Onorato, S. Geomechanical characterization of a rock cliff hosting a cultural heritage through ground and UAV rock mass surveys for its sustainable fruition. *Sustainability* **2021**, *13*, 924. [CrossRef]
102. RocScience DIPS. 2019. Available online: https://www.rocscience.com/software/dips (accessed on 30 October 2021).
103. Kallimogiannis, V.; Saroglou, C.; Zekkos, D.; Manousakis, J. 2D and 3D Back-analysis of a landslide in Egremnoi caused by the 17 November 2015 Lefkada earthquake. In Proceedings of the 2nd International Conference on Natural Hazards & Infrastructure, Chania, Greece, 23–26 June 2019; Volume 1, pp. 322–327.
104. Priest, S.D.; Hudson, J.A. Estimation of discontinuity spacing and trace length using scanline surveys. *Int. J. Rock Mech. Min. Sci.* **1981**, *18*, 183–197. [CrossRef]
105. Priest, S.D. The Collection and Analysis of Discontinuity Orientation Data for Engineering Design, with Examples. In *Rock Testing and Site Characterization. Principles, Practice and Projects*; Hudson, J.A., Ed.; Pergamon Press: Oxford, MI, USA, 1993; pp. 167–192.
106. Ferrero, A.M.; Forlani, G.; Roncella, R.; Voyat, H.I. Advanced geostructural survey methods applied to rock mass characterization. *Rock Mech. Rock Eng.* **2009**, *42*, 631–665. [CrossRef]
107. Hellmy, M.A.A.; Muhammad, R.F.; Shuib, M.K.; Fatt, N.T.; Abdullah, W.H.; Bakar, A.A.; Kugler, R. Rock Slope Stability Analysis based on Terrestrial LiDAR and Scanline Survey on Karst Hills in Kinta Valley Geopark, Perak, Peninsular Malaysia. *Sains Malaysiana* **2019**, *48*, 2595–2604. [CrossRef]
108. Pagano, M.; Palma, B.; Ruocco, A.; Parise, M. Discontinuity characterization of rock masses through terrestrial laser scanner and unmanned aerial vehicle techniques aimed at slope stability assessment. *Appl. Sci.* **2020**, *10*, 2960. [CrossRef]
109. Ampatzi, G.; Chatzigogos, N.; Makedon, M.; Papathanassiou, G.; Marinos, V. Application of terrestrial laser scanning (LIDAR) in rock slope stability. An example from Northern Greece. *Bull. Geol. Soc Greece* **2016**, *50*, 586–595. [CrossRef]

applied sciences

MDPI

Article

Seismological and Ground Deformation Study of the Ionian Islands (W. Greece) during 2014–2018, a Period of Intense Seismic Activity

Vassilis Sakkas [1,*], Vasilis Kapetanidis [1], George Kaviris [1], Ioannis Spingos [1], Spyridon Mavroulis [2], Michalis Diakakis [2], John D. Alexopoulos [1], Danai Kazantzidou-Firtinidou [1], Ioannis Kassaras [1], Spyridon Dilalos [1], Emmanuel Vassilakis [3], Evelina Kotsi [2], Gerasimos Tselentis [1,4], Efthymis Lekkas [2] and Nicholas Voulgaris [1]

1 Section of Geophysics—Geothermics, Department of Geology and Geoenvironment, National and Kapodistrian University of Athens, 15784 Athens, Greece; vkapetan@geol.uoa.gr (V.K.); gkaviris@geol.uoa.gr (G.K.); ispingos@geol.uoa.gr (I.S.); jalexopoulos@geol.uoa.gr (J.D.A.); dkazantzidou@geol.uoa.gr (D.K.-F.); kassaras@geol.uoa.gr (I.K.); sdilalos@geol.uoa.gr (S.D.); tselenti@noa.gr (G.T.); voulgaris@geol.uoa.gr (N.V.)
2 Section of Dynamic Tectonic Applied Geology, Department of Geology and Geoenvironment, National and Kapodistrian University of Athens, 15784 Athens, Greece; smavroulis@geol.uoa.gr (S.M.); diakakism@geol.uoa.gr (M.D.); ekotsi@geol.uoa.gr (E.K.); elekkas@geol.uoa.gr (E.L.)
3 Section of Geography & Climatology, Department of Geology and Geoenvironment, National and Kapodistrian University of Athens, 15784 Athens, Greece; evasilak@geol.uoa.gr
4 Institute of Geodynamics, National Observatory of Athens, 11810 Athens, Greece
* Correspondence: vsakkas@geol.uoa.gr

Citation: Sakkas, V.; Kapetanidis, V.; Kaviris, G.; Spingos, I.; Mavroulis, S.; Diakakis, M.; Alexopoulos, J.D.; Kazantzidou-Firtinidou, D.; Kassaras, I.; Dilalos, S.; et al. Seismological and Ground Deformation Study of the Ionian Islands (W. Greece) during 2014–2018, a Period of Intense Seismic Activity. *Appl. Sci.* **2022**, *12*, 2331. https://doi.org/10.3390/app12052331

Academic Editor: José A. Peláez

Received: 19 January 2022
Accepted: 17 February 2022
Published: 23 February 2022

Publisher's Note: MDPI stays neutral with regard to jurisdictional claims in published maps and institutional affiliations.

Abstract: Seismicity in the Ionian Sea (W. Greece) is mainly generated along the Cephalonia–Lefkada Transform Fault Zone (CLTFZ) in the central Ionian, and on the northwestern termination of the Hellenic subduction margin in the south. Joint pre-, co- and post-seismic ground deformation and seismological analysis is performed at the broad Ionian area, aiming to homogeneously study the spatiotemporal evolution of the activity prior to and after the occurrence of strong ($M > 6$) earthquakes during the period of 2014–2018. The 2014 Cephalonia earthquakes (Mw6.1 and Mw5.9) were generated on a faulting system adjacent to CLTFZ, causing local ground deformation. The post-seismic sequence is coupled in space and time with the 2015 Lefkada earthquake (Mw6.4), which occurred on the Lefkada segment of the CLTFZ. Co-seismic displacement was recorded in the broader area. Seismicity was concentrated along the CLTFZ, while its temporal evolution lasted for several months. The 2018 Zakynthos earthquake (Mw6.7) caused regional deformation and alterations on the near-velocity field, with the seismicity rate remaining above background levels until the end of 2021. In the northern Ionian, convergence between the Apulian platform and the Hellenic foreland occurs, exhibiting low seismicity. Seismic hazard assessment revealed high PGA and PGV expected values in the central Ionian.

Keywords: Ionian Islands; Cephalonia–Lefkada Transform Fault Zone; seismicity; ground deformation; GNSS; seismic hazard

1. Introduction

The area of the Ionian Islands in western Greece plays an important role in the kinematic processes of the eastern Mediterranean. This tectonically complex area is by far the most seismically active region in Greece and among the most seismogenic regions in Europe. It is characterized by the frequent occurrence of destructive large earthquakes and undergoes intense ground deformation. The central Ionian Islands constitute part of the Eastern Mediterranean lithosphere that is subducted beneath the Aegean lithosphere along the Hellenic Arc.

The Ionian Sea (Figure 1) hosts areas with different fault geometries and kinematics. The prevailing tectonic structure is the NNE–SSW-trending, right-lateral Cephalonia–Lefkada Transform Fault Zone (CLTFZ), which is the most seismically active structure not only in the Ionian Sea, but also in Greece. This zone is a major boundary in the kinematic field of the region, as it separates the Ionian Margin into two different areas. The North Ionian Islands, comprising the Diapontia Islands, Corfu, Paxoi and Antipaxoi Islands, move slowly northward and northwestward at rates lower than 5 mm/year with respect to Eurasia. The South Ionian Islands, comprising Lefkada, Cephalonia, Ithaca and Zakynthos, move rapidly southwestward with velocities of 6–30 mm/year [1–3].

The CLTFZ is composed of two segments: the 40 km-long, NE–SW-striking, ESE-dipping, right-lateral Lefkada segment, extending from the northwestern offshore part of Lefkada Island to the northern offshore part of Cephalonia Island [4,5], and the 90-km-long Cephalonia segment, close to the western offshore part of Cephalonia, with similar geometry and kinematic properties to the other segment [6,7].

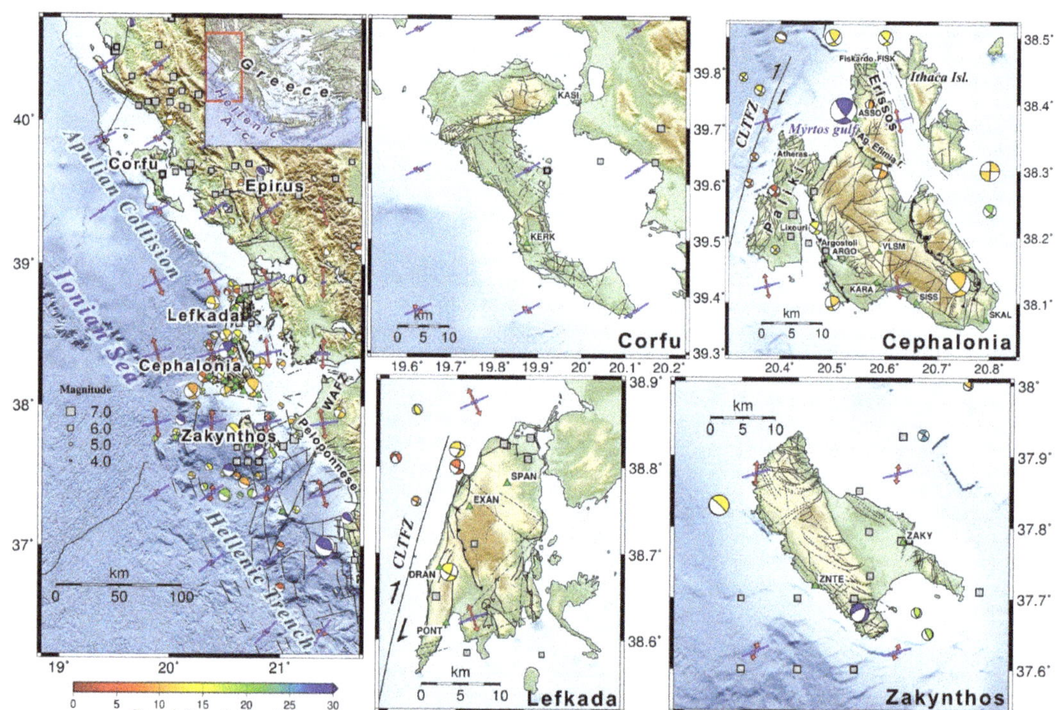

Figure 1. Seismotectonic background of the Ionian Islands region. Focal mechanisms of stronger events ($M \geq 5.5$) are presented (see Data Availability Statement for references). Historic seismicity is presented by gray squares (SHEEC database, [8,9]) Principal stress axes $\mathbf{S}_1$ (blue) and $\mathbf{S}_3$ (red) after [10]. Fault lines after [11–19]. CLTFZ: Cephalonia–Lefkada Transform Fault Zone, WAFZ: Western Achaia Fault Zone.

The South Ionian Islands (Zakynthos and Strofades) are located close to the northwesternmost tip of the Hellenic Arc, a few kilometers east of the Hellenic Trench in the southern Ionian Sea. The trench represents the active plate boundary where the eastern Mediterranean lithosphere is being subducted beneath the Aegean one. This subduction zone terminates against the Cephalonia segment of the CLTFZ.

The CLTFZ is not the only major right-lateral fault zone in the Ionian Sea. A few kilometers south of Zakynthos, another structure occurs. It constitutes the southward extension of the seismic NE–SW-striking, right-lateral Western Achaia Fault Zone (WAFZ), which extends from the northwestern part of the Peloponnese to offshore southern Zakynthos [20]. The epicenter of the 8 June 2008, Mw6.4 Andravida earthquake [21,22] and its aftershock sequence were distributed along the onshore part of the WAFZ, at the northwestern part of the Peloponnese. It is worth noting that the earthquake did not induce direct expression of primary surface faulting [23]. Furthermore, the onshore part of the WAFZ has no direct surficial morphotectonic or geological evidence onshore of the Western Peloponnese. On the contrary, its offshore extension is linked with an offshore pull-apart basin NE of the Strofades Islets [24,25].

The part of the Ionian Sea located south of Zakynthos constitutes a downthrown block of the External Hellenides at the northern end of the Hellenic Trench. Flat thrusts, strike-slip and normal faults are detected [25], with thrusting prevailing over strike-slip or normal faulting [7,26]. The prominent feature of this area is a 46 km-long, NW–SE-trending thrust system [27], responsible for the generation of the 1997 Mw6.6 [28] and the 26 October 2018 Mw6.7 [29] Zakynthos earthquakes.

The intense seismicity on the southern part of the Ionian Sea is attributed to the proximity of this area to the CLTFZ, the Hellenic Trench and many onshore faults in the South Ionian Islands. These islands have been formed either on the margins or within fault blocks. At Lefkada Island, typical cases are the fault zones in the western coastal part [15], along which extensive environmental effects were generated from the two strong Lefkada earthquakes on 14 August 2003 (Mw6.3) [30,31] and on 17 November 2015 (Mw6.4) [32], triggering many rockfalls and landslides in the western coastal part of the island [17,33]. In Cephalonia, several active faulting zones at the northern, western and southwestern parts of the island are susceptible to triggering of earthquake environmental effects, with the most typical example being the numerous rockfalls and slides due to the August 1953 earthquakes [34]. In Zakynthos, typical cases of active faults are the Volimes fault zone in the north part of the island, which has been formed in the transition from the Northern Zakynthos fault block to the Central Zakynthos one. Similar active faults are also located at the eastern and southern parts of Zakynthos Island [16].

Regarding the North Ionian Islands, a NNW–SSE-trending system, with a relatively extended shelf width represents the convergence (continental collision) between the Apulian Platform and the Hellenic foreland, with Corfu lying on the northwestern edge of the Hellenic Fold and Thrust Belt. The E–W-striking Southern Salerno–North Corfu fault zone is a major right-lateral structure that crosses Corfu from coast to coast [35] and which has resulted in the displacement of N–S-trending fold axes and thrusts [36]. This part of the Ionian Sea is characterized by lower seismicity than the southern one. The North Ionian Islands have suffered damage from earthquakes generated in adjacent areas, such as the 20 February 1743 Salento peninsula (Apulia, southern Italy) earthquake with Mw7.1 and I_0 = IX, which triggered landslides in Corfu [37] and severe structural damage, including building collapses in Corfu town [38], attributed to very efficient strong propagation with NW–SE preferential directivity [39].

The complex geotectonic status of the area resulted in the high seismic activity that occurs in the broad Ionian area. However, during the last decade, and mainly during the period between 2014 and 2018, increased seismicity was observed, and strong events ($M > 6.0$) shocked the central Ionian Islands. Early 2014, two earthquakes (Mw5.9 and Mw6.1) occurred on Cephalonia Island, while in November 2015 and in October 2018 two earthquakes of Mw6.4 and Mw6.7 took place at South Lefkada and offshore south of Zakynthos, respectively. The regional crustal motion along the entire Ionian Sea and western Greece, as well as the local deformation on the central Ionian Islands, have been studied and monitored with local dense Global Positioning System (GPS) networks [13,40,41] and continuous Global Navigation Satellite System (GNSS) stations [42,43]. Previous work focused on the seismological analysis and interpretation of each of these strong earthquakes [15,29,32,44–49], on the study and modeling of the co-seismic deformation during every event [50–58] and on the geological impact close to the epicentral areas [14,15,17,59,60].

The purpose of the present work is to study and present in a unified and homogeneous approach an overview of the seismicity and the ground deformation in the broad area of Ionian Sea, extending from Corfu Island to the north to Strofades Islet in the south. The time span of the data covers the period before, during and after the occurrence of the strong earthquakes of 2014–2018. The spatiotemporal evolution of the seismicity involving these significant events is analyzed. Pre-, co- and post-seismic deformation is quantitatively described, aiming to understand the pattern of the ground motion associated with the recorded seismicity. The regional implications of joint seismological and geodetic analysis are also considered and discussed with respect to the geotectonic status of the area. Seismological data from the Hellenic Unified Seismic Network (HUSN) and geodetic data from the commercial and institutional continuous GNSS networks in the area were used in the framework of this study.

2. Seismological Data and Results

In the Ionian Islands there are historical reports of 94 earthquakes from the years 358 to 1898 [9,61]. A significant number of historical earthquakes are located onshore, most possibly due to the limitations that arise from the use of historical sources. Instrumental seismicity, since 1900, is concentrated around the islands of Lefkada, Cephalonia and Zakynthos [28]. As described above, epicenters in Lefkada and Cephalonia are linked to local faults and mainly to the CLTFZ, whereas seismicity offshore and close to Zakynthos is related to the border between the Eurasian and African plates. Focal depths are limited to the upper 30–40 km of the crust, while foci at greater depths are related to the submerged African plate. The high seismicity rate is reflected in the high seismic hazard of the central Ionian Islands that belong to the highest category (Zone III) of the current Greek Building Code, with a Peak Ground Acceleration (PGA) value of 0.36 g for a return period of 475 years [62].

Since the implementation of the HUSN in 2007 [63], the monitoring of seismic activity in Greece has been enhanced, with increased detectability and improved location accuracy. This was particularly important for the region of the Ionian Islands, as it is situated at the western margins of the network and a significant part of its intense seismicity is located offshore. Herein, a catalogue of ~62,000 earthquakes was compiled. These events have occurred in the broad Ionian Islands area, including part of mainland western Greece. The seismological data were collected from the database of the Geodynamics Institute of the National Observatory of Athens (GI-NOA) for the period from February 2011 to May 2018. For the consecutive period from June 2018 to November 2021, the seismological data were extracted from the database of the Seismological Laboratory of the National and Kapodistrian University of Athens (SL-NKUA). Furthermore, in the present analysis, relocated seismicity catalogues were incorporated for the 2014 Cephalonia, 2015 Lefkada and 2018 Zakynthos aftershock sequences [29,32,64–67].

2.1. Seismicity Results

After visual inspection, the seismicity of the study area during the last decade was divided into 12 spatial groups (Figure 2), to enable the description of its spatiotemporal evolution. Seismicity marginal to the area of interest was placed in the miscellaneous group #13 (white). The magnitude of completeness is estimated at $M_c = 2.0 \pm 0.2$, with variations depending on the method used for its determination or the time period of a selected subset. At times, it reaches lower values, e.g., 1.5, whereas during the occurrence of significant earthquakes that produce series of large aftershocks, the magnitude of completeness may temporarily surpass 2.5. The cumulative number of events per spatial group (Figure 3a) and the spatiotemporal projection (Figure 3b) along the A–B profile of Figure 2a are drawn after the application of the magnitude threshold ($M \geq M_c = 2.0$). The herein referred magnitudes are M_L, unless otherwise stated.

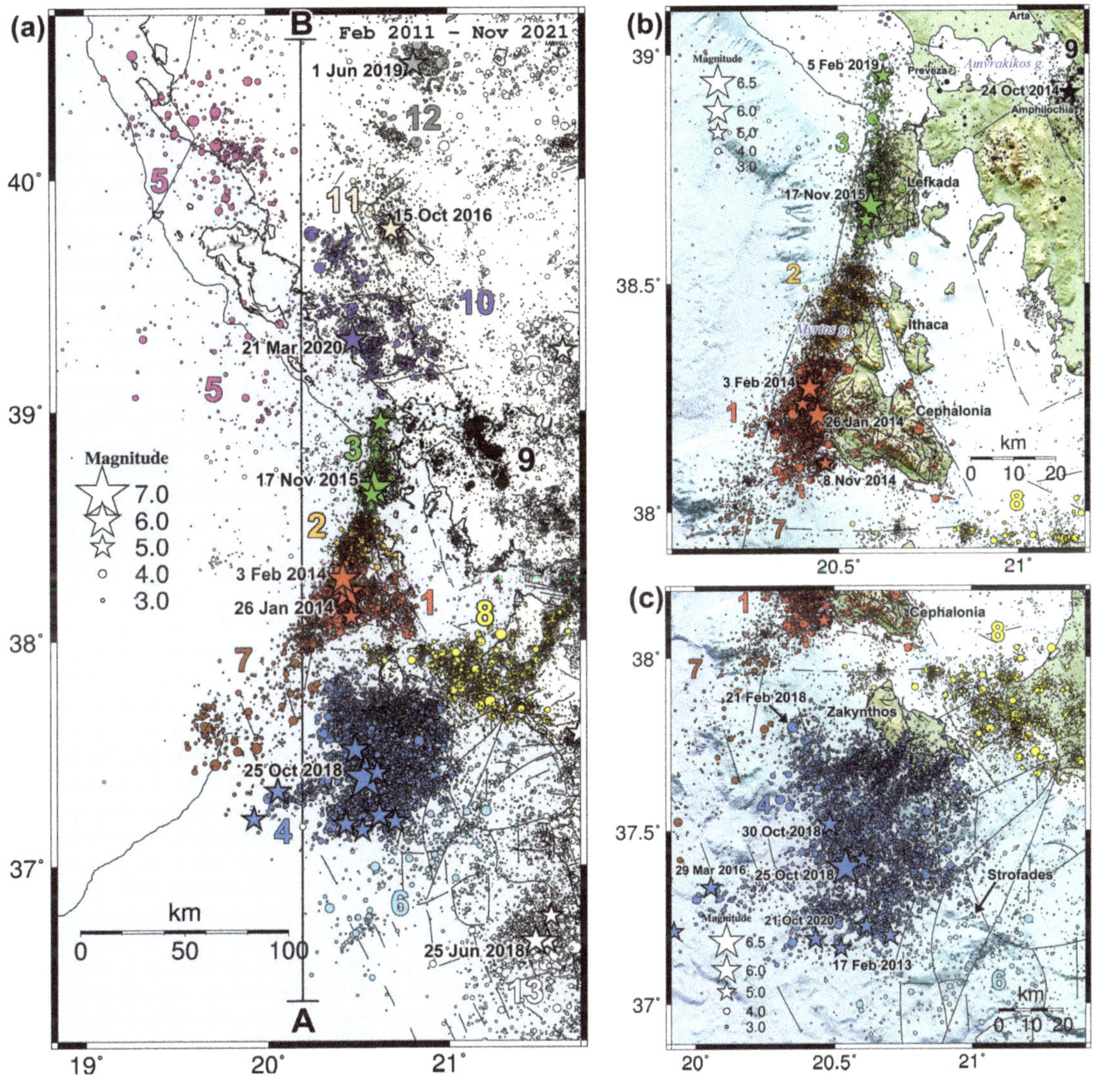

Figure 2. (**a**) Seismicity in the region of the Ionian Islands during the period February 2011–November 2021 from the databases of GI-NOA and SL-NKUA, including relocated catalogues for the three main aftershock sequences [29,32,64,65]. Colors and numbers correspond to different spatial groups. The north–south-oriented profile A–B is used for the spatiotemporal projection of Figure 3b. (**b**) Close-up of the seismicity in the region of Cephalonia–Ithaca–Lefkada Islands. (**c**) Close-up of the seismicity in the region of Zakynthos Island.

Starting in early 2011, scattered seismicity was observed throughout the Ionian Islands region. An M4.5 earthquake occurred near the northern tip of Cephalonia Island on 16 March 2011, an M4.1 event occurred just offshore south of the island on 5 December 2011 and an M4.2 event near the northwestern part of the Paliki peninsula (Atheras area) on 23 April 2013. The background seismicity rate with $M \geq 2.0$ was ~0.4 events/day, occurring mainly onshore of the island (group #1, red) and in Myrtos Gulf (group #2, orange). Overall, very few events are located near to or onshore of Ithaca Island during the study period. Background seismicity was also recorded along the SW extension of the CLTFZ, at a rate of 0.1–0.2 events/day. Further south, near Zakynthos Island, clustered seismicity was recorded following events of magnitude around 4 at group #4 (indigo; ~0.3 events/day) and group #8 (yellow; ~0.5 events/day), which extended from the area between Zakynthos and Cephalonia to the NW Peloponnese. An M5.0 event occurred at the westernmost end of group #4 on 19 July 2011, triggering seismicity along SW–NE-trending streaks. It should be noted that such delineations of epicenters in that area have been argued to be affected by the large azimuthal gap, causing location biases along this direction [29], so they should not be overinterpreted. The southernmost group #6 (cyan) near the Strofades Islets has a background rate of ~0.2 events/day. On Lefkada Island (group #3, green; with a background rate of 0.1–0.2 events/day), seismicity often occurs in spatiotemporal clusters, whereas the background seismicity on Cephalonia Island tends to be more randomly scattered. Significant earthquakes that took place during the study period also include an M4.5 event on 23 October 2012 offshore north of Lefkada and west of Preveza, and an M4.7 event on 23 May 2013 at the southern tip of the island.

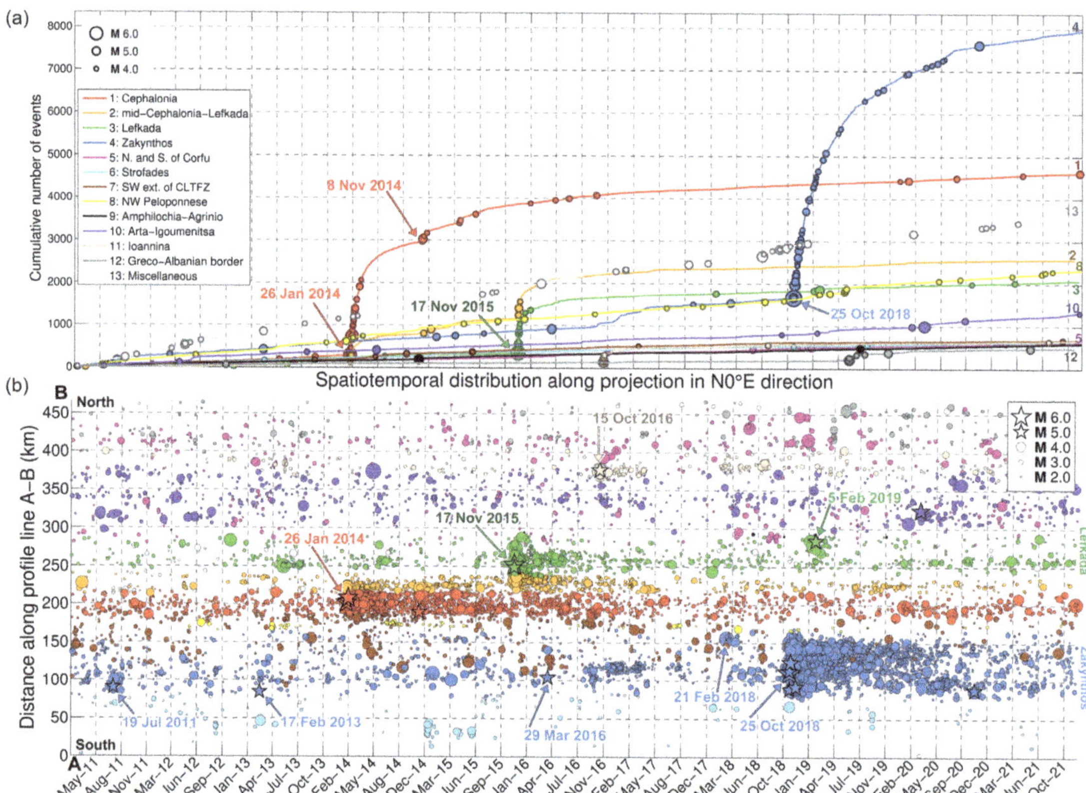

Figure 3. (**a**) Cumulative number of events per spatial group of Figure 2 (different colors), for events with $M \geq M_c = 2.0$. The larger events ($M \geq 4.0$) are marked by circles. (**b**) Spatiotemporal projection along the north–south-oriented profile A–B of Figure 2a, for events within a range of ±50 km from the profile and $M \geq M_c = 2.0$. Symbol size is proportional to magnitude. Events with $M \geq 5.0$ are marked with stars. Close-ups for the periods January 2014 to September 2017 and August 2016 to November 2021 are presented in Figures A1 and A2 of Appendix A, respectively.

Group #9 (black), near the eastern shore of Amvrakikos Gulf, with a background rate of ~0.1 events/day, contains an M5.2 event that occurred on 24 October 2014, producing a small cluster. Further north, Corfu Island is characterized by very low seismicity onshore throughout the study period. Sparse seismic activity was recorded offshore to the south and a bit denser distribution of epicenters is identified to the north, near the shores of Albania (group #5, purple; with a background rate of 0.1–0.2 events/day), related with events of magnitude $4.0 \leq M < 5.0$. West of Corfu Island, at NW Epirus, group #10 (blue; background rate of 0.2–0.4 events/day) included an $M = 5.6$ thrust event that occurred on 21 March 2020 near Kanallaki [68,69]. Group #11 (beige; background rate of 0.1 events/day) contains an M5.3 event that occurred on 15 October 2016 near Ioannina. At the northern edge of the study area, group #12 (gray; background rate of 0.1 events/day) near the Greco-Albanian border, contains an M5.3 event that occurred on 1 June 2019. The most significant bursts of earthquake activity at the Ionian Islands region during the study period comprise the 2014 Cephalonia (groups #1 and #2), the 2015 Lefkada (groups #3 and #2) and the 2018–2019 Zakynthos mainshock–aftershock sequences (group #4), which are described in more detail in the following subsections.

2.1.1. The 2014 Cephalonia Sequence

The 2014 Cephalonia sequence was initiated by an Mw6.1 mainshock that occurred on 26 January 2014, ~2 km NE of the city of Lixouri, on the Paliki peninsula [44–46,60,64]. About one week later, on 3 February 2014, another significant earthquake of magnitude Mw5.9 occurred at the NW part of Paliki. The seismicity on Cephalonia Island during 2014 was relocated using the double difference method (HypoDD; [70]) and a minimum 1D velocity model [64,65]. The aftershock distribution extended ~32 km in a N20° E direction, covering the entire Paliki peninsula (group #1), but seismic activity was also triggered in a spatially separated cluster inside Myrtos Gulf (group #2). The southern half of the onshore seismicity, related mostly to the first major earthquake of 26 January 2014 at a depth of 16 km, appeared more complex, being distributed between 5 and 17 km and extending to a width of 15 km in a N110° E direction, whereas the northern half, mostly related to the second major earthquake, was more linearly distributed (width ~5 km) and shallower, at focal depths between 5 and 12 km [65]. Seismicity after the first earthquake was mainly concentrated near its hypocenter during the first hours, but quickly spread throughout the aftershock zone. Some clusters further south, offshore Cephalonia, were triggered at a later stage. A significant cluster occurred south of group #1 after an M5.0 event on 8 November 2014, soon followed by clustered activity in Myrtos Gulf (group #2). The focal mechanisms of the two major events and most of the major aftershocks indicate dextral strike-slip faulting along SSW–NNE-trending, subvertical faults. However, seismicity inside the Myrtos Gulf is characterized by several smaller clusters trending E–W, i.e., transverse to the axis of the CLTFZ, which suggests antithetic sinistral faulting. The 2014 sequence temporarily increased the seismicity rate of the neighboring groups #7 and #8. The temporal evolution of the post-seismic activity in Cephalonia (group #1) shows a long relaxation period that lasted up to July 2017, considering a relatively high background seismicity level (0.3 events/day) for $M \geq M_c = 2.0$ (Figure 4). In the northern area of Cephalonia (group #2), with a smaller background seismicity rate (0.1 events/day prior to 2014), increased activity was observed up to September 2017, due to its reactivation after the 2015 Lefkada mainshock, which is presented in the following Section 2.1.2.

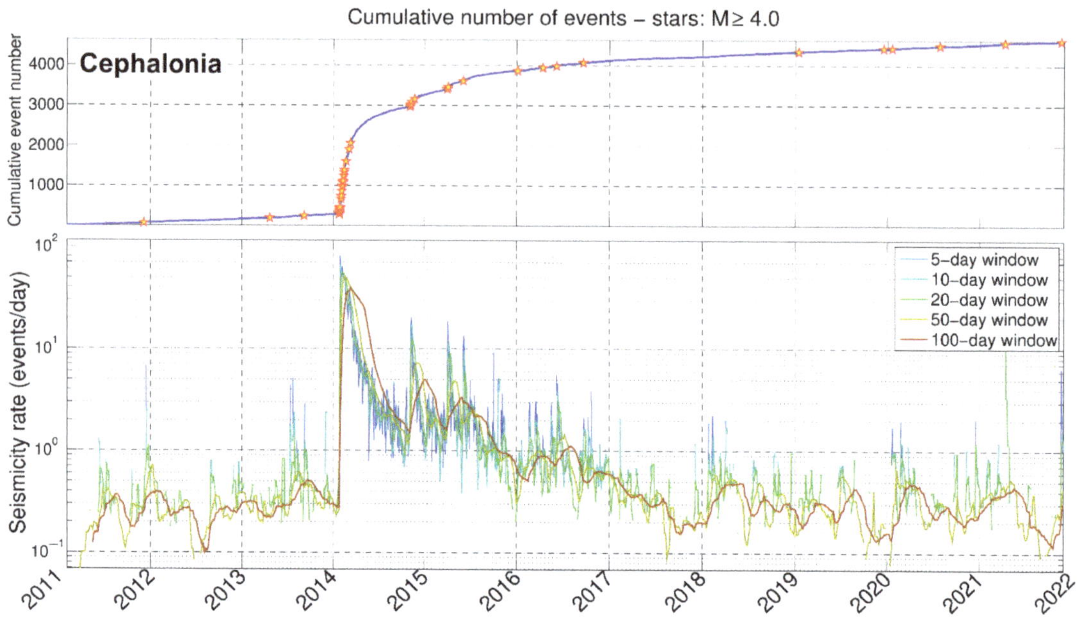

Figure 4. Temporal evolution of the seismic sequence on Cephalonia Island (group #1), for events with $M \geq M_c = 2.0$. The upper panel shows the cumulative number of events, with the occurrence of $M \geq 4.0$ events depicted by stars, while the lower panel describes the seismicity rate on a semi-logarithmic scale, measured for sliding windows of various lengths (5 days to 100 days).

2.1.2. The 2015 Lefkada Sequence

A few months after the Cephalonia events, on 17 November 2015, an Mw6.4 earthquake struck Lefkada Island, with its epicenter located near the mid-southern part of the western coast. The seismicity during the period from 17 November to 3 December 2015, relocated with the double-difference method [65], revealed a very different distribution than that of the 2014 Cephalonia sequence. The aftershocks were divided in distinct clusters, distributed at focal depths between 5 and 15 km. The aftershock zone extended to ~60 km in a N16°E direction, mainly onshore of the island. Very few events occurred south of the mainshock, suggesting that a major fault patch was ruptured, leaving only few small unbroken asperities. The largest aftershock was an M5.0 event that occurred on the same day as the mainshock and close to its vicinity. The densest cluster was located just north of the mainshock, in the same region that was activated during an M5.1 event that had occurred on 29 November 1994 [71], and also triggered after an Mw6.3 earthquake that struck the northern part of the island on 14 August 2003 [30,31]. Notably, the mainshock of 2015 at Lefkada Island, which mainly involved group #3, also triggered seismicity in group #2, between Cephalonia and Lefkada, and even inside the Myrtos Gulf. Furthermore, these offshore southern clusters presented similar characteristics as those in the gulf of Myrtos, i.e., east–west alignment, likely related to antithetic sinistral faulting, transverse to the CLTFZ. The earthquake rate of group #3 that increased since the occurrence of the 17 November 2015 earthquake, remained higher than the background level (0.1 events/day; for $M \geq M_c = 2.0$) as late as September–October 2017. To exclude possible biases due to the increase of event detectability, setting a higher magnitude threshold of 2.3 (Figure 5) indicated a return of the seismicity rate to its background level around October 2016, dropping to very low levels in May–July 2017.

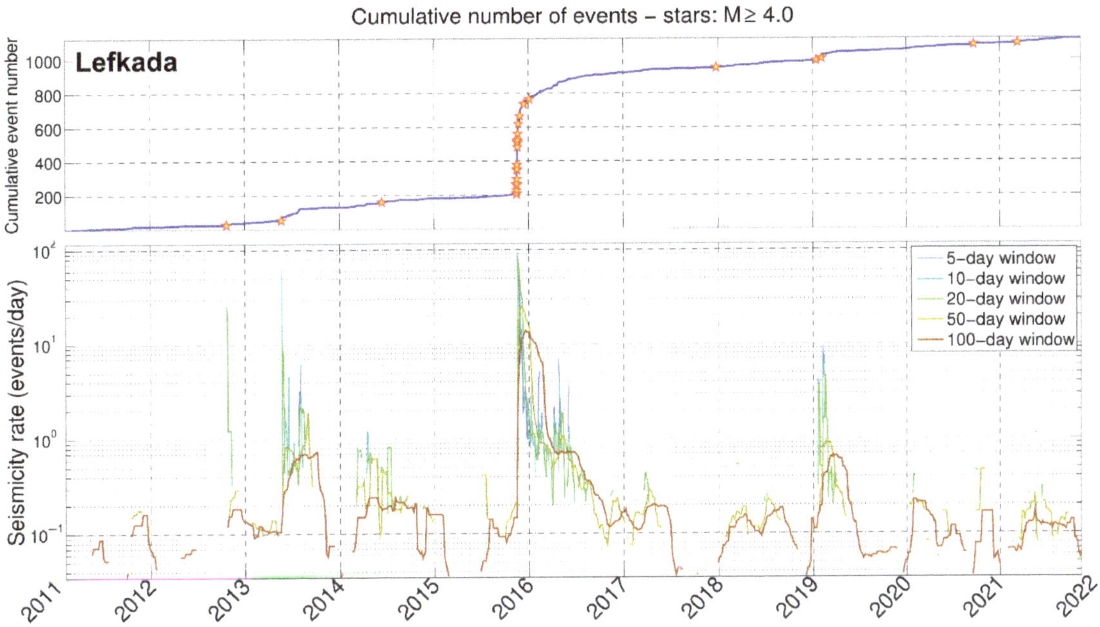

Figure 5. Temporal evolution of the seismic sequence on Lefkada Island (spatial group #3) for events with $M \geq 2.3$. Panels are similar to Figure 4.

2.1.3. The 2018–2019 Zakynthos Sequence

The most recent significant seismic occurrence in the region of the Ionian Islands during the study period is the 2018–2019 Zakynthos mainshock-aftershock sequence [29,49,72]. An Mw6.7 offshore earthquake occurred on 25 October 2018, 45 km SW of Zakynthos Island, at an estimated depth of ~12 km. The focal mechanism of the mainshock indicates dextral strike-slip on a low-angle (25°), east-dipping plane. The relocated catalogue of over 4000 aftershocks in the period between 25 October 2018 and 31 March 2019 reveals an extensive aftershock zone, ~80-km-long in a N110° E direction and ~55 km wide [29]. The largest aftershock was an Mw5.5 event that occurred on 30 October 2018, 15 km NNW of the mainshock, with a similar focal mechanism, albeit with a steeper dip angle. The distribution of epicenters permits the distinction of some sub-clusters within the seismicity cloud. A large group in the vicinity of the mainshock presented a smoothly diminishing seismicity rate. On the other hand, other clusters, particularly at the northern part of the aftershock zone, but also at the southern end, presented outbreaks, some related with the occurrence of significant aftershocks. The most persistent activity was observed at the northern subclusters of group #4, with intense activity that lasted until October 2019. A last burst of seismic activity occurred between March and July 2020 and was triggered at the southern end of the 2018 aftershock zone, after an M4.3 event on 26 March 2020, followed by a few more $M > 4$ events. A final significant M5.2 event occurred on 21 October 2020 in the same region, after a period of quiescence.

The 2018 Zakynthos mainshock–aftershock sequence was preceded by several foreshocks since the beginning of the year in the vicinity of the mainshock. An M4.8 event had occurred at the northern end of group #4 on 21 February 2018 and another Mw4.8 event was recorded half an hour before the mainshock. Earlier activity was also documented near the mainshock's region between September 2016 and March 2017, in at least three outbreaks, while a smaller one occurred in August 2017. Focal mechanisms of the 2018 Zakynthos aftershock sequence involved both low-angle and steeper strike-slip faulting [29]. The latter is mainly related with the northern subclusters, which also present a form of SW–NE-trending streaks, similar to the cluster at the southern end of group #4. The kinematics of the mainshock and many of the major aftershocks are consistent with SW–NE-trending compression occurring in this region that is found in a transition zone between strike-slip faulting in the north (CLTFZ) and to the east (WAFZ in the NW Peloponnese, e.g., the 2008 Andravida earthquake [21,22]) and the northwestern end of the Hellenic Arc subduction zone. The temporal evolution of the post-seismic activity indicates that even towards the end of 2021 the seismicity rate had still not returned to its background level, despite the magnitude threshold being set as high as 2.5, quite above the average M_c = 2.0 (Figure 6).

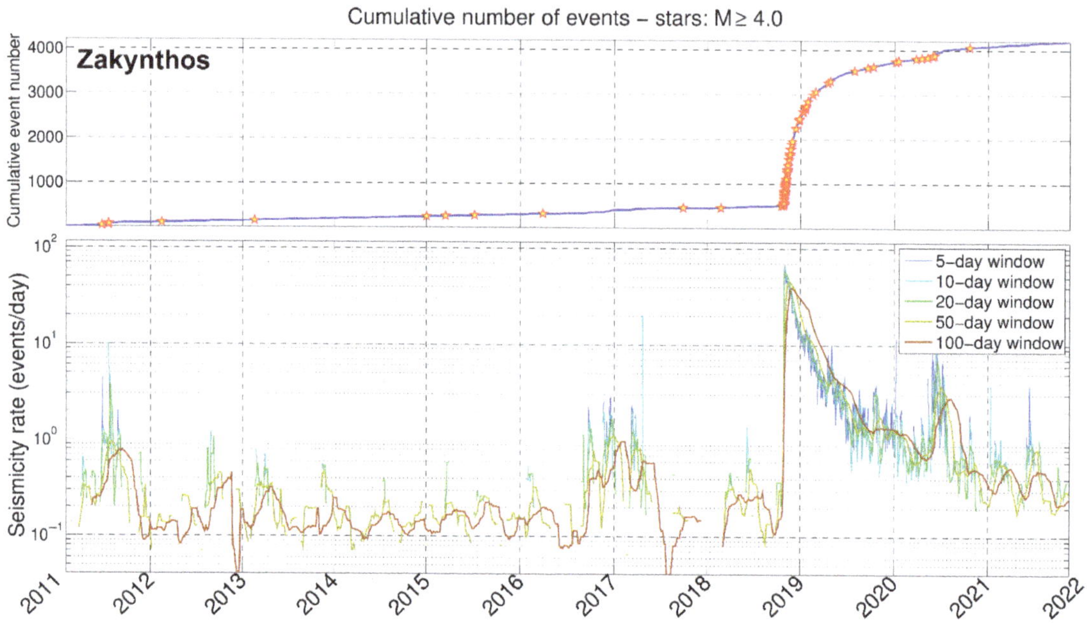

Figure 6. Same as Figure 4, but for the temporal evolution of seismicity south of Zakynthos Island (group #4), for events with $M \geq 2.5$.

2.2. Seismic Hazard

The intense and large magnitude seismic activity in the broad area of the Ionian Islands makes the assessment of seismic hazard crucial in the urban planning, as well as in the construction of critical infrastructure. Additionally, the tourism industry in the Ionian Islands is a major concern for the local and national economy that requires enhancing its earthquake resilience.

The strong earthquakes that occurred in the area have yielded high values of Peak Ground Acceleration (PGA). At Cephalonia, the two events of 2014 induced shakings reaching ~560 cm/s^2 and 735 cm/s^2, respectively [73]. At Lefkada, PGAs of ~412 cm/s^2 were recorded in the 2003 event [74] and 363 cm/s^2 during the 2015 earthquake [67]. Finally, the Mw6.7 earthquake of Zakynthos led to a recorded PGA of ~382 cm/s^2 [75], on the island.

On the framework of this study, the seismic hazard of the Ionian Islands was evaluated, aiming to estimate the maximum anticipated ground motion values in this highly active seismogenic area. A probabilistic seismic hazard assessment (PSHA) method has been applied. Based on the regional character of this study, a source-zone approach was selected. Zone boundaries were obtained from the SHARE project [76,77]. However, the b-value, earthquake rate and maximum magnitude for each area were re-estimated, from the catalogue compiled in the context of the current study. To account for time periods of incomplete data, the modified b-value (β) was determined using a maximum likelihood method [78]. The ground motion prediction equations (GMPEs) of [79], which have been successfully applied in seismic hazard studies in Greece [80–82], were used to obtain PGA values in the final PSHA model. A reverse/strike-slip focal mechanism type and a rock basement were considered in the GMPE, for all areas. Finally, the seismic hazard was assessed using the R-CRISIS software [83]. The software can calculate the expected hazard value (in this case, PGA and the Peak Ground Velocity; PGV) for estimated earthquake occurrence probabilities at specific timeframes, for the given distribution of sources and the event magnitudes within the limits of each zone. The PGA and PGV values were computed for a 10% exceedance probability level in 50 years (return period of 475 years) (Figure 7).

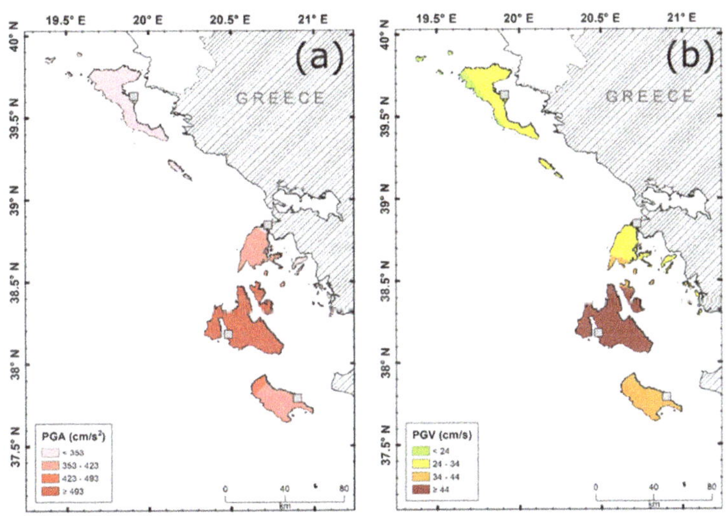

Figure 7. Maps showcasing the spatial distribution of PGA (**a**) and PGV (**b**) for a return period of 475 years. Major cities in the Ionian complex are marked by gray boxes.

The two hazard quantities follow similar distributions. Highest values are observed at Cephalonia and the nearby Ithaca Island. The high seismic activity on these islands fully complied with the obtained hazard values. Similarly, increased values are observed at the NW edge of Zakynthos. The hazard quantities decrease towards the central part of Zakynthos, in accordance with the lack of strong events in this area. The high hazard zone is terminated at the southern part of Lefkada Island. The lowest hazard values in the area are observed in the north Ionian Islands (Paxoi and Corfu); as expected, since no large-magnitude events ($M \geq 6.0$) have occurred in this area during the past 120 years [28].

The Greek Building Code [62] classifies the Ionian islands into two zones: Zone II (with PGA of 235 cm/s^2) and Zone III (with a value of 353 cm/s^2). Lefkada, Cephalonia, Ithaca and Zakynthos Islands are in Zone III, with Corfu and Paxoi classified in Zone II. While the code provides wide ranges, the current analysis indicates that there may be a need for a more detailed approach. The expected PGA at Corfu and Paxoi was found to be the lowest in the Ionian Islands, being less than 353 cm/s^2. However, it was observed that for Zone III islands, a much finer distribution could be extracted (Figure 7a). Moreover, Cephalonia and Ithaca seemed to feature PGA values much higher than that of the Seismic Code. Making this distinction by introducing a new Zone with stricter building guidelines could prove to be a useful approach to reduce seismic vulnerability—and therefore, risk—in a financially significant area of the Ionian prefecture. Revising the national code would also offer the opportunity to rethink local actions for improving the seismic resilience of older and historical buildings in cities and towns (e.g., Argostoli in Cephalonia) that teem with them. Other seismic hazard studies, albeit Greece-wide, also find PGA values that well exceed those proposed by the building code [80,84].

3. GNSS Data and Results

Daily GNSS data from stations located in the Ionian Islands and western mainland Greece and the Peloponnese were processed for the period from 2009 to 2021. The analysis intended to determine the crustal velocity field of the broad area, detect possible pre-seismic displacements and study co- and post-seismic deformation. The continuous GNSS stations extend from the northern Corfu Island to the Strofades Islets, south of Zakynthos. In western mainland Greece and the Peloponnese, the stations are located in the broad area of Patras Gulf and in the city of Pyrgos in the western Peloponnese (Figure 8). Sites KASI, SPAN, PONT, KIPO, VLSM, ZNTE, STRF, KTCH and RLSO belong to the National Observatory of Athens (NOA network) [42,43]. The commercial network of METRICA SA (HexagonSmartNet) [85] provided data for stations KERK, PAXO, ARGO, ZAKY, AGRI and PYRG. In Cephalonia Island, the sites SISS, KARA and SKAL are part of the National and Kapodistrian University of Athens network (NKUA net) [54]. Meanwhile stations EXAN and DRAN in Lefkada Island and FISK and ASSO in Cephalonia are part of the GNSS network of the Corinth Rift Laboratory (CRL GNSS net) [86], which is the only international Near-Fault Observatory (NFO) of the European Plate Observing System. Finally, PAT0 station in the city of Patras is a European Reference Frame (EUREF) station [87].

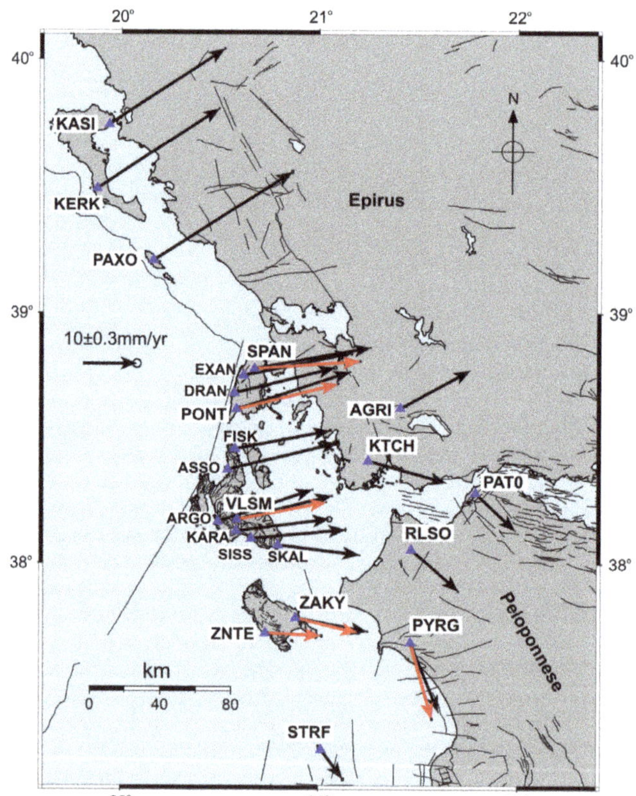

Figure 8. Horizontal velocity vectors (black arrows) for the continuous GNSS sites (blue triangles) of the broad Ionian Islands area. Red arrows indicate the horizontal velocity component estimated for the period after the strong ($M > 6$) earthquakes in the area (ITRF 2014).

The raw GNSS data were processed using the Bernese v5.2 software [88]. Several GNSS stations of the EUREF and IGS were included in the processing together with the local GNSS sites, while auxiliary files were introduced in the procedure. The absolute antenna phase center corrections were used, together with precise orbital solutions from the Center for Orbit Determination in Europe (CODE) and the Vienna Mapping Functions for the tropospheric modeling. The FES2004 model (http://holt.oso.chalmers.se/loading; accessed on 11 January 2022) was used for the tide-loading corrections. The precise double-difference method was used for static-mode solutions. Several ambiguity-resolution strategies were applied, based on the length of the formed baselines between the GNSS stations. The daily calculated coordinates for the GNSS stations were evaluated for the repeatability error on a weekly basis, and solutions were excluded in cases of large deviations from the weekly solution. The processing resulted in the estimation of high-precision station coordinates. Time series were formed, annual velocities were calculated (Figure 8; Tables 1–5) and co-seismic displacements were determined (Appendix B). The daily coordinates of the Ionian GNSS stations were estimated on the global ITRF2014 reference frame.

Table 1. Velocity components for the GNSS stations on Corfu and Paxoi Islands (ITRF 2014).

Site	Latitude (°)	Longitude (°)	Period	V_{East} (mm/Year)	V_{North} (mm/Year)	V_{Up} (mm/Year)
KASI	39.7464	19.9355	1 January 2013–6 May 2017 4.35 years	20.91 ± 0.04	13.75 ± 0.04	0.39 ± 0.09
KERK	39.4937	19.8734	1 March 2017–31 October 2021 4.67 years	22.15 ± 00.3	14.57 ± 0.03	−2.07 ± 0.11
PAXO	39.2108	20.1639	1 November 2019–31 October 2021 2.00 years	24.96 ± 0.07	16.08 ± 0.11	−3.1 ± 0.30

Table 2. Velocity components for the Lefkada GNSS stations (ITRF 2014).

Site	Latitude (°)	Longitude (°)	Period	V_{East} (mm/Year)	V_{North} (mm/Year)	V_{Up} (mm/Year)
SPAN	38.7813	20.6736	1 January 2009–16 November 2015 6.88 years	20.86 ± 0.02	3.83 ± 0.02	−0.50 ± 0.05
			18 November 2015–31 October 2021 5.94 years	18.97 ± 0.04	1.37 ± 0.03	−0.95 ± 0.07
EXAN	38.7540	20.6186	25 November 2015–16 May 2021 5.48 years	19.46 ± 0.07	4.06 ± 0.12	−5.90 ± 0.51
DRAN	38.6838	20.5746	24 November 2015–17 June 2018 2.56 years	18.10 ± 0.13	4.59 ± 0.11	−5.06 ± 0.36
PONT	38.6190	20.5852	1 January 2009–16 November 2015 6.88 years	20.45 ± 0.03	6.81 ± 0.02	−0.71 ± 0.08
			18 November 2015–6 August 2016 0.72 years	5.28 ± 0.81	−17.31 ± 0.65	18.52 ± 2.06
			6 August 2016–31 October 2021 5.24 years	18.27 ± 0.05	4.61 ± 0.04	−0.63 ± 0.13

Table 3. Velocity components for the Cephalonia GNSS stations (ITRF 2014).

Site	Latitude (°)	Longitude (°)	Period	V_{East} (mm/Year)	V_{North} (mm/Year)	V_{Up} (mm/Year)
FISK	38.4597	20.5771	15 March 2017–17 August 2020 3.49 years	17.14 ± 0.08	3.32 ± 0.07	1.01 ± 0.28
ASSO	38.3778	20.5417	15 March 2017–18 May 2021 4.18 years	19.81 ± 0.07	4.79 ± 0.04	−2.03 ± 0.19
VLSM	38.1768	20.5886	1 January 2009–25 January 2014 5.07 years	17.10 ± 0.04	4.27 ± 0.03	−0.70 ± 0.09
			8 February 2014–4 September 2016 2.57 years	14.07 ± 0.13	−1.49 ± 0.07	0.44 ± 0.25
			5 September 2016–31 October 2021 5.16 years	16.10 ± 0.03	3.15 ± 0.03	−1.20 ± 0.09
ARGO	38.1690	20.4925	1 May 2016–31 Ocotber 2021 5.50 years	17.08 ± 0.02	5.78 ± 0.02	−2.27 ± 0.27
KARA	38.1308	20.5843	6 May 2014–16 November 2015 1.53 years	16.78 ± 0.19	2.17 ± 0.18	2.48 ± 0.71
SISS	38.1009	20.6594	6 February 2016–31 October 2021 5.73 years	16.97 ± 0.03	1.59 ± 0.0.03	−3.57 ± 0.10
SKAL	38.0746	20.7937	4 March 2015–13 August 2021 4.45 years	14.69 ± 0.02	−1.80 ± 0.02	−1.52 ± 0.07

Table 4. Velocity components for the Zakynthos—Strofades GNSS stations (ITRF 2014).

Site	Latitude (°)	Longitude (°)	Period	V_{East} (mm/Year)	V_{North} (mm/Year)	V_{Up} (mm/Year)
ZAKY	37.7792	20.8850	2 December 2013–25 October 2018 4.90 years	12.83 ± 0.03	−2.5 ± 0.03	−0.31 ± 0.08
			27 October 2018–5 July 2019 0.69 years	2.37 ± 0.42	−1.34 ± 0.42	6.40 ± 1.24
			6 July 2019–31 October 2021 2.32 years	11.21 ± 0.10	−2.68 ± 0.09	−0.88 ± 0.32
ZNTE	37.7176	20.7308	23 November 2018–31 May 2019 0.52 years	−8.30 ± 0.83	20.61 ± 0.51	−3.39 ± 2.18
			1 June 2019–31 October 2021 2.42 years	10.17 ± 0.08	−0.48 ± 0.08	0.10 ± 0.025
STRF	37.2454	21.0156	6 October 2016–25 October 2018 2.05 year	3.87 ± 0.08	−5.31 ± 0.09	−0.46 ± 0.24

3.1. *Corfu–Paxoi Islands*

At the North Ionian Islands, three GNSS stations were processed (Table 1): two in Corfu, KASI and KERK; and the PAXO station located at Paxoi Island. The linear type of motion in all three components for the whole-time span is evident in these stations. There was no static deformation due to the seismic activity that occurred in the south, i.e., related to the 2015 Lefkada or the 2014 Cephalonia events (Figure S1). The calculated velocity vectors are similar to the ones that are presented in [68,89]. For the two Corfu stations (KASI and KERK), the horizontal velocity components are almost parallel, with the southern site exhibiting a slightly increased eastward motion. The most noticeable aspect is the significant subsidence that occurs at the KERK and PAXO stations.

Table 5. Velocities of GNSS stations in western Greece and the NW Peloponnese (ITRF 2014).

Site	Latitude (°)	Longitude (°)	Period	V_{East} (mm/Year)	V_{North} (mm/Year)	V_{Up} (mm/Year)
AGRI	38.6240	21.4090	5 January 2011–12 February 2019 8.11 years	12.51 ± 0.01	6.96 ± 0.01	−0.30 ± 0.04
KTCH	38.4116	21.2469	20 November 2013–31 October 2021 7.95 years	14.12 ± 0.01	−3.94 ± 0.01	−1.25 ± 0.03
PAT0	38.2837	21.7868	27 January 2009–31 October 2021 12.77 years	7.20 ± 0.01	−6.82 ± 0.01	−0.46 ± 0.02
RLSO	38.0558	21.4647	3 June 2011–26 February 2017 5.74 years	8.45 ± 0.03	−7.53 ± 0.03	0.34 ± 0.07
			1 March 2019–31 October 2021 2.67 years	10.03 ± 0.08	−10.95 ± 0.06	−2.27 ± 0.16
PYRG	37.6788	21.4622	2 December 2013–25 October 2018 4.90 years	5.03 ± 0.02	−12.51 ± 0.02	−0.40 ± 0.06
			27 October 2018–12 July 2019 0.70 years	−6.47 ± 0.29	−17.74 ± 0.29	−0.91 ± 1.03
			13 July 2019–31 October 2021 2.30 years	3.86 ± 0.06	−13.79 ± 0.08	−0.94 ± 0.23

3.2. *Lefkada Island*

On central Lefkada Island, data from two GNSS stations (SPAN in the north and PONT in the south) were processed for the period from 2009 to 2021 (Figure S2). Partial data from two other sites, EXAN and DRAN, were available after the strong (Mw6.4) 2015 earthquake (Table 2). For the two stations—SPAN and PONT—which operated prior, during and after the November 2015 event, a linear type of motion was observed for the pre-seismic period. There was no clear evidence of any kinematic changes or any alterations on the pattern of motion, in any of the three components, prior to the earthquake. Baseline changes between the two sites calculated for the pre-seismic period (2009 to 2015) indicated shortening of the distance, with a rate of V = (−2.50 ± 0.03) mm/year. The latter is indicative of the compressional status that occurred along the island during the pre-earthquake period.

The co-seismic displacements at the two sites (SPAN and PONT) were quite significant, mainly in the southern PONT station, which is located in the near vicinity of the epicenter (Appendix B). Static co-seismic displacements, which were smaller but not negligible, were also observed at stations located in Cephalonia (VLSM, KARA and SKAL), in Zakynthos (ZAKY) and in western Greece (AGRI) (Appendix B).

During the post-seismic period, the kinematic status of the area was characterized by a long relaxation process, expressed as a non-linear motion on the stations. This post-seismic kinematic behavior was observed at the two Lefkada stations, but mainly in PONT [90], as well as at sites of northern Cephalonia (FISK, ASSO). The slow relaxation was also evident at the two new sites that were established on the island, EXAN and DRAN (CRL GNSS net). The relaxation period on the PONT station was estimated to last up to August 2016, exhibiting a non-linear type of motion during that period. The estimation was performed with an average moving window of 5–60 days in length. The relaxation process was characterized by a significant southward motional component and a strong uplifted tension, following the significant co-seismic subsidence. For the subsequent period (from 2016 to 2021), the time series of the two stations showed an analogous motion to the pre-seismic period, but with slightly smaller horizontal amplitudes (Figure 8). Similar velocity vectors for the post-seismic period were also estimated for the EXAN and DRAN GNSS sites. Calculation of the baseline changes between SPAN and PONT sites after August 2016 shows that the shortening of the distance between these two sites remains, as was the case for the pre-seismic period, but with a slightly decreased amplitude (-1.81 ± 0.05 mm/year).

3.3. *Cephalonia Island*

Regarding the island of Cephalonia, most GNSS stations were established after the 2014 intense seismic activity. For the period prior to 2014, data were available only from the VLSM station. The time series of this station (Table 3) showed an ENE linear type of motion with small seasonal fluctuations, mainly in the north component, and a subsiding pattern (Figure 9).

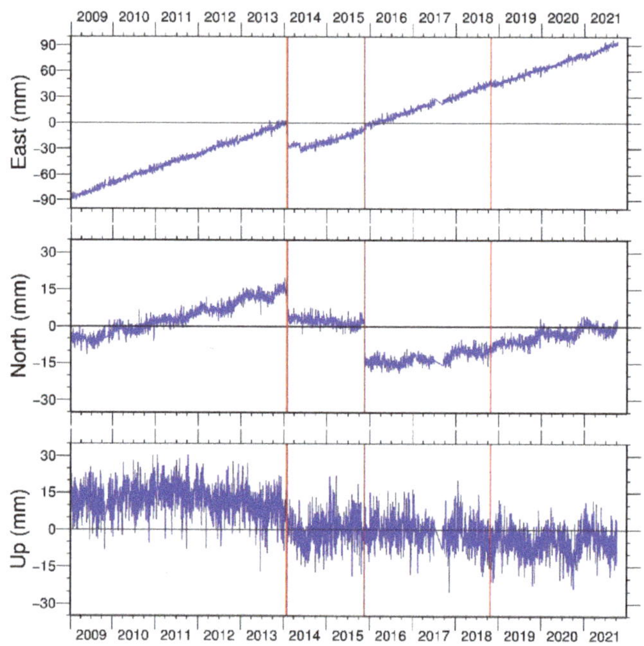

Figure 9. Time Series of VLSM Cephalonia GNSS station. Red lines mark the dates of the strong earthquakes in the Ionian Islands.

The two strong events of Mw6.1 in late January and Mw5.9 in early February 2014, recorded on the VLSM and KIPO stations, caused measurable static displacement (Appendix B). KIPO station, located on the western part of the Paliki peninsula, closer to the earthquakes' epicenters, revealed the most significant displacement. VLSM station, to the southeast of the epicentral area, situated on the limestone unit of central Aenos Mountain, exhibited smaller co-seismic motion. The co-seismic displacement vectors in these two GNSS sites indicated a right-lateral activated faulting zone [50]. Nevertheless, these events have not caused any displacement to GNSS stations located at the neighboring islands or in western Greece (Appendix B).

The GNSS velocity vectors for the sites located at Cephalonia Island after the seismic activity of 2014 exhibited a gradual rotation on the horizontal component from ENE in the north to ESE motion to the southern part of the island (SKAL station) (Figure 8). Most stations showed linear type of motion since 2014, with a subsiding vertical component (Table 3). The motional behavior of the VLSM station is, however, also worth noting. The time series of VLSM (Figure 9) show irregular motion, in both the north and vertical components, for the period from February 2014 to September 2016. During that period, the station shows a non-linear ESE horizontal motion and uplifting behavior. The station acquired the pre-seismic ENE and subsiding motional character in the subsequent period (from September 2016 to present). This long-lasting irregular behavior—which may be described as a post-seismic relaxation period and also encompassed the 2015 Lefkada earthquake—has not been observed at any other site in the Ionian Islands. Calculating for the post-seismic period (from 2017 to 2020) the baseline changes between the stations in Cephalonia, extension occurred between the northernmost (FISK) and the southernmost (SKAL) stations (V = 4.66 ± 0.10 mm/year). Meanwhile, in the central part of the island the distance between the ARGO and VLSM stations remained almost stable (V = −0.44 ± 0.04 mm/year), for a similar period (from September 2016 to October 2021).

3.4. *Zakynthos–Strofades Islands*

In the southern part of the Ionian Islands, GNSS data from three stations were processed (Table 4). For the sites of ZAKY and STRF, located in the city of Zakynthos and at the islet of Strofades, respectively, data were available prior to the strong 2018 earthquake (Mw6.7). Both stations exhibited a linear type of ground motion. ZAKY station revealed an ESE horizontal motion, with a negligible subsiding pattern, while STRF station showed an SE motion with lower amplitude (Figure 8). Baseline change calculation between these two sites for the pre-earthquake period showed a lengthening of the distance with an estimated velocity of V = 3.50 ± 0.10 mm/year. ZAKY station's pre-seismic kinematic behavior indicated shortening relative to the GNSS station PYRG, located in the western Peloponnese (V = −5.12 ± 0.09 mm/year). Likewise, the baseline distance between the STRF and PYRG stations decreased with a similar velocity, V = −4.52 ± 0.08 mm/year. These observations revealed a compressional regime between the Peloponnese and the South Ionian Islands, while extension occurred between Zakynthos and Strofades.

The 2018 earthquake, southwest of Zakynthos, caused strong ground displacement in both Zakynthos and Strofades. Regarding GNSS stations at the island of Cephalonia, significant ground displacement was recorded at the central ARGO and the southern SISS and SKAL stations (Appendix B), but not at the VLSM site. Moreover, static ground displacement was observed at stations located in the western (PYRG) and northwestern (PAT0) Peloponnese, even in the southwestern part of mainland Greece (KTCH).

After the occurrence of the mainshock, a new continuous GNSS station was established in the island (station ZNTE). The two sites that operated in Zakynthos after the Mw6.7 event, together with the PYRG station, revealed an almost eight-month-long post-seismic relaxation period. During this period, the kinematic behavior of ZAKY and PYRG stations showed a significant alteration of motion with respect to the pre-seismic one (Figure 10). A distinctive alteration of the east motional component was observed at PYRG station, where the eastward pre-seismic pattern reversed to a westward post-seismic motion. The post-seismic relaxation period was estimated to last up to early July 2019. During the following period (from July 2019 to the end of 2021), both ZAKY and PYRG stations again presented a pre-seismic motion pattern, with small alterations in the horizontal components (Figure 10). Similarly, the ZNTE station indicated a velocity vector significantly different compared to that of the post-seismic period, with a pattern of motion consistent to that of ZAKY.

3.5. *Neigboring Stations–Overall Image*

Together with the continuous GNSS stations in the Ionian Islands, stations located in western mainland Greece (AGRI, KTCH) and western and northwestern Peloponnese (PAT0, RLSO, PYRG) were also processed (Table 5). The goal was to study the velocity field, ground deformation and differential motions in the broad region of the Ionian Sea, focusing on the area close to the highly activated central Ionian Islands.

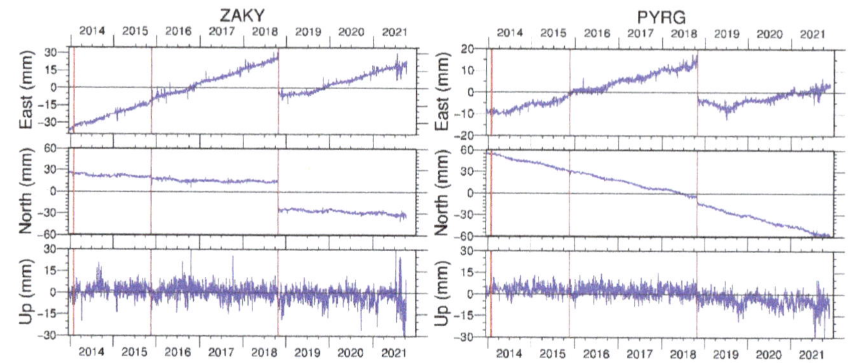

Figure 10. Time series for the two GNSS stations in Zakynthos (ZAKY) and the western Peloponnese (PYRG). Red lines are the same as Figure 9.

The deduced velocity field of the area shows the differentiation of the motion between western Greece and the Peloponnese, resulting in the opening of the Patras Gulf with an estimated rate of about 8.0 ± 0.5 mm/year. As previously described, the seismic activity on Cephalonia has not caused any displacement to the broad area, as co-seismic deformation was observed only at local sites. The Lefkada 2015 event affected only the northern AGRI station, located in western Greece. It was only the strong 2018 Zakynthos event that caused noticeable static displacement in both western Greece and the Peloponnese (Appendix B). The most distinct co- and post-seismic deformation was recorded at PYRG station. A small displacement was also observed at KTCH station (Figure S3). The strong events in the Ionian Islands did not appear to affect the velocity field of these stations, which exhibited a linear type of motion throughout the processed period. The only noticeable alterations were observed in the vertical component of PAT0 station (NW Peloponnese) (Figure S3). In this station, the vertical component shows a non-linear motion. Initially, up to the end of 2015, an uplift pattern was observed, which since the beginning of 2016 has reversed to a subsiding motion. A similar pattern also was observed at the neighboring RLSO site.

The overall image of the velocity field in the broad Ionian Islands area shows a gradual rotation of the horizontal velocity vectors (Figure 8). GNSS stations located to the north exhibited a NW horizontal motion, while towards the south, the horizontal vector component shifted to a SW motional direction. The latter has been combined with a decrease in the amplitude of the horizontal velocity from ~26 mm/year in the northern Ionian Islands, to ~13 mm/year in Zakynthos Island and ~7 mm/year in Strofades Islet, respectively. The transition from a northward to southward motion pattern occurs in Cephalonia Island. A similar rotation pattern is also observed at the stations located in western Greece and the NW Peloponnese, with the transition occurring between AGRI and KTCH stations, roughly at the same latitude as Cephalonia.

Analyzing the baseline-length changes for GNSS stations located on different Ionian Islands, as well as between stations in Ionian and western Greece, information about the tectonic status of the area can be extracted (Appendix C). In the northern Ionian Sea, the distance between Corfu and Paxoi has remained almost unchanged ($V_{KERK\text{-}PAXO} = -0.60 \pm 0.11$ mm/year). Intense lengthening was calculated between the northern islands (Corfu, Paxoi) and Lefkada (i.e., $V_{PAXO\text{-}SPAN} = 7.55 \pm 0.11$ mm/year). The latter indicates an extensional regime between the northern and the central Ionian Islands. Extension with a lower rate is also observed by the baseline changes between the stations located in the central and southern Ionian Islands. Fluctuations in the baseline velocity rates were detected before and after the strong seismic events that occurred in the area during the years 2014–2018. According to the baseline velocity between stations on the islands of Lefkada and Cephalonia, lengthening was taking place during the period before the 2014 seismic events at Cephalonia (i.e., $V_{PONT\text{-}VLSM} = 2.91 \pm 0.04$ mm/year). For the period following the 2015 Lefkada earthquake, up to late 2016, the observed intense shortening ($V_{PONT-VLSM} = -13.61 \pm 0.73$ mm/year) could be attributed to the post-seismic process. Re-establishment of the pre- 2014 lengthening character then gradually took place. Negative baseline velocity changes between the central Ionian Islands and stations located in western Greece and the Peloponnese imply compressional forces in this area which were not affected by the intense seismic activity. The latter is supported by the small variations on the velocity values before and after the strong earthquakes (Appendix C).

4. Discussion

Three major tectonic features in the Ionian Islands define the regional kinematic field and control the seismic activity: the convergence of the Apulian Platform and the Hellenic foreland in the north; the long NNE–SSW Cephalonia–Lefkada Transform Fault Zone offshore and west of the respective islands in the central part; and the northwestern tip of the Hellenic Arc in the southern Ionian Sea. These major structures, together with smaller local faults, create a complex tectonic environment generating intense seismic activity and strong ground deformation. In the present study, seismological and geodetic data from the area of the Ionian Islands were combined for the periods before, during and after the occurrence of strong earthquakes in 2014–2018. In the following, the results of the joint analysis are discussed in terms of the overall tectonic status of the study area.

4.1. North Ionian Sea

The northern Corfu and Paxoi Islands located at the Hellenic foreland present the largest amplitude regarding the horizontal velocity vectors (with respect to ITRF), but the recorded seismicity is by far the lowest in the region. The seismicity in this area is located mainly west of Corfu and Paxoi Islands, in mainland Greece and north of Corfu, close to the shores of Albania. The estimated horizontal velocities for the local GNSS stations match with the ones calculated by [68], although the datasets cover slightly different time periods. Considering that the horizontal velocity amplitude of the Apulian Platform is ~31.1 mm/year ($V_{East} = 25.1$ mm/year, $V_{North} = 18.4$ mm/year) [68], the Paxoi Island shows a similar horizontal amplitude of ~29.7 mm/year, while Corfu stations have smaller average velocity ~25.8 mm/year. The latter may indicate that, kinematically, Paxoi is controlled by the Apulian motion, as [68] has suggested, although the baseline change between stations KERK and PAXO is estimated to be rather small (about -0.6 ± 0.1 mm/year). The subsiding pattern observed in the southern Corfu and Paxoi stations (KERK, PAXO), though they could be attributed to local geological setting, may also reflect the continental collision that occurs in the area. The absence of significant seismic activity along the NNW–SSE-trending boundary of the Apulian Platform and Hellenic foreland—not just in the limited period of the present study but also during the past (Figure A3 in Appendix A)—indicates a nearly aseismic convergence process, mainly owed to the relatively low motion rate (~5.5 ± 0.7 mm/year; that is, the average converge velocity of the two Corfu stations towards Apulian Platform, similar to the velocity estimated by [89]).

4.2. Central Ionian Sea

The two strong earthquakes in early 2014 on Cephalonia Island triggered a long-lasting aftershock sequence, partially overlapping with that of the 2015 Lefkada mainshock. Prior to the 2014 events, the background seismicity occurred along the regional CLTFZ and onshore at the northern part of the island, likely associated with the local major faulting features. The baseline change between the VLSM station and the southern PONT site in Lefkada indicated that extension occurred between the two islands. The 2014 events occurred on a faulting system adjacent and parallel to the CLTFZ on the Paliki peninsula. Strong co-seismic deformation occurred mainly at the western part of the island, as has been detected from local GPS network measurements [50]. However, ground displacement was observed only on Cephalonia; none of the continuous GNSS stations in the near vicinity of the island, either on Lefkada or western Greece, showed any deformation, proving the local character of the activated seismogenic faulting zone. The post-seismic activity expanded mainly along Paliki, towards the north, but seismic clusters were also observed at the southern part of the activated zone. Seismicity in the area of the Myrtos Gulf indicated activation of structures transverse to the axis of the CLTFZ, oriented with local faults, e.g., the Agia Efimia fault (Figure 1), as the major thrust fault that separates the northern Erissos peninsula from the main island is called. It is important to notice that a similar cluster, on the northern part of Cephalonia, was also formed after the Mw6.3 2003 northern Lefkada earthquake [13,30]. The seismicity rate remained at a high level (above the background level) for the whole period until the occurrence of the November 2015 event on Lefkada. This long post-seismic period is correlated with the ground deformation observed on the GNSS VLSM station, where different pattern of motion was exhibited with respect to the pre-seismic era. Analytically, a southward motion with uplift tension occurred, and increased lengthening of the baseline distance between PONT and VLSM was observed.

The late 2015 Mw6.4 Lefkada earthquake was the second strong event to occur in less than two years in the central Ionian Islands. However, this earthquake occurred on the Lefkada segment of the major CLTFZ or a sub-parallel structure. The regional character of the event reflected on the ground deformation observed in the broad area, extending to Zakynthos Island to the south and to AGRI station in western Greece. The post-seismic activity expanded onshore of Lefkada Island, but also triggered clustered activity on the northern part of Cephalonia, in an area previously activated after the 2003 Lefkada event and, more recently, after the 2014 Cephalonia events. The increased seismicity rate (above the background level) after the strong earthquakes, marks a long relaxation period (up to October 2017), indicative of the activated area (Figures 4 and 5).

The recorded motion at the PONT GNSS site, near the epicentral area, as well as to other neighboring GNSS sites at Lefkada and Cephalonia, confirmed the long relaxation period after the earthquake [90]. Based on the geodetic data, the kinematic field of the area returned to its pre-seismic pattern during August–September 2016, for both Lefkada and Cephalonia. However, the amplitude of the velocity vector at the sites on Cephalonia and Lefkada show small but noticeable differentiation (~15%) with respect to the pre-seismic period from 2009 to 2014.

Regarding the possibility that the 2014 seismic sequence caused or triggered the 2015 earthquake, previous studies [48,91] claimed that the southern Lefkada event was a continuation of the seismic process of the 2003 earthquake in the northern part of CLTFZ. The seismological data in the present study reveal that after the occurrence of the 2014 earthquakes, seismicity propagated northwards, while it has been shown [48] that the 2014 main events caused Coulomb stress changes in the epicentral area of 2015, where the 2003 earthquake also induced stress transfer. The GNSS data, for the inter-seismic period from February 2014 to November 2015, showed motional changes on the VLSM station (southward pattern), while the PONT Lefkada site maintained its kinematic status. The latter resulted in increased velocity (lengthening character) of the baseline change between these two stations, doubled during that period (Appendix C). Assessing the spatiotemporal expansion of the 2014 Cephalonia seismic sequence, together with the geodetic evidence presented in this study, it may be argued that the occurrence of the anticipated event after the 2003 earthquake in southern Lefkada Island may have been accelerated—and even triggered—by the 2014 Cephalonia events. It is likely that the cause of the activity, in both the Cephalonia and Lefkada epicentral areas, was the 2003 event on the major CLTFZ structure [48], expressed as a local-scale activity in Cephalonia, and as a regional one in Lefkada.

4.3. South Ionian Sea

The last strong earthquake (Mw6.7) in the Ionian Sea occurred south of Zakynthos Island, three years after the major event at Lefkada. The source of this earthquake was located on the northwestern end of the active plate boundary between the eastern Mediterranean lithosphere and the Aegean one. Geodetic data from the broad area prior to the earthquake show extensional status between Cephalonia and Zakynthos, as well as between Zakynthos and Strofades. Compression occurred between Zakynthos–Strofades and the Peloponnese, as depicted by the baseline change between the GNSS sites (Appendix C). The earthquake was preceded by a number of intermediate events (~M5.0) in the near vicinity of the epicentral area. Detailed seismological and geodetic work by [58] has suggested slow-slip events in the area, with transient signals on the surface deformation. This earthquake resulted in the longest and most intense seismic sequence of the study period (Figure 6), accompanied by significant ground deformation and motional alterations in the affected area.

The Mw6.7 event caused strong ground deformation on a broad area, extending to Cephalonia Island, to Patras city (NW Peloponnese) and to western mainland Greece (KTCH GNSS station). The extent of the co-seismically deformed area highlights the regional character of this event. The post-seismic activity is concentrated offshore, north and east of the epicenter. It is worth noting the limited activity on the northern onshore part of the island. The latter observation, which may indicate a differentiation between the northern and southern parts of the island, has also been observed on the kinematic field detected by local GPS measurements [41].

Post-seismic evolution of the ground deformation suggests a significantly shorter period until the area returns to the pre-seismic kinematic pattern. This period extended up to July 2019, as revealed by the horizontal component of the three closer sites to the epicenter. The orientation of the seismic clusters on the northern extent of the Zakynthos sequence (group #4; Figure 2c) in a SW–NE-trending direction, coincide with the direction of the 2008 activated zone in NW Peloponnese [21]. This is an area where the geodetic data show a differential motion between the PAT0, RLSO and the PYRG GNSS stations.

4.4. The Broad Ionian Sea Area

The overall image of the seismic activity in the broad area of Ionian Islands is dominated by the three seismic sequences generated after the large-magnitude events ($M > 6$) in Cephalonia, Lefkada and Zakynthos Islands (Figures 2 and 3). The Lefkada sequence is mainly aligned and concentrated along the north Lefkada segment of the regional CLTFZ structure, expanding offshore and west of the northern part of Cephalonia. The 2014 Cephalonia sequence occurred on a local faulting zone, associated with the Cephalonia segment of the CLTFZ but not comprising part of it. It is distributed spatially on a wider zone, along the western Paliki peninsula, and extends offshore, west of the northern Erissos peninsula. These two sequences appear to be coupled to each other in both time and space and linked with the earlier 2003 Lefkada earthquake. The southern Zakynthos sequence also has regional characteristics and is spatially distributed in a region much larger than anticipated for the magnitude of the mainshock [29]. This is an area affected by the transition of the northwestern end of the Hellenic Arc subduction zone to the CLTFZ major faulting structure.

The geodetic data clearly depicted the kinematic pattern of the broad area, with respect to major tectonic formations and seismic activity (Figure 11). Comparing the regional velocity field on the northern Ionian Islands with the one of the central islands, a clear kinematic boundary emerges between these two areas. The northern islands show a uniform motional pattern, while the velocity field at the central and southern Ionian Sea shows a clockwise rotation around an axis located on Cephalonia Island (calculating the horizontal velocities with respect to VLSM station). Based on the results of this study, a rotation rate of 6.9 ± 2.1 degrees/Myr was estimated. This agrees with previous regional studies [92,93], as well as with local geodetic studies on Cephalonia [13]. Baseline changes between several GNSS sites reveal the extensional regime along the Ionian Islands, which is gradually reduced from north to south, while compression occurs between the central Ionian Islands, and western Greece and the Peloponnese. Extension is also observed between western Greece and the Peloponnese, as well as across the NW Peloponnese. The strain regime alternates the periods after the strong earthquakes, marking the post-seismic relaxation process. The spatial extension of the co-seismic displacements reflects the regional or local character of the activated area. The Lefkada and Zakynthos earthquakes caused static displacement in the broad area, while the Cephalonia events affected only the local GNSS sites.

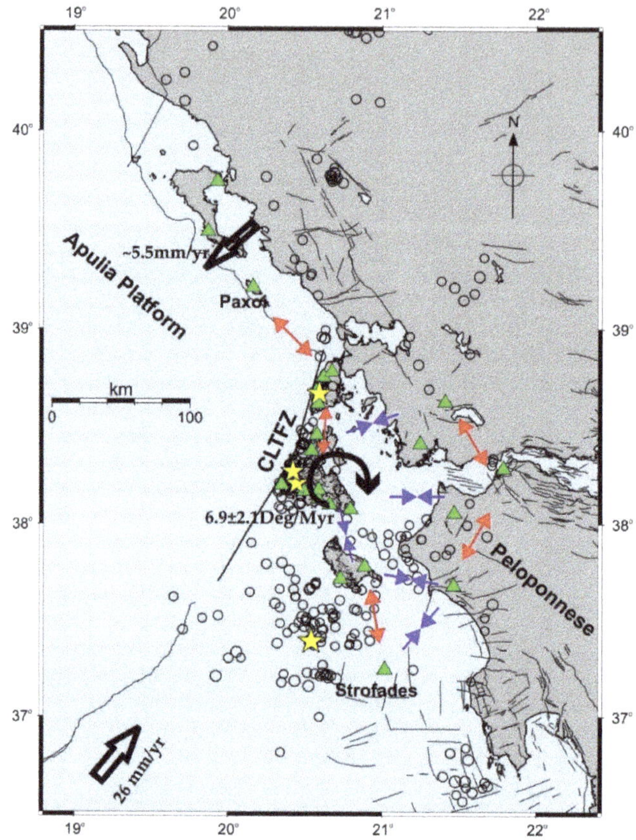

Figure 11. Overview of the Ionian Islands area showing the main tectonic structures together with seismic activity ($M > 4$) (black circles), and strong ($M > 6$) events (yellow stars). Black arrows show the relative Eurasia–Africa motion in the SW [94], and the motion of the Hellenic foreland with respect to the Apulian Platform (estimated herein) in the north. The extensional (red arrows) and compressional (blue arrows) pattern calculated based on baseline changes between GNSS stations (green triangles) is qualitatively presented.

The temporal and spatial joint analysis of the seismological and geodetic data revealed that the 2014 and 2015 seismic sequences are connected, while the Zakynthos one is part of the tectonic process in the western termination of the Hellenic subduction margin. The expansion of the seismic activity after the 2014 and 2015 events offshore of northern Cephalonia is the spatial link. In this area, previous work [52] identified a seismic gap, indicating that a segment of the CLTFZ has not ruptured yet and may generate a strong event in the future. The present analysis shows that seismicity in this region is clustered in a large number of small segments oriented almost in an E–W direction, transverse to the axis of the CLTFZ. The latter may indicate the transition between the Cephalonia and Lefkada segments of CLTFZ, and the increased seismicity may relieve the accumulated stress. Thus, the occurrence of a strong event in this region seems less likely, but cannot be excluded. Meanwhile, the area between Cephalonia and Zakynthos is characterized by low seismic activity. It is a region whose kinematic status is affected by the CLTFZ in the north and the subduction process in the south. The last major earthquake in the area dates to 1983, with an Mw6.8 earthquake [52] occurring offshore SW of Cephalonia and NW of Zakynthos. It may be argued that this is an area that has the potential to generate another strong event. However, the geodetic data show small changes in the baseline length between the SKAL GNSS site in southern Cephalonia and Zakynthos station, ZAKY (Appendix C), and a similar pattern of motion, in both direction and amplitude. Another area where low seismic activity was observed during the study period is Ithaca Island. Although the general seismic activity in this area is relatively low (e.g., Figure 2b), a few strong earthquakes have occurred in the past (Figure A3). Previous GPS campaign results [54] showed a differential motion between the northern and southern parts of the island, and compression between Cephalonia and Ithaca, compatible with the regional field [93]. The area is considered to have the potential to generate a significant earthquake in the future, taking into account the strong events in its vicinity, but more data are required to assess such a scenario.

The strong earthquakes and large co-seismic ground deformation recorded in the study area clearly exhibit the necessity for seismic hazard analysis in such a seismically active area. As expected, large earthquakes yield violent shaking that coincides well with the expected PGA from the probabilistic analysis. The latter confirms the definition of the SHARE seismic source zones. However, future studies in the area should focus on much smaller scales and identify potential hazards from specific active fault zones in an area as tectonically complex as the broad Ionian Sea. The process of seismogenesis, as described by statistical laws in any PSHA approach, should also be revised according to fault-specific knowledge. Finally, a detailed survey of local site conditions would be crucial in better refining PSHA results and offer greater accuracy of the estimations.

5. Conclusions

Seismological analysis and geodetic results from continuous GNSS stations were combined to study the spatiotemporal evolution of seismic activity and ground deformation in the Ionian Islands during the period from 2014 to 2018, characterized by the occurrence of strong earthquakes in Cephalonia, Lefkada and Zakynthos Islands.

Low seismicity was observed in the north Ionian Sea, where the collision between the Apulian platform and the Hellenic foreland is taking place. The velocity field of the area, deduced by stations on Corfu and Paxoi Islands, shows horizontal NW motion, transverse to the collision front. The vertical component reveals a subsiding pattern, compatible with the convergence process in the area.

The CLTFZ feature offshore of the central Ionian Islands of Cephalonia and Lefkada is the prevailing structure, generating the strong and intense seismic activity observed in the region. The 2014 Mw6.1 and Mw5.9 earthquakes occurred on an adjacent local faulting zone, causing strong ground displacement on the island but not in the broader area. The seismic sequence following these two events expanded in the vicinity of the activated area along the Paliki peninsula, and offshore northwards. Few months later, on 17 November 2015, another strong earthquake (Mw6.4) occurred on or parallel to the Lefkada segment of the CLTFZ. Co-seismic displacement was recorded in almost all the southern Ionian Islands, highlighting the regional character of the seismogenic source. The post-seismic activity expanded along the CLTFZ and southwards in the same area as the 2014 sequence. The temporal evolution of the post-seismic activity and the ground deformation show a long relaxation period. However, the motional pattern obtains its pre-seismic character earlier (August 2016) compared to the respective seismic excitation. The seismic activity reaches its background levels (0.1–0.2 events/day) after a longer period, between late 2016 and mid-2017. The two seismic sequences are linked in space and time, with the 2014 activity likely accelerating the 2015 event.

The October 2018 Mw6.7 Zakynthos earthquake occurred close to the northwestern tip of the Hellenic Arc. Strong ground displacement took place on the near-field GNSS stations, as well as on Cephalonia Island and the Peloponnese. The seismic sequence expanded in the vicinity of the epicentral area, but mainly offshore south and west of Zakynthos. The motion field of the stations in Zakynthos and the western Peloponnese showed significant alterations with respect to the pre-seismic period, on both north and east motional components, which regained its anticipated pattern about eight months after the mainshock. However, the seismicity rate has remained at higher values, compared to the background activity (0.2 events/day), for more than three years.

Probabilistic seismic hazard assessment was performed in the Ionian Islands to estimate PGA and PGV for a return period of 475 years. The smallest values were obtained, as anticipated, for the northern Ionian Islands, corelating with the low seismicity of the area. However, the calculated PGA values in Cephalonia and Ithaca are significantly higher and well exceed the ones proposed by the current Greek Building Code. Based on the herein-performed PSHA analysis, a finer, more detailed zonation of the area and re-evaluation of the PGA values provided by the Building Code must be considered in the future.

Supplementary Materials: The following supporting information can be downloaded at: https://www.mdpi.com/article/10.3390/app12052331/s1, Figure S1: Time series for station KERK on Corfu Island, and PAXO on Paxoi Island; Figure S2: Time series for station SPAN and PONT on the northern and southern part of Lefkada, respectively; Figure S3: Time series for station KTCH in Western Greece and PAT0 on the NW Peloponnese, in the city of Patras.

Author Contributions: Conceptualization, V.S. and G.K.; methodology, V.S., V.K., G.K. and I.S.; software, V.S., V.K. and I.S.; validation, V.S., V.K., G.K. and S.M.; formal analysis, V.S., V.K., G.K. and I.S.; investigation, M.D., J.D.A., D.K.-F., I.K., S.D., E.V. and E.K.; resources, G.T., E.L. and N.V.; data curation, V.S., V.K. and G.K.; writing—original draft preparation, V.S., V.K., G.K., I.S. and S.M.; writing—review and editing, V.K., G.K., I.S., S.M., J.D.A., I.K., S.D., E.V. and E.K.; visualization, V.S., V.K., I.S. and S.M.; supervision, G.K., J.D.A., I.K., E.V., G.T., E.L. and N.V.; project administration, N.V.; funding acquisition, N.V. All authors have read and agreed to the published version of the manuscript.

Funding: This research was funded by the project "Telemachus Innovative Seismic Risk Management Operational System of the Ionian Islands" (MIS 5007986) which is part of the Regional Operational Programme «Ionian Islands 2014 2020» and is co-financed by the European Regional Development Fund (ERDF) (National Strategic Reference Framework NSRF 2014 20).

Informed Consent Statement: Not applicable.

Data Availability Statement: Focal mechanisms presented in Figure 1a are compiled from various sources, including [5,95–106]. Seismological data from GI-NOA and SL-NKUA are available at https://bbnet.gein.noa.gr (accessed on 18 January 2022) and http://www.geophysics.geol.uoa.gr/stations/gmapv3_db/ (accessed on 18 January 2022), respectively. GNSS data from NOA-network and for station SISS (NKUA-network) are available at http://geodesy.gein.noa.gr:8000/nginfo/ (accessed on 18 January 2022).

Acknowledgments: The authors would like to thank the personnel of the Hellenic Unified Seismological Network (http://eida.gein.noa.gr/; accessed on 18 January 2022) for the installation and operation of the seismological stations used in the current study. Continuous GNSS data were also provided by METRICA SA (HexagonSmartNet). Some of the figures were made using the Generic Mapping Tools software [107].

Conflicts of Interest: The authors declare no conflict of interest.

Appendix A

Close-up diagrams of the spatiotemporal distribution in the areas of Cephalonia–Lefkada Islands and Zakynthos Island.

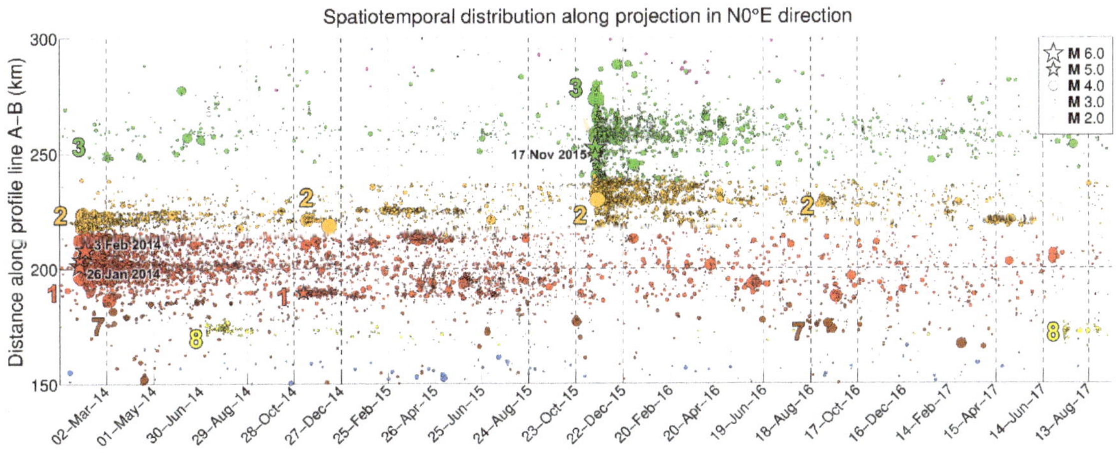

Figure A1. Close-up of Figure 3b in the region of Cephalonia–Lefkada between January 2014 and 12 September 2017.

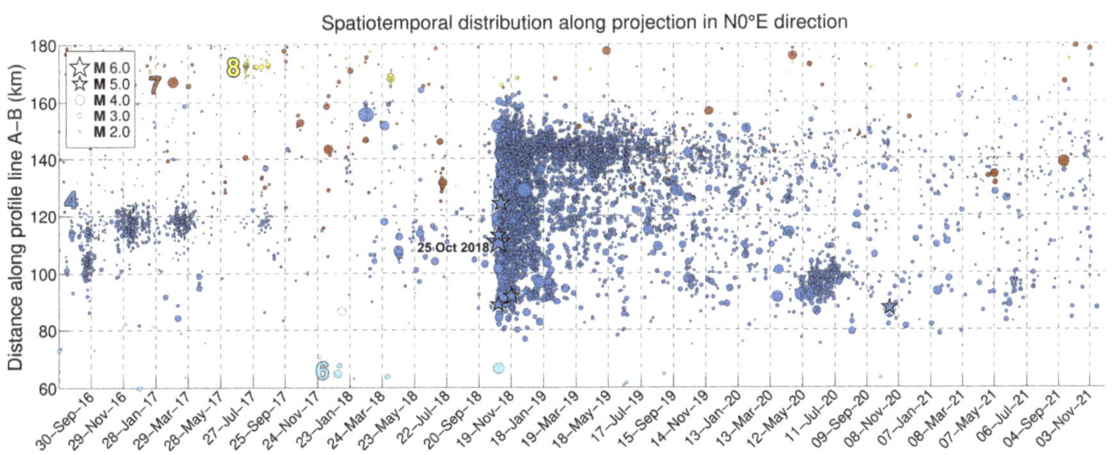

Figure A2. Close-up of Figure 3b in the region of Zakynthos between August 2016 and November 2021.

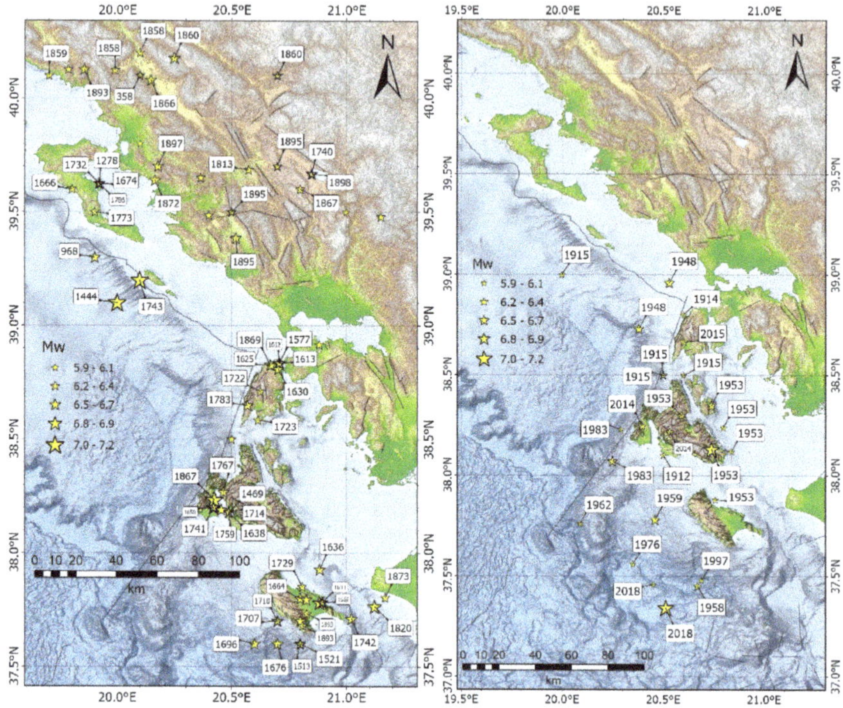

Figure A3. (**Left**) Historic seismicity [8,9] and (**right**) significant earthquakes ($M_w \geq 5.9$) during the instrumental era (1900–present) (from [28] and the databases of GI-NOA and SL-NKUA) in the broad region of the Ionian Islands.

Appendix B

The co-seismic displacements on the continuous GNSS stations processed in this study are presented in Table A1. The co-seismic displacement and its error were estimated based on the average value of each coordinate seven days prior to the earthquake and one to seven days after its occurrence, depending on the proximity of each site to the epicentral area. Results are presented only for the stations where the co-seismic displacement was larger than its error, otherwise it is considered that no displacement occurred.

Table A1. Co-seismic displacements on the continuous GNSS stations.

	Site	Component	Cephalonia Eqs January–February 2014 Mw6.1 & 5.9	Lefkada Eq November 2015 Mw6.4	Zakynthos Eq October 2018 Mw6.6
Lefkada Island	SPAN	DEast (mm) DNorth (mm) Dup (mm)	No displacement	−76.90 ± 2.47 −56.44 ± 2.62 −2.83 ± 5.53	No displacement
	PONT	DEast (mm) DNorth (mm) Dup (mm)	No displacement	−218.01 ± 1.93 −370.07 ± 2.85 −55.71 ± 6.24	No displacement
Cephalonia Island	KIPO	DEast (mm) DNorth (mm) Dup (mm)	1.89 ± 2.42 73.91 ± 2.42 55.23 ± 6.19	No data	
	VLSM	DEast (mm) DNorth (mm) Dup (mm)	−25.49 ± 2.22 −12.01 ± 2.14 −7.07 ± 5.19	4.17 ± 1.36 −16.61 ± 1.43 2.67 ± 5.44	No displacement
	ARGO	DEast (mm) DNorth (mm) Dup (mm)	No data		−2.34 ± 0.93 0.91 ± 1.91 −3.77 ± 5.00
	KARA	DEast (mm) DNorth (mm) Dup (mm)	No data	1.37 ± 1.54 −12.81 ± 1.57 −3.14 ± 6.78	No data
	SISS	DEast (mm) DNorth (mm) Dup (mm)	No data		−4.22 ± 2.65 −2.28 ± 2.44 0.62 ± 6.88
	SKAL	DEast (mm) DNorth (mm) Dup (mm)	No data	4.12 ± 1.58 −9.17 ± 1.62 2.27 ± 7.59	−4.55 ± 1.57 −4.16 ± 2.00 −2.91 ± 6.59
Zakynthos –Strofades	ZAKY	DEast (mm) DNorth (mm) Dup (mm)	No data	2.43 ± 3.49 5.20 ± 2.78 −0.60 ± 3.83	−30.86 ± 1.34 −38.68 ± 1.69 −4.12 ± 3.14
	STRF	DEast (mm) DNorth (mm) Dup (mm)	No data		22.21 ± 1.05 −42.47 ± 0.63 5.03 ± 2.84
Western Greece –Peloponnese	AGRI	DEast (mm) DNorth (mm) Dup (mm)	No displacement	−6.70 ± 2.15 0.56 ± 1.80 3.94 ± 4.75	No displacement
	KTCH	DEast (mm) DNorth (mm) Dup (mm)	No displacement	No displacement	3.70 ± 1.40 −5.63 ± 1.48 1.20 ± 5.63
	PAT0	DEast (mm) DNorth (mm) Dup (mm)	No displacement	No displacement	−4.05 ± 1.60 −1.43 ± 1.29 −2.73 ± 5.29
	PYRG	DEast (mm) DNorth (mm) Dup (mm)	No displacement	No displacement	−17.24 ± 0.88 −7.46 ± 0.95 −2.04 ± 3.93

Appendix C

The baseline change between two stations was estimated based on the daily coordinates of the selected sites. Baselines were formed only when data were available for both sites and for a period exceeding 24 h. The following Table A2 lists some characteristic formed baselines between the Ionian Islands, as well as between GNSS stations in Ionian Sea and western Greece and the Peloponnese.

Table A2. Baseline velocity between continuous GNSS stations.

Baseline	Period	Velocity (mm/Year)
KASI–SPAN	1 January 2013–15 November 2015	8.00 ± 0.10
	18 November 2015–6 May 2017	11.97 ± 0.22
KERK–PAXO	1 November 2019–31 October 2021	−0.60 ± 0.11
KERK–SPAN	1 March 2017–31 October 2021	8.26 ± 0.04
PAXO–SPAN	1 November 2019–31 October 2021	7.55 ± 0.11
PONT–VLSM	1 January 2009–25 January 2014	2.91 ± 0.04
	8 February 2014–16 November 2015	6.21 ± 0.18
	18 November 2015–4 September 2016	−13.61 ± 0.73
	5 September 2016–31 October 2021	1.64 ± 0.05
VLSM–ZAKY	8 February 2014–16 November 2015	2.67 ± 0.16
	17 November 2015–4 September 2016	−1.05 ± 0.465
	5 September 2016–25 October 2018	0.42 ± 0.10
	27 October 2018–31 October 2021	1.82 ± 0.07
SKAL–ZAKY	4 March 2015–16 November 2015	−0.72 ± 0.51
	18 November 2015–25 October 2018	−1.31 ± 0.08
	27 October 2018–13 August 2021	0.32 ± 0.08
ZAKY–STRF	6 October 2016–25 October 2018	3.46 ± 0.09
STRF–PYRG	6 October 2016–25 October 2018	−4.52 ± 0.07
SPAN–AGRI	5 January 2011–16 November 2015	−8.65 ± 0.04
	18 November 2015–28 April 2018	−5.46 ± 0.14
SPAN–KTCH	20 November 2013–16 November 2015	−0.97 ± 0.15
	18 November 2015–31 October 2021	0.51 ± 0.02
PONT–KTCH	23 November 2013–15 November 2015	−3.29 ± 0.22
	17 November 2015–31 October 2021	0.28 ± 0.04
VLSM–KTCH	22 November 2013–31 October 2021	−4.24 ± 0.05
VLSM–PAT0	27 January 2009–25 January 2014	−10.41 ± 0.05
	8 February 2014–31 October 2021	−10.60 ± 0.03
VLSM–RLSO	1 January 2009–25 01 2014	−5.32 ± 0.08
	8 February 2014–31 October 2021	−5.68 ± 0.03
ZAKY–PYRG	2 February 2013–25 October 2018	−5.12 ± 0.03
	27 October 2018–5 July 2019	−0.91 ± 0.18
	6 July 2019–31 October 2021	−4.47 ± 0.09
ZAKY–RLSO	8 February 2014–27 February 2017	−5.54 ± 0.11
	28 02 2019–31 October 2021	−4.93 ± 0.11
PAT0–AGRI	5 January 2011–12 February 2019	7.62 ± 0.02
PAT0–KTCH	20 November 2013–31 October 2021	−5.91 ± 0.02
PAT0–PYRG	2 December 2013–25 October 2018	6.29 ± 0.04
	27 October 2018–31 October 2021	8.66 ± 0.08

References

1. Anzidei, M.; Baldi, P.; Casula, G.; Crespi, M.; Riguzzi, F. Repeated GPS surveys across the Ionian Sea: Evidence of crustal deformations. *Geophys. J. Int.* **1996**, *127*, 257–267. [CrossRef]
2. Hollenstein, C.H.; Geiger, A.; Kahle, H.-G.; Veis, G. CGPS time-series and trajectories of crustal motion along the West Hel-lenic. *Arc. Geophys. J. Int.* **2006**, *164*, 182–191. [CrossRef]
3. Hollenstein, C.; Müller, M.; Geiger, A.; Kahle, H.-G. Crustal motion and deformation in Greece from a decade of GPS measurements, 1993–2003. *Tectonophys* **2008**, *449*, 17–40. [CrossRef]
4. Underhill, R. Triassic evaporites and Plio-Quaternary diapirism in western Greece. *J. Geol. Soc. Lond.* **1988**, *145*, 26–282. [CrossRef]
5. Louvari, E.; Kiratzi, A.A.; Papazachos, B.C. The Cephalonia Transform Fault and its extension to western Lefkada Island (Greece). *Tectonophys* **1999**, *308*, 223–236. [CrossRef]

6. Sachpazi, M.; Hirn, A.; Clément, C.; Haslinger, F.; Laigle, M.; Kissling, E.; Charvis, P.; Hello, Y.; Lépine, J.-C.; Sapin, M.; et al. Western Hellenic subduction and Cephalonia Transform: Local earthquakes and plate transport and strain. *Tectonophys* **2000**, *319*, 301–319. [CrossRef]
7. Kokinou, E.; Kamberis, E.; Vafidis, A.; Monopolis, D.; Ananiadis, G.; Zelilidis, A. Deep seismic reflection data from offshore western greece: A new crustal model for the ionian sea. *J. Pet. Geol.* **2005**, *28*, 185–202. [CrossRef]
8. Stucchi, M.; Rovida, A.; Capera, A.A.G.; Alexandre, P.; Camelbeeck, T.; Demircioglu, M.B.; Gasperini, P.; Kouskouna, V.; Musson, R.M.W.; Radulian, M.; et al. The SHARE European Earthquake Catalogue (SHEEC) 1000–1899. *J. Seismol.* **2013**, *17*, 523–544. [CrossRef]
9. Papazachos, B.C.; Papazachou, C.B. *The Earthquakes of Greece*; Ziti Editions: Thessaloniki, Greece, 2003; 286p.
10. Kapetanidis, V.; Kassaras, I. Contemporary crustal stress of the Greek region deduced from earthquake focal mechanisms. *J. Geodyn.* **2019**, *123*, 55–82. [CrossRef]
11. Lekkas, E.L.; Danamos, G.; Mavrikas, G. Geological Structure and Evolution of Kefallonia and Ithaki islands. *Bull. Geol. Soc. Greece* **2001**, *34*, 11. [CrossRef]
12. Lekkas, E.L.; Danamos, G.; Lozios, S.G. Neotectonic structure and evolution of Lefkada island. *Bull. Geol. Soc. Greece* **2001**, *34*, 157. [CrossRef]
13. Lagios, E.; Sakkas, V.; Papadimitriou, P.; Parcharidis, I.; Damiata, B.N.; Chousianitis, K.; Vassilopoulou, S. Crustal defor-mation in the Central Ionian Islands (Greece): Results from DGPS and DInSAR analyses (1995–2006). *Tectonophysics* **2007**, *444*, 119–145. [CrossRef]
14. Lekkas, E.L.; Mavroulis, S.D. Earthquake environmental effects and ESI 2007 seismic intensities of the early 2014 Cephalonia (Ionian Sea, western Greece) earthquakes (January 26 and February 3, Mw 6.0). *Nat. Hazards* **2015**, *78*, 1517–1544. [CrossRef]
15. Lekkas, E.; Mavroulis, S.; Carydis, P.; Alexoudi, V. The 17 November 2015 Mw 6.4 Lefkas (Ionian Sea, Western Greece) Earthquake: Impact on Environment and Buildings. *Geotech. Geol. Eng.* **2018**, *36*, 2109–2142. [CrossRef]
16. Mavroulis, S.; Stanota, E.-S.; Lekkas, E. Evaluation of environmental seismic intensities of all known historical and recent earthquakes felt in Zakynthos Island, Greece using the Environmental Seismic Intensity (ESI 2007) scale. *Quat. Int.* **2019**, *532*, 1–22. [CrossRef]
17. Rondoyanni, T.; Sakellariou, M.; Baskoutas, J.; Christodoulou, N. Evaluation of active faulting and earthquake secondary effects in Lefkada Island, Ionian Sea, Greece: An overview. *Nat. Hazards* **2012**, *61*, 843–860. [CrossRef]
18. Ganas, A. NOAFAULTS KMZ layer Version 3.0 (2020 update) (V3.0) [Data set]. *Zenodo* **2020**. [CrossRef]
19. Lekkas, E.; Diakakis, M.; Mavroulis, S.; Kotsi, E. *Work Package 1.3: Compilation of Neotectonic Maps. Report, "Telemachus—Innovative Operational Seismic Risk Management System of the Ionian Islands" Project, Priority Axis "Environmental Protection and Sustainable Development" of the Operational Program "Ionian Islands 2014–2020"*; National and Kapodistrian University of Athens: Athens, Greece, 2020; p. 150.
20. Makris, J.; Papoulia, J. The backstop between the Mediterranean Ridge and western Peloponnese, Greece: Its crust and tec-tonization. An active seismic experiment with ocean bottom seismographs. *Boll. Geofis. Teor. Appl.* **2014**, *55*, 249–279.
21. Ganas, A.; Serpelloni, E.; Drakatos, G.; Kolligri, M.; Adamis, I.; Tsimi, C.; Batsi, E. The Mw 6.4 SW-Achaia (Western Greece) Earthquake of 8 June 2008: Seismological, Field, GPS Observations, and Stress Modeling. *J. Earthq. Eng.* **2009**, *13*, 1101–1124. [CrossRef]
22. Karakostas, V.; Mirek, K.; Mesimeri, M.; Papadimitriou, E.; Mirek, J. The Aftershock Sequence of the 2008 Achaia, Greece, Earthquake: Joint Analysis of Seismicity Relocation and Persistent Scatterers Interferometry. *Pure Appl. Geophys.* **2017**, *174*, 151–176. [CrossRef]
23. Mavroulis, S.D.; Fountoulis, I.G.; Skourtsos, E.N.; Lekkas, E.L.; Papanikolaou, I.D. Seismic intensity assignments for the 2008 Andravida (NW Peloponnese, Greece) strike-slip event (June 8, Mw = 6.4) based on the application of the Environmental Seismic Intensity scale (ESI 2007) and the European Macroseismic scale (EMS-98). *Ann. Geophys.* **2014**, *56*, 0681. [CrossRef]
24. Camera, L.; Mascle, J.; Wardell, N.; Accettella, D. The SEAHELLARC Team. The Peloponnese continental margin from Zak-ynthos Island to Pylos: Morphology and recent sedimentary processes. *Boll. Geofis. Teor. Appl.* **2014**, *55*, 325–342.
25. Wardell, N.; Camera, L.; Mascle, J.; Nicolich, R.; Marchi, M.; Barison, E. The structural framework of the Peloponnese conti-nental margin from Zakynthos to Pylos from seismic reflection and morpho-bathymetric data. *Boll. Geofis. Teor. Appl.* **2014**, *55*, 343–367.
26. Kokinou, E.; Papadimitriou, E.; Karakostas, V.; Kamberis, E.; Vallianatos, F. The Kefalonia Transform Zone (offshore Western Greece) with special emphasis to its prolongation towards the Ionian Abyssal Plain. *Mar. Geophys. Res.* **2006**, *27*, 241–252. [CrossRef]
27. SEAHELLARC Working Group. A new seismogenic model for the Kyparissiakos Gulf and western peloponnese (SW Hellenic Arc). *Boll. Geofis. Teor. Appl.* **2014**, *55*, 405–432.
28. Makropoulos, K.; Kaviris, G.; Kouskouna, V. An updated and extended earthquake catalogue for Greece and adjacent areas since 1900. *Nat. Hazards Earth Syst. Sci.* **2012**, *12*, 1425–1430. [CrossRef]
29. Papadimitriou, P.; Kapetanidis, V.; Karakonstantis, A.; Spingos, I.; Pavlou, K.; Kaviris, G.; Kassaras, I.; Sakkas, V.; Voulgaris, N. The 25 October 2018 Zakynthos (Greece) earthquake: Seismic activity at the transition between a transform fault and a subduction zone. *Geophys. J. Int.* **2021**, *225*, 15–36. [CrossRef]
30. Papadimitriou, P.; Kaviris, G.; Makropoulos, K. The MW=6.3 2003 Lefkada earthquake (Greece) and induced stress transfer changes. *Tectonophys* **2006**, *423*, 73–82. [CrossRef]

31. Kassaras, I.; Kalantoni, D.; Benetatos, C.; Kaviris, G.; Michalaki, K.; Sakellariou, N.; Makropoulos, K. Seismic damage sce-narios in Lefkas old town (W. Greece). *Bull. Earth. Engin.* **2015**, *13*, 3669–3711. [CrossRef]
32. Papadimitriou, P.; Karakonstantis, A.; Kapetanidis, V.; Bozionelos, G.; Kaviris, G.; Voulgaris, N. The 2015 Lefkada earth-quake sequence (W. Greece). In Proceedings of the 35th General Assembly of the European Seismological Commission, Trieste, Italy, 4–10 September 2016; p. ESC2016-139.
33. Margaris, B.; Papaioannou, C.; Theodulidis, N.; Savvaidis, A.; Anastasiadis, A.; Klimis, N.; Makra, K.; Demosthenous, M.; Karakostas, C.; Lekidis, V.; et al. *Preliminary Observations on the 14 August 2003 Lefkada Island (Western Greece) Earth-Quake. EERI Special Earthquake Report; Joint Report by Institute of Engineering Seismology and Earthquake Engineering, National Technical University of Athens and University of Athens*; EERI: Oakland, CA, USA, 2003; pp. 1–12.
34. Mavroulis, S.; Lekkas, E. Revisiting the Most Destructive Earthquake Sequence in the Recent History of Greece: Environmental Effects Induced by the 9, 11 and 12 August 1953 Ionian Sea Earthquakes. *Appl. Sci.* **2021**, *11*, 8429. [CrossRef]
35. Lekkas, E.; Danamos, G.; Antoniou, V.; Bosinakou, G.; Vassilakis, E. *Neotectonic Map of Greece, Kerkyra Island sheet 1:100,000 scale (in Greek)*; Applied Research Project, National and Kapodistrian University of Athens: Athens, Greece, 1995.
36. Caputo, R. The Neogenic dextral trascurrent system of Corfu (Central Mediterranean). *Ann. Geol. Pays Hell.* **1988**, *33*, 327–335.
37. Guidoboni, E.; Ferrari, G.; Mariotti, D.; Comastri, A.; Tarabusi, G.; Valensise, G. CFTI4Med, Catalogue of Strong Earthquakes in Italy (461 B.C.-1997) and Mediterranean Area (760 B.C.-1500). INGV-SGA. 2007. Available online: https://www.researchgate.net/publication/271078106_CFTI4Med_Catalogue_of_strong_earthquakes_in_Italy_461_BC-1997_and_Mediterranean_area_760_BC-1500 (accessed on 17 January 2022).
38. Galli, P.; Naso, G. The "taranta" effect of the 1743 earthquake in Salento (Apulia, southern Italy). *Boll. Geofis. Teor. Appl.* **2008**, *49*, 177–204.
39. Nappi, R.; Gaudiosi, G.; Alessio, G.; De Lucia, M.; Porfido, S. The environmental effects of the 1743 Salento earthquake (Apulia, southern Italy): A contribution to seismic hazard assessment of the Salento Peninsula. *Nat. Hazards* **2017**, *86*, 295–324. [CrossRef]
40. Lagios, E.; Papadimitriou, P.; Novali, F.; Sakkas, V.; Fumagalli, A.; Vlachou, K.; Del Conte, S. Combined Seismicity Pattern Analysis, DGPS and PSInSAR studies in the broader area of Cephalonia (Greece). *Tectonophys* **2012**, *524-525*, 43–58. [CrossRef]
41. Sakkas, V.; Novali, F.; Lagios, E.; Vassilopoulou, S.; Damiata, B.N.; Fumagalli, A. Ground deformation of Zakynthos Island (Western Greece) observed by PSI and DGPS. In Proceedings of the 2014 IEEE Geoscience and Remote Sensing Symposium, Quebec City, QC, Canada, 13–18 July 2014; pp. 4792–4795.
42. Ganas, A.; Drakatos, G.; Rontogianni, S.; Tsimi, C.; Petrou, P.; Papanikolaou, M.; Argyrakis, P.; Boukouras, K.; Melis, N.; Stavrakakis, G. NOANET: The new permanent GPS network for Geodynamics in Greece. *Geophys. Res. Abs.* **2008**, *10*, EGU2008-A-04380.
43. Chousianitis, K.; Papanikolaou, X.; Drakatos, G.; Tselentis, G.-A. NOANET: A Continuously Operating GNSS Network for Solid-Earth Sciences in Greece. *Seism. Res. Lett.* **2021**, *92*, 2050–2064. [CrossRef]
44. Karakostas, V.; Papadimitriou, E.; Mesimeri, M.; Gkarlaouni, C.; Paradisopoulou, P. The 2014 Kefalonia Doublet (MW6.1 and MW6.0), Central Ionian Islands, Greece: Seismotectonic Implications along the Kefalonia Transform Fault Zone. *Acta Geophys.* **2015**, *63*, 1–16. [CrossRef]
45. Karastathis, V.K.; Mouzakiotis, E.; Ganas, A.; Papadopoulos, G.A. High-precision relocation of seismic sequences above a dipping Moho: The case of the January–February 2014 seismic sequence on Cephalonia island (Greece). *Solid Earth* **2015**, *6*, 173–184. [CrossRef]
46. Papadopoulos, G.; Karastathis, V.K.; Koukouvelas, I.; Sachpazi, M.; Baskoutas, I.; Chouliaras, G.; Agalos, A.; Daskalaki, E.; Minadakis, G.; Moshou, A.; et al. The Cephalonia, Ionian Sea (Greece), sequence of strong earthquakes of January–February 2014: A first report. *Res. Geophys.* **2014**, *4*. [CrossRef]
47. Sokos, E.; Kiratzi, A.; Gallovič, F.; Zahradnik, J.; Serpetsidaki, A.; Plicka, V.; Janský, J.; Kostelecký, J.; Tselentis, A. Rupture process of the 2014 Cephalonia, Greece, earthquake doublet (Mw6) as inferred from regional and local seismic data. *Tectonophysics* **2015**, *656*, 131–141. [CrossRef]
48. Papadimitriou, E.; Karakostas, V.; Mesimeri, M.; Chouliaras, G.; Kourouklas, C. The Mw6.5 17 November 2015 Lefkada (Greece) Earthquake: Structural Interpretation by Means of the Aftershock Analysis. *Pure Appl. Geophys.* **2017**, *174*, 3869–3888. [CrossRef]
49. Sokos, E.; Gallovič, F.; Evangelidis, C.P.; Serpetsidaki, A.; Plicka, V.; Kostelecký, J.; Zahradník, J. The 2018 Mw 6.8 Zakynthos, Greece, earthquake: Dominant strike-slip faulting near subducting slab. *Seismol. Res. Lett.* **2020**, *91*, 721–732. [CrossRef]
50. Sakkas, V.; Lagios, E. Fault modelling of the early-2014 ~M6 Earthquakes in Cephalonia Island (W. Greece) based on GPS measurements. *Tectonophys* **2015**, *644-645*, 184–196. [CrossRef]
51. Chousianitis, K.; Konca, A.O.; Tselentis, G.-A.; Papadopoulos, G.A.; Gianniou, M. Slip model of the 17 November 2015 Mw = 6.5 Lefkada earthquake from the joint inversion of geodetic and seismic data. *Geophys. Res. Lett.* **2016**, *43*, 7973–7981. [CrossRef]
52. Ganas, A.; Elias, P.; Bozionelos, G.; Papathanassiou, G.; Avallone, A.; Papastergios, A.; Valkaniotis, S.; Parcharidis, I.; Briole, P. Coseismic deformation, field observations and seismic fault of the 17 November 2015 M = 6.5, Lefkada Island, Greece earthquake. *Tectonophys* **2016**, *687*, 210–222. [CrossRef]
53. Boncori, J.P.M.; Papoutsis, I.; Pezzo, G.; Tolomei, C.; Atzori, S.; Ganas, A.; Karastathis, V.; Salvi, S.; Kontoes, C.; Antonioli, A. The February 2014 Cephalonia Earthquake (Greece): 3D Deformation Field and Source Modeling from Multiple SAR Techniques. *Seism. Res. Lett.* **2015**, *86*, 124–137. [CrossRef]

54. Sakkas, V.; Lagios, E. Ground deformation effects from the ~M6 earthquakes (2014–2015) on Cephalonia–Ithaca Islands (Western Greece) deduced by GPS observations. *Acta Geophys.* **2017**, *65*, 207–222. [CrossRef]
55. Saltogianni, V.; Taymaz, T.; Yolsal-Çevikbilen, S.; Eken, T.; Moschas, F.; Stiros, S. Fault model for the 2015 Leucas (Aegean arc) earthquake: Analysis based on seismological and geodetic observations. *Bull. Seismol. Soc. Am.* **2017**, *107*, 433–444. [CrossRef]
56. Saltogianni, V.; Moschas, F.; Stiros, S. The 2014 Cephalonia Earthquakes: Finite Fault Modeling, Fault Segmentation, Shear and Thrusting at the NW Aegean Arc (Greece). *Pure Appl. Geophys.* **2018**, *175*, 4145–4164. [CrossRef]
57. Ganas, A.; Briole, P.; Bozionelos, G.; Barberopoulou, A.; Elias, P.; Tsironi, V.; Valkaniotis, S.; Moshou, A.; Mintourakis, L. The 25 October 2018 Mw=6.7 Zakynthos earthquake (Ionian Sea, Greece): A low-angle fault model based on GNSS data, relocated seismicity, small tsunami and im-plications for the seismic hazard in the west Hellenic Arc. *J. Geodyn.* **2020**, *137*, 101731.
58. Mouslopoulou, V.; Bocchini, G.M.; Cesca, S.; Saltogianni, V.; Bedford, J.; Petersen, G.; Gianniou, M.; Oncken, O. Earthquake swarms, slow slip and fault interactions at the western-end of the Hellenic subduction system precede the moment Mw 6.9 Zakynthos earthquake, Greece. *Geochem. Geophys. Geosystems* **2020**, *21*, e2020GC009243. [CrossRef]
59. Chousianitis, K.; Konca, A.O. Intraslab Deformation and Rupture of the Entire Subducting Crust During the 25 October 2018 Mw 6.8 Zakynthos Earthquake. *Geophys. Res. Lett.* **2019**, *46*, 14358–14367. [CrossRef]
60. Valkaniotis, S.; Ganas, A.; Papathanassiou, G.; Papanikolaou, M. Field observations of geological effects triggered by the January–February 2014 Cephalonia (Ionian Sea, Greece) earthquakes. *Tectonophysics* **2014**, *630*, 150–157. [CrossRef]
61. Kouskouna, V.; Makropoulos, K. Historical earthquake investigations in Greece. *Ann. Geophys.* **2004**, *47*, 723–731. [CrossRef]
62. EAK. Greek Seismic Code edited by: Earthquake Planning & Protection Organization. *Athens Greece* **2003**, *72*, 7. (In Greek)
63. Evangelidis, C.P.; Triantafyllis, N.; Samios, M.; Boukouras, K.; Kontakos, K.; Ktenidou, O.-J.; Fountoulakis, I.; Kalogeras, I.; Melis, N.S.; Galanis, O.; et al. Seismic Waveform Data from Greece and Cyprus: Integration, Archival, and Open Access. *Seism. Res. Lett.* **2021**, *92*, 1672–1684. [CrossRef]
64. Papadimitriou, P.; Voulgaris, N.; Kouskouna, V.; Kassaras, I.; Kaviris, G.; Pavlou, K.; Karakonstantis, A.; Bozionelos, G.; Kapetanidis, V. The Kefallinia Island earthquake sequence January–February 2014. In Proceedings of the Second European Conference on Earthquake Engineering and Seismology (2ECEES), Istanbul, Turkey, 24–29 August 2014.
65. Kapetanidis, V. Spatiotemporal Patterns of Microseismicity for the Identification of Active Fault Structures Using Seismic Waveform Cross-Correlation and Double-Difference Relocation. Ph.D. Thesis, Department of Geophysics-Geothermics, Faculty of Geology and Geoenvironment, University of Athens, Athens, Greece, 2017.
66. Papadimitriou, P.; Kapetanidis, V.; Karakonstantis, A.; Spingos, I.; Pavlou, K.; Kaviris, G.; Kassaras, I.; Voulgaris, N. The 25th October, 2018 Zakynthos Earthquake. *Bull. Geol. Soc. Greece* **2019**, *7*, 259–260.
67. Bonatis, P.; Akinci, A.; Karakostas, V.; Papadimitriou, E.; Kaviris, G. Near-fault Broadband Ground Motion Simulation Ap-plications at the Central Ionian Islands, Greece. *Pure Appl. Geophys.* **2021**, *178*, 3505–3527. [CrossRef]
68. Valkaniotis, S.; Briole, P.; Ganas, A.; Elias, P.; Kapetanidis, V.; Tsironi, V.; Fokaefs, A.; Partheniou, H.; Paschos, P. The Mw = 5.6 Kanallaki Earthquake of 21 March 2020 in West Epirus, Greece: Reverse Fault Model from InSAR Data and Seismotec-tonic Implications for Apulia-Eurasia Collision. *Geosciences* **2020**, *10*, 454. [CrossRef]
69. Lekkas, E.; Mavroulis, S.; Carydis, P.; Skourtsos, E.; Kaviris, G.; Paschos, P.; Ganas, A.; Kazantzidou-Firtinidou, D.; Par-charidis, I.; Gatsios, T.; et al. The March 21, 2020, Mw 5.7 Epirus (Greece) Earthquake. Newsletter of the Postgraduate Studies Program "Environmental Disasters & Crises Management Strategies" of the National and Kapodistrian University of Athens. Issue No. 17. 2020. Available online: https://www.researchgate.net/publication/346039612_The_March_21_2020_Mw_57_Epirus_Greece_Earthquake (accessed on 17 January 2022).
70. Waldhauser, F. hypoDD—A Program to Compute Double-Difference Hypocenter Locations. *U. S. Geol. Surv. Open File Rep.* **2001**, *113*. [CrossRef]
71. Makropoulos, K.; Diagourtas, D.; Kassaras, J.; Kouskouna, V.; Papadimitriou, P.; Ziazia, M. The November–December 1994 Lefkas (W. Greece) earthquake sequence: Results from in situ seismological survey. In Proceedings of the XXV General Assembly of ESC, Reykavik, Iceland, 1996; p. 108, Book of abstracts.
72. Karakostas, V.; Kostoglou, A.; Chorozoglou, D.; Papadimitriou, E. Relocation of the 2018 Zakynthos, Greece, aftershock sequence: Spatiotemporal analysis deciphering mechanism diversity and aftershock statistics. *Acta Geophys.* **2020**, *68*, 1263–1294. [CrossRef]
73. Kassaras, I.; Papadimitriou, P.; Kapetanidis, V.; Voulgaris, N. Seismic site characterization at the western Cephalonia Island in the aftermath of the 2014 earthquake series. *Int. J. Geo-Eng.* **2017**, *8*, 1251. [CrossRef]
74. Papathanassiou, G.; Pavlides, S.; Ganas, A. The 2003 Lefkada earthquake: Field observations and preliminary microzonation map based on liquefaction potential index for the town of Lefkada. *Eng. Geol.* **2005**, *82*, 12–31. [CrossRef]
75. Cirella, A.; Romano, F.; Avallone, A.; Piatanesi, A.; Briole, P.; Ganas, A.; Theodoulidis, N.; Chousianitis, K.; Volpe, M.; Bozionellos, G.; et al. The 2018 Mw 6.8 Zakynthos (Ionian Sea, Greece) earthquake: Seismic source and local tsunami characteriza-tion. *Geophys. J. Int.* **2020**, *221*, 1043–1054. [CrossRef]
76. Giardini, D.; Wössner, J.; Danciu, L. Mapping Europe's Seismic Hazard. *Eos* **2014**, *95*, 261–262. [CrossRef]
77. Woessner, J.; Laurentiu, D.; Giardini, D.; Crowley, H.; Cotton, F.; Grünthal, G.; Valensise, G.; Arvidsson, R.; Basili, R.; Demircioglu, M.B.; et al. The 2013 European Seismic Hazard Model: Key components and results. *Bull. Earthq. Eng.* **2015**, *13*, 3553–3596. [CrossRef]
78. Weichert, D. Estimation of the Earthquake Recurrence Parameters for Unequal Observation Periods for Different Magnitudes. *Bull. Seismol. Soc. Am.* **1980**, *70*, 1337–1340. [CrossRef]

79. Danciu, L.; Tselentis, G.-A. Engineering Ground-Motion Parameters Attenuation Relationships for Greece. *Bull. Seism. Soc. Am.* **2007**, *97*, 162–183. [CrossRef]
80. Tselentis, G.-A.; Danciu, L. Probabilistic seismic hazard assessment in Greece-Part 1: Engineering ground motion parameters. *Hazards Earth Syst. Sci.* **2010**, *10*, 25–39. [CrossRef]
81. Pavlou, K.; Kaviris, G.; Kouskouna, V.; Sakkas, G.; Zymvragakis, A.; Sakkas, V.; Drakatos, G. Minor seismic hazard changes in the broader area of Pournari artificial lake after the first filling (W. Greece). *Results Geophys. Sci.* **2021**, *7*, 100025. [CrossRef]
82. Kaviris, G.; Zymvragakis, A.; Bonatis, P.; Sakkas, G.; Kouskouna, V.; Voulgaris, N. Probabilistic Seismic Hazard Assessment for the Broader Messinia (SW Greece) Region. *Pure Appl. Geophys.* **2022**, 1–17. [CrossRef]
83. Ordaz, M.; Salgado-Gálvez, M.A.; Giraldo, S. R-CRISIS: 35 years of continuous developments and improvements for proba-bilistic seismic hazard analysis. *Bull. Earthq. Eng.* **2021**, *19*, 2797–2816. [CrossRef]
84. Burton, P.W.; Xu, Y.; Tselentis, G.A.; Sokos, E.; Aspinall, W. Strong ground acceleration seismic hazard in Greece and neigh-boring regions. *Soil Dyn. Earthq. Eng.* **2003**, *23*, 159–181. [CrossRef]
85. HexagonSmartNet METRICA S.A. Available online: https://hxgnsmartnet.com (accessed on 17 January 2022).
86. Corinth Rift Laboratory (CRL) GNSS network. Available online: https://nfo.crlab.eu/gnss (accessed on 17 January 2022).
87. EUREF Permanent GNSS Network—ETRF/ITRF Transformation. Available online: http://www.epncb.oma.be (accessed on 24 April 2020).
88. Dach, R.; Lutz, S.; Walser, P.; Fridez, P. *Bernese GNSS Software Version 5.2*; User manual; Astronomical Institute, University of Bern, Bern Open Publishing: Bern, Switzerland, 2015.
89. Briole, P.; Ganas, A.; Elias, P.; Dimitrov, D. The GPS velocity field of the Aegean. New observations, contribution of the earthquakes, crustal blocks model. *Geophys. J. Int.* **2021**, *226*, 468–492. [CrossRef]
90. Vallianatos, F.; Sakkas, V. Multiscale Post-Seismic Deformation Based on cGNSS Time Series Following the 2015 Lefkas (W. Greece) Mw6.5 Earthquake. *Appl. Sci.* **2021**, *11*, 4817. [CrossRef]
91. Sokos, E.; Zahradník, J.; Gallovič, F.; Serpetsidaki, A.; Plicka, V.; Kiratzi, A. Asperity break after 12 years: The Mw6.4 2015 Lefkada (Greece) earthquake. *Geophys. Res. Lett.* **2016**, *43*, 6137–6145. [CrossRef]
92. Chousianitis, K.; Ganas, A.; Evangelidis, C.P. Strain and rotation rate patterns of mainland Greece from continuous GPS data and comparison between seismic and geodetic moment release. *J. Geophys. Res. Solid Earth* **2015**, *120*, 3909–3931. [CrossRef]
93. D'Agostino, N.; Métois, M.; Koci, R.; Duni, L.; Kuka, N.; Ganas, A.; Georgiev, I.; Jouanne, F.; Kaludjerovic, N.; Kandić, R. Active crustal deformation and rotations in the southwestern Balkans from continuous GPS measurements. *Earth Planet. Sci. Lett.* **2020**, *539*, 116246. [CrossRef]
94. Pérouse, E.; Sébrier, M.; Braucher, R.; Chamot-Rooke, N.; Bourlès, D.; Briole, P.; Sorel, D.; Dimitrov, D.; Arsenikos, S. Transition from collision to subduction in Western Greece: The Katouna–Stamna active fault system and regional kinematics. *Geol. Rundsch.* **2017**, *106*, 967–989. [CrossRef]
95. Baker, C.; Hatzfeld, D.L.; Lyon-Caen, H.; Papadimitriou, E.; Rigo, A. Earthquake mechanisms of the Adriatic Sea and Western Greece: Implications for the oceanic subduction-continental collision transition. *Geophys. J. Int.* **1997**, *131*, 559–594. [CrossRef]
96. Constantinescu, L.; Ruprechtová, L.; Enescu, D. Mediterranean-Alpine Earthquake Mechanisms and their Seismotectonic Implications. *Geophys. J. Int.* **1966**, *10*, 347–368. [CrossRef]
97. Drakopoulos, J.; Delibasis, N. The Focal Mechanisms of Earthquakes in the Major Area of Greece for the Period 1947–1981. *Seismol. Lab. Univ. Athens Publ. Athens Greece* **1982**, *2*, 1–72.
98. Jackson, J.; Haines, J.; Holt, W. The horizontal velocity field in the deforming Aegean Sea region determined from the moment tensors of earthquakes. *J. Geophys. Res.* **1992**, *97*, 17657. [CrossRef]
99. Kiratzi, A.; Louvari, E. Focal mechanisms of shallow earthquakes in the Aegean Sea and the surrounding lands determined by waveform modelling: A new database. *J. Geodyn.* **2003**, *36*, 251–274. [CrossRef]
100. Main, I.G.; Burton, P.W. Moment—magnitude scaling in the Aegean area. *Tectonophysics* **1990**, *179*, 273–285. [CrossRef]
101. Papadimitriou, E. Focal mechanism along the convex side of the Hellenic arc. *Boll. De Geofis. Teor. Ed Appl.* **1993**, *XXXV*, 401–426.
102. Papadopoulos, G.; Kondopoulou, D.; Leventakis, G.-A.; Pavlides, S. Seismotectonics of the Aegean region. *Tectonophys* **1986**, *124*, 67–84. [CrossRef]
103. Papazachos, B.; Delibasis, N. Tectonic stress field and seismic faulting in the area of Greece. *Tectonophys* **1969**, *7*, 231–255. [CrossRef]
104. Papazachos, B.; Kiratzi, A.; Papadimitriou, E. Regional focal mechanisms for earthquakes in the Aegean area. *Pure Appl. Geophys.* **1991**, *136*, 405–420. [CrossRef]
105. Ritsema, A. The earthquake mechanics of the Balkan region. *R. Netherl. Meteorol. Inst. De Bilt. Sci. Rep.* **1974**, 4–74.
106. Wickens, A.J.; Hodgson, J.H. *Computer Re-Evalutation of Earthquake Mechanism Solutions*; Departement of Energy, Mines and Resources: Ottawa, ON, Canada, 1967.
107. Wessel, P.; Smith, W.H.F.; Scharroo, R.; Luis, J.; Wobbe, F. Generic Mapping Tools: Improved Version Released. *EOS Trans. Am. Geophys. Union* **2013**, *94*, 409–410. [CrossRef]

applied sciences

MDPI

Article

Investigation of the Thiva 2020–2021 Earthquake Sequence Using Seismological Data and Space Techniques

George Kaviris [1,*], Vasilis Kapetanidis [1], Ioannis Spingos [1], Nikolaos Sakellariou [1], Andreas Karakonstantis [1], Vasiliki Kouskouna [1], Panagiotis Elias [2], Andreas Karavias [3], Vassilis Sakkas [1], Theodoros Gatsios [1,3], Ioannis Kassaras [1], John D. Alexopoulos [1], Panayotis Papadimitriou [1], Nicholas Voulgaris [1] and Issaak Parcharidis [3]

1 Section of Geophysics-Geothermics, Department of Geology and Geoenvironment, National and Kapodistrian University of Athens, 15784 Athens, Greece; vkapetan@geol.uoa.gr (V.K.); ispingos@geol.uoa.gr (I.S.); nsakel@geol.uoa.gr (N.S.); akarakon@geol.uoa.gr (A.K.); vkouskouna@geol.uoa.gr (V.K.); vsakkas@geol.uoa.gr (V.S.); theogat@geol.uoa.gr (T.G.); kassaras@geol.uoa.gr (I.K.); jalexopoulos@geol.uoa.gr (J.D.A.); ppapadim@geol.uoa.gr (P.P.); voulgaris@geol.uoa.gr (N.V.)

2 Institute of Astronomy, Astrophysics, Space Applications and Remote Sensing, National Observatory of Athens, 15236 Penteli, Greece; pelias@noa.gr

3 Department of Geography, Harokopio University of Athens, 17676 Kallithea, Greece; gp219309@hua.gr (A.K.); parchar@hua.gr (I.P.)

* Correspondence: gkaviris@geol.uoa.gr

Abstract: We investigate an earthquake sequence involving an M_w = 4.6 mainshock on 2 December 2020, followed by a seismic swarm in July–October 2021 near Thiva, Central Greece, to identify the activated structures and understand its triggering mechanisms. For this purpose, we employ double-difference relocation to construct a high-resolution earthquake catalogue and examine in detail the distribution of hypocenters and the spatiotemporal evolution of the sequence. Furthermore, we apply instrumental and imaging geodesy to map the local deformation and identify long-term trends or anomalies that could have contributed to stress loading. The 2021 seismic swarm was hosted on a system of conjugate normal faults, including the eastward extension of the Yliki fault, with the main activated structures trending WNW–ESE and dipping south. No pre- or coseismic deformation could be associated with the 2021 swarm, while Coulomb stress transfer due to the M_w = 4.6 mainshock of December 2020 was found to be insufficient to trigger its nucleation. However, the evolution of the swarm is related to stress triggering by its major events and facilitated by pore-fluid pressure diffusion. The re-evaluated seismic history of the area reveals its potential to generate destructive M_w = 6.0 earthquakes; therefore, the continued monitoring of its microseismicity is considered important.

Keywords: seismic swarm; seismology; geodesy; double-difference relocation; Coulomb stress transfer; Thiva; Greece

Citation: Kaviris, G.; Kapetanidis, V.; Spingos, I.; Sakellariou, N.; Karakonstantis, A.; Kouskouna, V.; Elias, P.; Karavias, A.; Sakkas, V.; Gatsios, T.; et al. Investigation of the Thiva 2020–2021 Earthquake Sequence Using Seismological Data and Space Techniques. *Appl. Sci.* **2022**, *12*, 2630. https://doi.org/10.3390/app12052630

Academic Editor: Valerio Comerci

Received: 31 January 2022
Accepted: 28 February 2022
Published: 3 March 2022

Publisher's Note: MDPI stays neutral with regard to jurisdictional claims in published maps and institutional affiliations.

1. Introduction

Greece is located at the southeastern margin of Europe, a region with rich localized seismotectonic phenomena, a result of the convergence between the Eurasian and African tectonic plates. These processes are expressed through intense deformation and high seismicity along thrust, strike-slip, and normal faults [1,2]. Geodetic, geological, and seismological surveys have revealed that the N–S to NE–SW extensional tectonic regime of Central Greece in the Quaternary is a result of the back-arc extension between the Hellenic arc and the transtensional tectonic regime of the North Aegean [3–5].

Thiva (Central Greece) is located at the transition zone between two major WNW–ESE- and NW–SE-striking rifts: the Corinth Gulf in the south and the Euboekos Gulf in the east (Figure 1). The Gulf of Corinth is an area characterized by high seismicity [6,7] expressed through the frequent occurrence of seismic swarms [8–12].

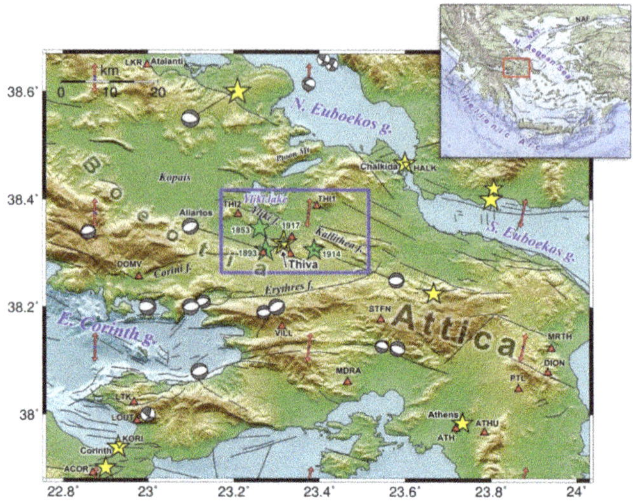

Figure 1. Seismotectonic map of part of Central Greece showcasing Thiva (Boeotia) and the Corinth and Euboekos gulfs. The study area is marked by the blue rectangle. Focal mechanisms of significant earthquakes obtained from the catalogue of the Seismological Laboratory of the National and Kapodistrian University of Athens (SL-NKUA) and the compilation in [13] are shown, along with the locations of seismological stations of the Hellenic Unified Seismic Network (HUSN; red triangles) and GNSS stations (inverse yellow triangles). Green stars indicate epicenters whose parameters were estimated through macroseismic data in the present study. Yellow stars are epicenters of events adopted by the SHEEC catalogue [14]. Faults, after [15,16], are also shown (black lines). Inset: Greece and its main geotectonic features. The area of the larger map is marked by a red rectangle. NAT: North Aegean Trough; NAF: North Anatolian Fault.

The broad Thiva area is characterized by normal faulting (Figure 1). It is worth noting that recent strong and destructive earthquakes in Greece have occurred on normal faults, such as the Lesvos 2017 [5,17,18], Kos 2017 [5,19,20], Samos 2020 [21–24], Thessaly 2021 [25–30], and Arkalochori (Crete) 2021 [31,32] earthquakes. Fault segmentation is common near Thiva, with the presence of a multitude of smaller normal faults all around the area. One of these segments is the 6 km long Kallithea fault, striking between N 80° E and N 115° E, with a probable eastward extension [33]. North of Thiva, a denser fault network with smaller south-dipping ruptures dominates the area around Yliki Lake. Due to the small length of the mapped faults, the occurrence of an earthquake with a magnitude (M) of at least 6.0 seems unlikely [34,35].

On 2 December 2020, 10:54:56 UTC, an M_w = 4.6 earthquake occurred east of Thiva [36]. Moment tensor inversion yielded a normal focal mechanism with strike N106° E, dip 31°, and rake −81° [36]. The analysis of both geodetic and seismological data indicated an uncommonly shallow depth of approximately 2 km for the main event. The mainshock and its aftershock sequence, which lasted until 3 January 2021, were triggered by ruptures on or adjacent to the Kallithea fault [36]. A few kilometers to the west and 7 months later, a new earthquake sequence was initiated, characterized by high productivity, lasting for over 6 months. The whole activity led to the recording of over 2500 earthquakes.

In this study, we present a detailed analysis of the seismic crises in the Thiva area between late November 2020 and mid-October 2021 through high-resolution double-difference relocation. The spatiotemporal evolution of the various outbreaks composing the 2021 seismic swarm is presented, along with the activated structures at depth and their relation to surface traces of mapped faults. The triggering mechanisms of the 2021 Thiva seismic swarm are investigated, such as stress redistribution due to the 2 December 2020 mainshock or possible aseismic factors, including fluid-induced pore-pressure diffusion, through the migration of seismicity, as well as deformation transients. The deformation study resulted from the joint analysis of local GNSS data and satellite aperture radar interferometry (InSAR) products.

2. Past Earthquakes

The study area has had a mild-to-moderate seismic footprint since antiquity. There are ancient reports of various incidents in the broad area of Thiva that could be interpreted as geotectonic in nature [37]. Limited archaeological evidence suggests earthquake activity, which has had an impact on Thiva between 1350 and 1230 BCE without any substantial proof regarding the epicentral area [38]. Many centuries later, in 1321–1323 CE, the eyewitness Jordanus reports collapse and damage of many houses in Thiva [14,39]. In [40] it is noted that, in general, the areas of Boeotia, Locris, Fokis, and Fthiotis have suffered from catastrophic earthquakes, such as the 427 BCE Orchomenos earthquake sequence, which shook Boeotia, Euboea, and Athens and was perhaps a forerunner of the large Maliakos Gulf 426 BCE earthquake [39].

After another long period of seismic quiescence, significant activity was observed in the 19th century. On 13 December 1833, [41] reports a local earthquake was felt in Thiva. On 18 August and 29 September 1853, two destructive earthquakes struck Thiva, affecting the area between it and Yliki Lake, a few kilometers to the north. The first shock destroyed the northern outskirts and caused considerable damage to the rest of the urban environment, leading to 11 fatalities and 60 injuries. The second one caused the collapse of the already-damaged houses and ruined the rest, causing one casualty. At least 3 felt aftershocks occurred up to 1854 [14,39]. Thiva, with a population of 4000, was rebuilt by 1860, with wider spaces between houses. On 26 September 1872, a rather strong event occurred, preceded by a few foreshocks, causing nonstructural damage [39]. The earthquake of 23 May 1893 was also preceded by damaging foreshocks, ruining Thiva and nearby places [39]. The earthquake of 17 October 1914 occurred without a foreshock and caused heavy damage in Thiva and nearby villages. A strong aftershock, a few hours later, extended the damage area with a westward migration of seismicity [14,39,42,43]. This earthquake was recorded by the two seismographs of the National Observatory of Athens, of Mainka and Agamemnone type, followed by a series of aftershocks. All recorded events were located at a distance of 45 km from Athens, assuming their epicenters approximately at 6 km SE of Thiva, in agreement with the epicenter of the mainshock derived from the isoseismals drawn from a significant number of direct macroseismic observations [40,44]. The *Annales de l'Observatoire National d'Athènes* reported another damaging earthquake near Thiva on 23 September 1917, preceded by two foreshocks and followed by at least 37 recorded aftershocks lasting up to the end of the year. This information is passed on to the catalogues of Galanopoulos [45], Kárnik [46], and Shebalin et al. [47]. However, the event is not reported in the catalogue of Makropoulos et al. [6].

For the three events with I > 8, information on coseismic effects is also reported. More specifically, regarding the 18 August 1853 event, in the S. Euboekos gulf successive waves were observed and the sea flooded the coast without any damage. In Kopais Lake (artificially drained in 1880, approximately 25 km NW of Thiva) the banks were flooded. Rockfalls fell from a mountain near Thiva and from Mt. Ptoon (18 km to the north). Liquefaction near Atalanti was also reported. No ground cracks or surface ruptures were observed [39]. For the 23 May 1893 earthquake, ground cracks were reported near Thiva and Mulki, a settlement near Aliartos (Figure 1). Rockfalls and ground cracks, triggered by landslides, were reported from localities in the vicinity of Thiva. No information about ground deformations was reported [39]. For the 17 October 1914 earthquake, no ground cracks or any kind of ground ruptures was found [40].

For the pre-1900 events, macroseismic intensities were collected from [48], whereas for those in the 20th century, intensities were assigned using EMS98 (European macroseismic scale) [49] from the reports of the *Annales de l'Observatoire National d'Athènes* [40,44], as well as from Ambraseys's unpublished material. The latter was donated to SL-NKUA and consists of notes, earthquake lists, slides, maps, and books. Having this information, we compiled the seismic history of Thiva, that is, a distribution of EMS98 intensities, I, over time, for events in the study area between 1893 and 1934 (Figure 2a). The apparent clustering of damaging ($I \geq 6$) and felt earthquakes is associated with two periods (1914–1917 and 1923–1927) of the 20th century, the first related to the two most damaging shocks in this period (1914 and 1917). The activity in 1914–1917 is almost continuous, including fore- and aftershocks of the major events. However, at least five subclusters can be identified in the year 1915 (Figure 2b). This activity resembles the recent 2020–2021 activity in the vicinity of Thiva regarding its duration. It is noted, however, that the 1915 events represent the long aftershock activity of the 1914 event, a fact also mentioned in [40,44].

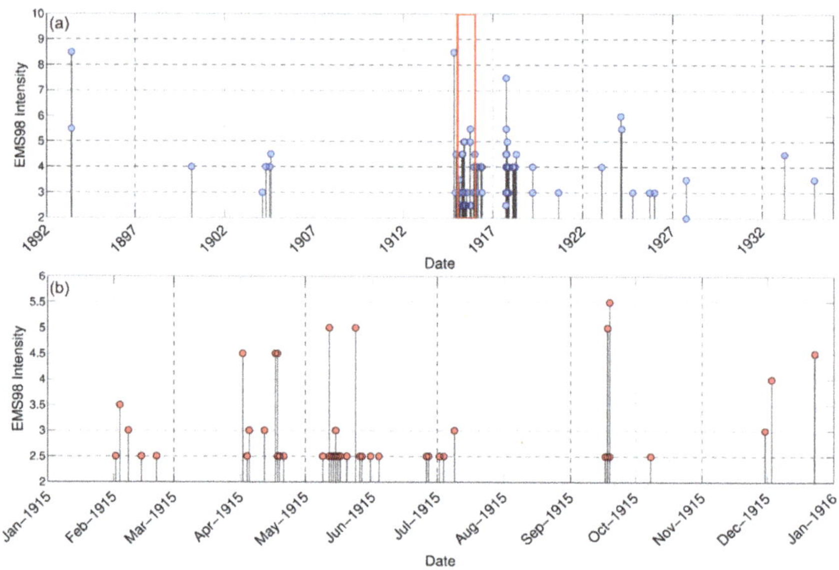

Figure 2. Seismic histories of the area (**a**) between 1893–1934 and (**b**) 1915 in terms of EMS98 (macroseismic scale). The red rectangle in panel (**a**) marks the window presented in panel (**b**).

The significant number of available intensity data points allowed for the macroseismic estimation of earthquake parameters (MEEP) using the MEEP software [50], described in detail in [51]. The software calculates four types of epicenters (Centroid, MEEP, Bakun–Wentworth, and Pairwise). The first is based on the [52] "boxer" method, while the second is the MEEP procedure with iterations adopting the point of lowest RMS [50]. The other two procedures are the methods in [53,54], respectively. The uncertainties of the parameters are calculated via a bootstrap resampling routine of the intensity data points (IDP) data set [55], repeating the procedure 1000 times, from which a standard deviation is calculated. However, the final uncertainties concerning both the epicenter coordinates and the magnitude are obviously significantly larger compared with the respective ones obtained through analysis of recent instrumental data, depending on the number and distribution of the available IDPs. Table 1 and Figure 1 (green stars) present the results of MEEP software applied to the four most damaging earthquakes in the study area, with updated, in comparison with previous studies, parameters due to the increased number of IDPs (Table 1). It is observed that, within a period of 65 years, Thiva has experienced three events of equivalent moment magnitude, M_w, ranging from 6.0 to 6.2. Figure A1 presents the spatial distribution of macroseismic intensities assigned in this study and the calculated macroseismic epicenter of the 1914 earthquake derived from 62 IDPs. Our analysis yielded an epicenter between Thiva and Kallithea (site of the December 2020 activity). For the 1917 event, the epicenter is located in Thiva, where the sole maximum intensity is observed.

Table 1. Estimated macroseismic parameters in comparison with previous studies for the four damaging earthquakes in Thiva (Figure 1, green stars). I_0: epicentral intensity; φ: latitude of estimated epicenter; λ: longitude of estimated epicenter; U_{epi}: epicentral uncertainty; M_w: equivalent moment magnitude (macroseismic); D: focal depth; IDPs: number of intensity data points.

Source	Date (yyyy/mm/dd)	I_0	φ (°N)	λ (°E)	U_{epi} (km)	M_w	D (km)	IDPs
This study	1853/08/18	9	38.347	23.261	8.5	6.2 ± 0.1	-	41
[14]		9–10	38.319	23.317	20	6.7 ± 0.3	-	8
This study	1893/05/23	9	38.306	23.275	5.7	6.2 ± 0.3	5	13
[14]		8	38.310	23.250	5	6.2 ± 0.3	-	-
This study	1914/10/17	8–9	38.306	23.388	6.7	6.0 ± 0.1	8 ± 1	62
[6]		-	38.20	23.50	-	5.9	8	-
This study	1917/09/23	7–8	38.319	23.317	0.0	4.5	2	3
[47]		7	38.200	23.500	33	4.5	4	1

3. Seismological Data and Methods

Here, we study the seismicity in the area of Thiva between 25 November 2020 and 16 October 2021. During that time, P- and S-phase arrivals for over 2000 earthquakes were manually determined at the Seismological Laboratory of the National and Kapodistrian University of Athens (SL-NKUA). The catalogue data were enriched with additional events and manually picked arrival times from the database of the Geodynamics Institute of the National Observatory of Athens (GI-NOA). The compiled catalogue contains a total of ~2800 events for the study period. The sequence was recorded by the regional Hellenic Unified Seismic Network [56], complemented by the deployment of temporary local stations, which play a significant role in constraining the focal depths and increasing the detection threshold. Specifically, two stations (AMPE and THIV; Figure 3) were installed by SL-NKUA, and three stations (THVA, THI1, and THI2; Figures 1 and 3) were installed by GI-NOA in the immediate area of Thiva.

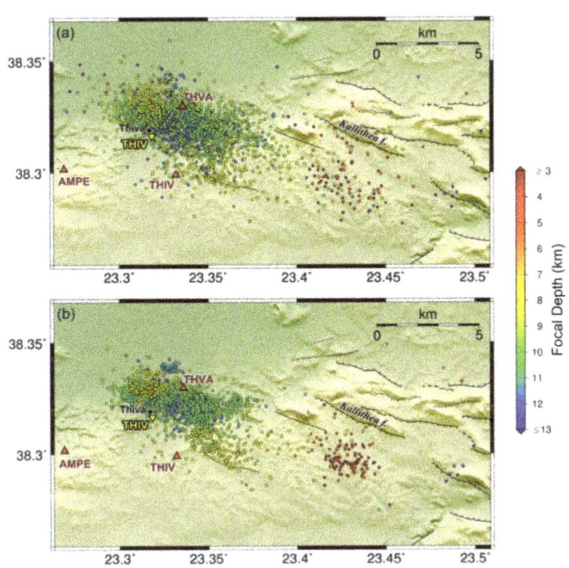

Figure 3. Seismicity in the area of Thiva between 25 November 2020 and 16 October 2021. (**a**) Initial locations with HypoInverse and (**b**) after double-difference relocation with HypoDD. Fault lines are

from [35] and the NOAfaults database [15,16]. Local seismological and GNSS stations are marked with red triangles and a yellow inverted triangle, respectively.

3.1. Location and Relocation of Seismicity

Initial locations for the earthquake sequence were acquired using the HypoInverse code [57], employing the P-wave velocity model of [58] for the neighboring Eastern Gulf of Corinth, as well as for Boeotia, obtained from recordings of the local CORNET network [59–61]. Other velocity models that were tried out (e.g., regional models for Central Euboea or Atalanti; [62]) provided similar results and comparable uncertainties; however, the [58] model was preferred for consistency with previous results for the 2020 Kallithea sequence, where the same model was used [36]. The importance of local velocity models for earthquake location has recently been demonstrated in the case of the 2019 Constance Lake (Central Europe) seismic sequence [63]. Regarding the present study, the V_p/V_s ratio was confirmed to be ~1.79 with the Chatelain [64] method (Figure A2), as in the original model of [58]. The initial locations are presented in Figure 3a. The average root mean square (RMS) travel-time residual for the dataset is 0.19 s, with average horizontal and vertical location uncertainties, as reported by HypoInverse, at 0.38 and 0.80 km, respectively (Figure A3). The average focal depth is 9.4 km, although much shallower hypocenters (a median focal depth of 4.2 km) are resolved for the eastern group, related to the 2020 Kallithea sequence, previously presented by [36]. The magnitude of completeness for the dataset is $M_c = 1.6$, determined by the entire magnitude range (EMR) method [65].

As this sequence involves a large number of events that occurred in a series of outbreaks clustered in space and time within a limited area, it is important to reduce the relative location uncertainties in order to better distinguish the activated structures. For this purpose, we employed the double-difference relocation code, HypoDD [66]. This algorithm minimizes the double difference between calculated and observed P- and S-wave travel times for pairs of neighboring events, resulting in the reduction of location uncertainties due to unmodeled velocity structure. Waveform cross-correlation data are also incorporated in the relocation scheme to enhance the hypocentral distribution of strongly correlated events by reducing relative location uncertainties due to arrival-time picking inconsistencies.

To facilitate the relocation procedure, the dataset was divided into subsets, each one with a smaller number of events. This was performed after examining the initial locations and separating the sequence into groups, roughly based on their spatiotemporal distribution. Several outbreaks were distinguished in time related to both temporal and spatial clustering of the occurring earthquakes. The Thiva earthquake sequence was divided into 9 temporal groups, each covering a different period and beginning with an abrupt increase in the seismicity rate.

The event waveforms were initially examined in terms of waveform similarity using reference stations located in the vicinity of the epicentral area and operational during the whole study period. For this purpose, waveforms of the station VILL (Figure 1) at a distance of ~16 km from Thiva were mainly employed, while additional data were also used from the temporary station THVA (Figure 3), deployed by GI-NOA at Thiva, 1 day after the 2 December 2020 $M_w = 4.6$ Kallithea event, and the station STFN at an epicentral distance of ~16 km from the 2020 Kallithea sequence (Figure 1). The full event waveforms (both P and S waves) were cropped and band-pass-filtered between 2 and 8 Hz for the stations VILL and STFN and in the range of 2–15 Hz for the station THVA. The waveforms of all combinations of event pairs were cross-correlated, and the RMS values of cross-correlation maxima from each component were registered in a correlation matrix. Nearest neighbor linkage was applied, and multiplets were formed with an optimal threshold value of $C_{th} = 0.88$, which maximized the difference between the size of the largest cluster and the sum of clustered events [67]. A total of 1948 events were classified in 163 multiplets. Next, for each of the temporal groups, the P- and S-wave windows of events belonging to the same multiplets were cross-correlated at each station with available arrival-time data, and their cross-correlation maxima with their respective time lags were registered as input for HypoDD.

The double-difference relocation procedure was performed separately for each one of the 9 temporal groups. The parameterization of the procedure for each group varies slightly due to the different numbers of events, travel-time data, and clustering properties. However, a similar scheme was employed for all cases, favoring stronger a priori weights for catalogue data during the first steps, followed by stronger weights for cross-correlation data during the last stage. A total of 2700 events were successfully relocated, representing 96.4% of the initial catalogue (Figure 3b).

3.2. Spatiotemporal Evolution of the Sequence

To examine the spatiotemporal distribution of the 2020–2021 Thiva sequence in detail, the dataset was divided into spatial groups, permitted by the achieved reduction of relative location uncertainties with the application of the double-difference relocation. Ward's linkage was applied to the matrix of 3D hypocentral distances between all combinations of relocated event-pairs, and a large number of spatial clusters were formed. After visual examination, some smaller clusters were merged, and a final subdivision into 12 spatial groups was adopted (Figure 4). An exception to this procedure is group #1, which is related to the 2 December 2020 event, but is also covering all seismicity that occurred during the first period of the seismic crisis, that is, before 10 July 2021, which mainly took place at the eastern part of the study area. In the following, we examine the spatiotemporal evolution of the sequence with reference to the 12 spatial groups, distinguished by different colors and numerical labels (Figure 4).

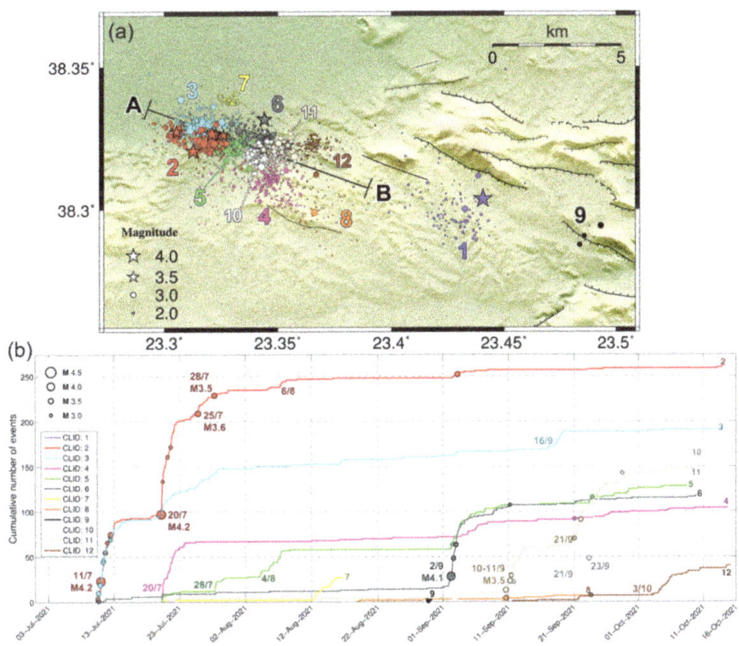

Figure 4. (**a**) Map of relocated epicenters with different colors and numbers representing the 12 spatial groups to which the sequence was divided. Events with $M \geq 3.5$ are depicted as stars. The profile line A–B is used for the spatiotemporal projection of Figure 5. (**b**) Cumulative number of events with $M \geq M_c = 1.6$ per spatial group represented by different colors and numbers (at the right end of each curve) for the period between July and 16 October 2021. The occurrence of events with $M \geq 3.0$ is marked with circles with size proportional to magnitude. Major events related to outbreaks at certain

groups are labeled with their date and largest event magnitude. Group #1 is presented in an alternate version of this plot in Figure A4, covering the entire study period, starting 25 November 2020.

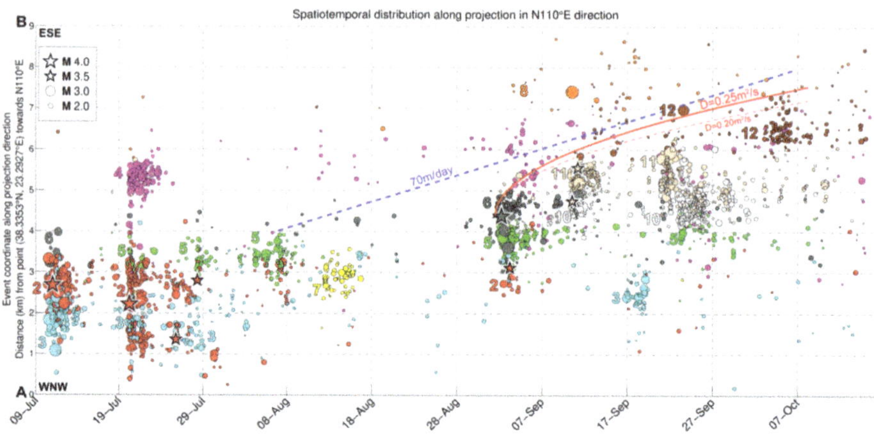

Figure 5. Spatiotemporal projection of relocated epicenters along the N 110° E-oriented profile A–B in Figure 4a for the period between 9 July and 16 October 2021, with colors and numbers corresponding to different spatial groups. Events with $M \geq 3.5$ are depicted as stars. The red lines indicate possible fluid-driven seismicity triggering fronts with a hydraulic diffusivity of 0.20 m^2/s (dashed line) or 0.25 m^2/s (solid line). The dashed blue line is drawn for reference, representing a seismicity migration speed of 70 m/day towards ESE.

The sequence commenced in late November 2020, with a small cluster of earthquakes occurring near Kallithea (eastern group #1, blue), including an M_L = 3.0 on 25 November. Then, on 2 December 2020, 10:54:56 UTC, an M_w = 4.6 earthquake struck near Kallithea. This earthquake apparently had a very shallow focus at a focal depth of 1.5 km after relocation in agreement with the results in [36]. The shallow focal depth of this event is evidenced by the fact that, despite its small magnitude, the deformation field was clearly identified by InSAR, exploiting Copernicus Sentinel-1 data in both ascending and descending orbital directions [36]. This sequence was short-lived, as it practically ceased after 17 December 2020, with a few earthquakes happening until January 2021, and was followed by a long period of quiescence (Figure A4 in Appendix B).

The seismic activity at the western part of the sequence started on 10 July 2021, with several events of $3.1 \leq M_L \leq 3.4$ (red group #2 and cyan group #3), which compose the western half of the 2021 swarm, also initiating seismic activity at group #6 (dark gray) at the midnorthern part with a rate of 0.3–0.2 events/day. A stronger event of M_w = 4.2 occurred on 11 July, 00:00:17 UTC, in the group #2 area, followed by another event of similar magnitude on 20 July in the same area. The latter earthquake increased the seismicity rate of groups #2 and #3, but also triggered seismicity towards other areas, mainly group #4 (purple) and, secondarily, group #5 (green), while a few events were located at groups #7 (yellow) and #11 (beige). Minor outbreaks, causing abrupt increase in the seismicity rate, occurred on 25 July (group #2, including an M_L = 3.6 event, and group #3), 28 July (including an M_L = 3.5 event at group #2), 4 August (group #5), 6 August (groups #2 and #5), and 12 August (group #7). A small isolated cluster of deeper events (~12 km; group #9) occurred on 29–30 August, ~4 km ESE of group #1, at the eastern end of the study area, initiated by an M_L = 3.2 earthquake.

On 1 September 2021, an increase in the seismicity rate was observed in the group #6 area (dark gray), followed by an M_w = 4.1 event on 2 September in the same area. The latter also caused an increase in the seismicity rate, mainly at groups #2, #4, and #5 and less at groups #3 and #11, while events began occurring at group #10 (light gray). An abrupt increase in the seismicity rate of group #11 followed the occurrence of two M_L = 3.5 events on 10 and 11 September in the same group, which also triggered events at group #10. A small cluster was triggered between 16 and 19 September at group #3, including four events with $2.5 \leq M_L \leq 2.8$. Another outbreak at group #11 occurred on 20 September, including an M_L = 3.2 event on 21 September and an M_L = 3.4 event on 22 September. A significant burst of seismicity at group #10 was triggered by an M_L = 3.4 event on 23 September. Lastly, group #12 (brown) at the eastern end of the main part of the 2021 sequence increased seismic activity after 3 October without any major event. The strongest event since then was an M_L = 2.8 earthquake on 10 October 2021 at group #11.

The spatiotemporal projection in Figure 5, along the WNW–ESE-oriented profile A–B in Figure 4a, roughly parallel to the main direction at which the epicenters are aligned, reveals the complex evolution of the 2021 Thiva earthquake sequence. The swarm was initiated on 10 July 2021 at the western end of the activated zone, spanning along groups #2 and #3, with a few events belonging to the neighboring group #6 at the eastern edge of the activated patches. On the outbreak of 20 July, the major M_w = 4.2 event occurred at group #2, but seismicity was also triggered eastwards to group #4, while seismic activity began at group #5, which presented some more bursts between 28 July and 8 August. The small isolated cluster #7, north of the main seismic cloud and slightly to the west of group #5, was activated between 12 and 15 August. A 2-week period of relative quiescence followed. Then in September, the outbreak at group #6 began with some foreshocks, followed by an M_w = 4.1 event on 2 September, mainly triggering aftershocks to the eastern half of the activated zone, including group #5 and a few events at group #2, at which the activity ceased later. Seismicity continued to spread eastwards, with few events at group #4. The next seismic burst occurred at groups #10 and #11 between 10 and 13 September. A last temporal cluster at group #3 took place between 16 and 19 September, followed by reactivation of groups #11 on 20 September and #10 on 23 September. Lastly, cluster #12 occurred at the easternmost end of the 2021 Thiva swarm.

Overall, the seismicity presents a general tendency for spatiotemporal migration towards ESE. This is not straightforward, as the early outbreaks are mainly confined at the western half, between groups #2 and #3, then spread suddenly to group #4 and later group #5. Since September 2021, however, seismicity mostly occurs at the eastern half of the activated zone. Foreshock seismicity near the origin of the M_w = 4.1 event of group #6 spreads towards WNW at a distance of ~1.5 km within 1 day, where it triggered an M_L = 3.5 event at group #2. However, a slower seismicity triggering front can also be observed, spreading towards the east, including some events at group #4, then passing by the eastern edges of group #11 and later group #12, at a migration speed of ~70 m/day (Figure 5). Considering the possibility that this part of the seismicity is induced by the fluid-driven diffusion of pore pressure through the fracture network, a parabolic seismicity triggering front can be drawn according to the relation of [68]:

$$r(t) = \sqrt{4\pi Dt} \tag{1}$$

where $r(t)$ is the distance from the beginning of fluid injection at the origin and t the time elapsed since then, while D is the hydraulic diffusivity. This front would begin on ~1 September near the origin of the M_w = 4.1 event at group #6 and spread eastwards through mid-October, with a hydraulic diffusivity value between 0.20 and 0.25 m^2/s (Figure 5). It can be observed that most of the eastward migrating seismicity is located behind the proposed parabolic triggering-front envelope, although some sparse events, mainly those of groups #4 and #8, are located on the outside, possibly triggered due to pore-fluid diffusion following paths of higher hydraulic diffusivity values.

3.3. Focal Mechanisms and Activated Structures at Depth

Focal mechanism data for the 4 larger events of the sequence (M_w > 4.0) are available from routine waveform modelling and moment tensor inversion analysis performed at

SL-NKUA and GI-NOA. The fault-plane solution of the 2 December 2020 $M_w = 4.6$ Kallithea event is adopted from [36]. In addition, first-motion polarities (FMPs) were manually picked for 20 significant events of smaller magnitude ($2.8 \leq M_L \leq 3.6$) from visual inspection of their recordings at local and regional stations of HUSN (Table A1 in Appendix B and Excel File S2 in the Supplementary Material). For each event, taking into account the relocated foci and the respective azimuth and angle of incidence corresponding to each station, a grid search was applied in the whole range of strike, dip, and rake (φ, δ, λ) values to find solutions compatible with the observed FMPs. In cases where no such solution was available, solutions with a lower percentage of compatibility with the FMPs were sought, considering the analyst's quality estimate of each FMP, with a lower weight applied to measurements of emergent P arrivals or uncertain polarities. Each individual solution compatible with a minimum percentage of polarities was converted into a unitary double-couple moment tensor from the respective φ, δ, and λ parameters, and their sum provided an average focal mechanism for each event. Then, the RMS angular difference between average and individual solutions was measured in terms of their Kagan [69] angle, and solutions with an RMS < 25° were retained. The accepted focal mechanisms have at least 13 FMP measurements with over 90% compatibility.

The fault plane solutions of the major events, presented in Figure 6, indicate dominant normal faulting in an E–W to WNW–ESE direction, with few events presenting oblique-normal kinematics. The spatial distribution of the relocated hypocenters and the acquired focal mechanisms were examined in a series of cross sections, oriented N200° E (Figure 7). This direction is roughly perpendicular to the general trend of the mapped faults on the surface, which are in line with the focal mechanisms and with the main axis of the distribution of epicenters (i.e., N 110° E-oriented profile line A–B in Figure 4a). Starting from the westernmost cross section, a_1–a_2, groups #2 and #3 appear to correspond to the deeper and shallower parts, respectively, of a steep, south-dipping structure. This is represented by the red dashed line, which is the cross section of a least-squares plane determined from the joint distribution of groups #2, #3, #5, and #10 (see also the 3D model of Interactive Matlab Figure S1 in the Supplementary Material). The same pattern persists in section b_1–b_2, where, in addition, a few events of group #5 appear below group #2. The small isolated cluster #7 is also presented, with a dashed line showing the cross section of the least-squares north-dipping plane determined from this group and extrapolated a few km updip.

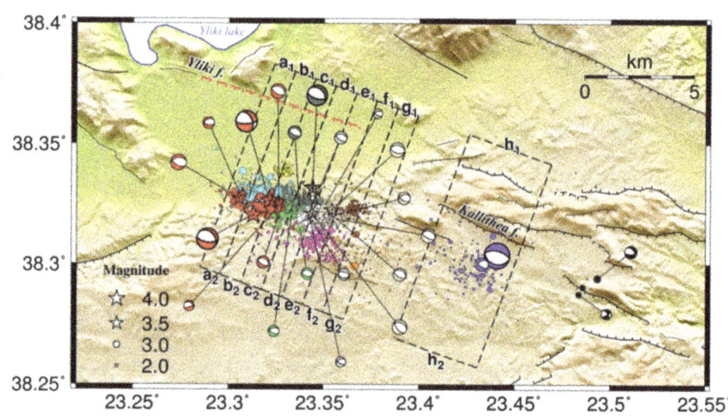

Figure 6. Map of relocated epicenters of the 2020–2021 Thiva earthquake sequence and focal mechanisms of the major events. The labeled (a_1–a_2 to h_1–h_2) dashed rectangles indicate the boundaries and orientation (N 200° E) of the respective cross sections of Figure 7. The red dashed line is the proposed eastward extension of the Yliki fault.

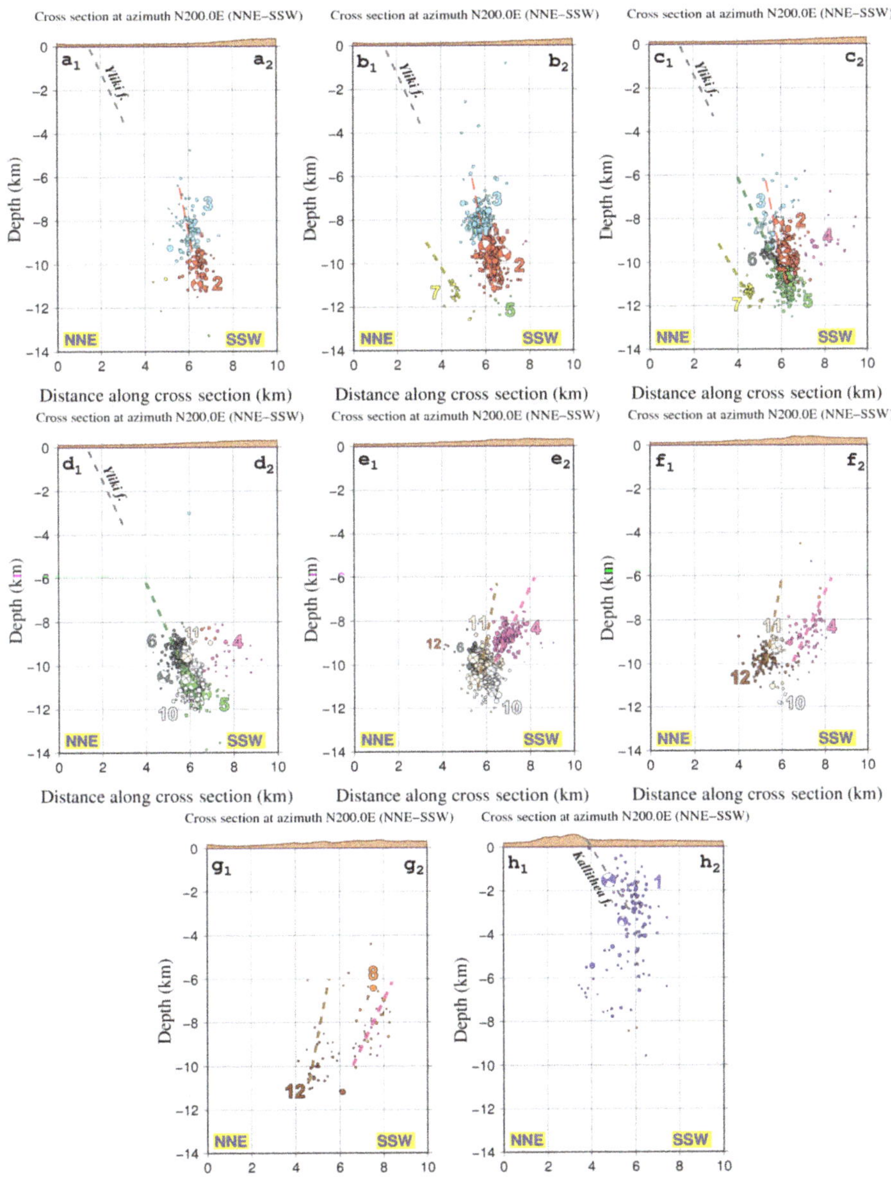

Figure 7. Vertical cross sections drawn in a N 200° E direction along the profiles in Figure 6. Dashed lines at depth are inferred planes from the distribution of relocated hypocenters, whereas dashed lines at the top indicate the planes of known faults on the surface, that is, the eastward extension of Yliki fault (red dashed line in Figure 6) and Kallithea fault. See also Interactive Matlab Figure S1 in the Supplementary Material for a 3D model of the earthquake sequence and the activated structures.

In the cross section c_1–c_2, the distribution of hypocenters becomes more complex, with the introduction of more spatial groups. The dark green dashed line, at a lower dip angle than the red one, represents a least-squares plane determined from the distribution of groups #5, #6, and #10. This plane seems to be consistent with the eastward extension of the Yliki fault when extrapolated downdip at an angle of 65°. Sparse seismicity belonging to group #4 can be observed in this slice. The cross section d_1–d_2 marks the beginning of the eastern half of the 2021 sequence, mainly activated after September. The dashed green line is consistent with the distribution of groups #6 and #10, as well as the remaining events of group #5, whereas groups #2 and #3 are not present in this slice. In the cross section e_1–e_2, a dense cluster of group #4, activated on 20 July 2021, can be observed, apparently belonging to a north-dipping structure (purple dashed line). The brown dashed line is the cross section of a least-squares plane determined from the distribution of groups #11 and #12, also north-dipping, but at a steeper dip angle. The latter structures can also be observed in the cross sections f_1–f_2 and g_1–g_2. The plane determined from groups #11 and #12 is oblique to the direction of the cross sections, increasing its distance from the plane of group #4 toward the east.

Finally, the cross section h_1–h_2 shows the distribution of group #1, which is related to the shallow 2 December 2020 M_w = 4.6 event associated with the south-dipping Kallithea fault. The relocated epicenters of the mainshock and most aftershocks (Figure 6) are clustered south of the Kallithea fault trace with the hypocentral distribution at depths shallower than 4 km being roughly subvertical but apparently favoring a south-dipping structure when considered together with the mainshock in agreement with [36]. The absence of a local seismological network in the area during the occurrence of the 2020 sequence limits our capacity to constrain the hypocentral depths for this group; however, the shallow depth of the rupture (or deformation source) is supported by the observed surface displacements given that almost two fringes are visible in Sentinel-1 coseismic interferograms [36].

4. Coulomb Stress Transfer Due to the 2 December 2020 Mainshock

In the following, we calculate the effect of the 2 December 2020 M_w = 4.6 earthquake at Kallithea on the redistribution of stress in the study area to examine possible stress load to the zone that was activated after 10 July 2021 near Thiva. For this purpose, we adopt the geodesy-favored fault model in [36], determined through modeling of the observed surface deformation together with seismological data, which associated the earthquake with a shallow rupture of the Kallithea fault. Taking into account the abovementioned fault geometry and kinematics, with strike = 120°, dip = 48°, rake = −74°, and net slip of 240 mm, homogeneously distributed on the fault plane with a rupture top at 0.70 km and bottom at 1.44 km (fault length L = 2 km, width W = 1 km), we calculate the Coulomb stress transfer using the Coulomb 3.3 code [70] with an effective coefficient of friction μ' = 0.4.

The Coulomb stress transfer model, determined for optimally oriented normal faults according to the regional stress field, which is dominantly extensional with a subhorizontal $\mathbf{S}_3$ axis oriented ~N186° E [71], is presented in Figure 8. The calculations indicate that a stress load greater than +0.1 bar is mainly confined at depths shallower than 6 km, whereas no significant stress transfer can be inferred at the seismogenic depths of the 2021 seismic swarm, which are between 7 and 12 km (Figure 7). Even at shallower depths, where the main rupture occurred, the +0.1 bar contour hardly reaches the eastern edge of the 2021 seismic sequence epicentral area. Furthermore, the nucleation of the 2021 sequence occurred at its western edge, that is, groups #2 and #3 (e.g., white star with blue outline in Figure 8), located at an even further distance from the Kallithea fault. This suggests that static stress transfer due to the 2 December 2020 mainshock could not have possibly triggered the 2021 seismic swarm.

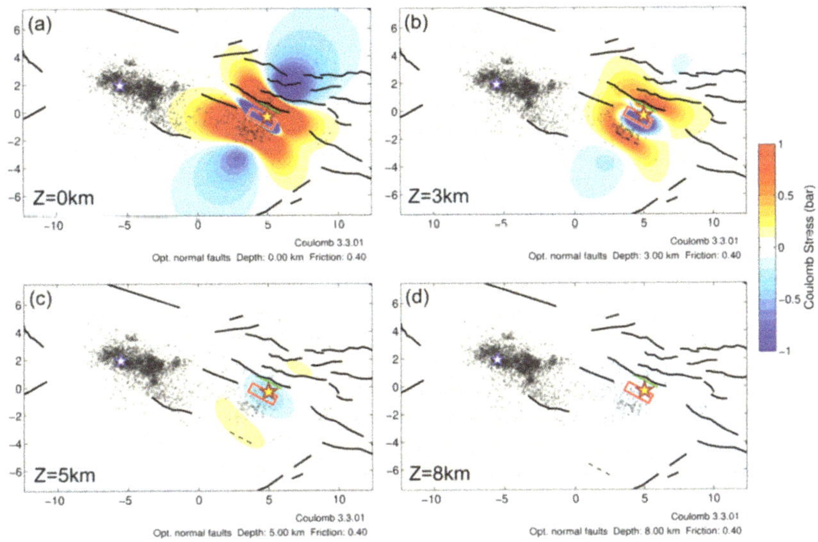

Figure 8. Coulomb stress transfer due to the 2 December 2020 mainshock at Kallithea (yellow star with red outline), following the fault model in [36], marked with a red rectangle, for normal faults optimally oriented to the regional stress field, at different depths Z: (**a**) Z = 0 km, (**b**) Z = 3 km, (**c**) Z = 5 km, and (**d**) Z = 8 km. Gray circles represent the relocated epicenters of the 2020–2021 Thiva seismic sequence, with the white star with blue outline marking the location of the 11 July 2021 M_W = 4.2 event. The dashed line shows the trace of the fault plane (extended downdip) at the depth of the respective horizontal slice. Black lines are fault traces after [35], the NOAfaults database [15,16], and the eastward extension of Yliki fault (Figure 6, dashed red line).

5. Instrumental and Imaging Geodesy to Map Local Deformation

Over the past 30 years, Interferometric Synthetic Aperture Radar (InSAR) and in particular Multi-temporal InSAR (MT-InSAR), as well as Global Navigation Satellite System (GNSS) networks, have been established as methodologies capable of monitoring surface deformations in urban and rural environments with millimeter resolution [72–77]. However, a GNSS network provides information only about a few points, but is characterized as a dynamic real-time monitoring system of high precision in all three motional components. On the other hand, spaceborne SAR interferometry provides a large number of measured points over long time periods. Moreover, GNSS systems have a big economic cost. In addition, SAR interferometry has the capability to remotely monitor areas much wider than traditional surveying techniques without the necessity to install in situ ground instrumentation.

The pre- and coseismic deformation of the Thiva area was investigated based on MT-InSAR results using interferometric data from the Sentinel-1 A and B satellites of the Copernicus program. Copernicus Sentinel-1 A and B SAR scenes are freely available from the Sentinel Hub portal [78] and can provide improved SAR SLC (Single Look Complex) data, ensuring (a) continuous, all-weather, day and night imagery; (b) rapid revisit period in the same imaging mode (6 days); (c) constant and regular acquisition to build a large global archive; and (d) wide area coverage, thanks to the 250 km image swath width. Satellite geodetic data (GNSS) were used to calibrate and validate the interferometric results, while the location of the geodetic device acted as a reference point during the interferometric processing.

5.1. GNSS Geodetic Data to Map Local Deformation

GNSS data from continuous stations (Figure 1, inverted yellow triangles) were processed for the period of January 2017 to November 2021. The available data cover the time prior to the outbreak of the seismicity, as well as the coseismic period. One of the analyzed stations, THIV (Figure 3), belonging to the HexagonSmartNet commercial network by Metrica S.A. [79], is located at Thiva and was operating continuously during the 2020–2021 earthquake sequence. Partial data from two more sites in Thiva were processed for the year 2018 and for the period 2019 to 2020 (belonging to the Uranus network by TreeComp [80]). The processing of the raw GNSS data was performed using the precise double-difference method with the Bernese v5.2 GNSS software [81]. The precise orbital solutions from the Center for Orbit Determination in Europe (CODE) were introduced in the estimation of the station's coordinates. The time series of all three components of the daily coordinates of the station THIV on the global ITRF2014 reference frame area are presented in Appendix C (Figure A6). The velocity vector of the THIV station was estimated to be $V_{\text{North}} = -8.97 \pm 0.04$ mm/yr, $V_{\text{East}} = 9.29 \pm 0.04$ mm/yr, and $V_{\text{Up}} = 1.16 \pm 0.08$ mm/yr. For the two other GNSS sites in Thiva, a similar velocity vector was estimated, although they refer to a slightly different time period. These velocity values are in accordance with previous estimations for the station THIV [5,82]. Based on the continuous daily solutions of the site THIV, it appears that the increased seismicity in the area in 2021 has not caused any observable static ground deformation or alteration of the velocity vector. The only noticeable change that was identified was a small alteration (changing point of the rate) on the vertical component since May 2019. More specifically, a slightly increased uplift motion was detected, from 0.09 ± 0.23 mm/yr prior to May 2019 to 1.18 ± 0.39 mm/yr, for the rest of the period. Moreover, a distinctive seasonal signal is emerged in the vertical component that could be associated with the seasonal fluctuations of the groundwater horizon. Calculation of the baseline changes between the THIV GNSS site and stations located in Chalkida (~30 km NE from Thiva) and Corinth (~54 km SW of Thiva), with both sites belonging to HexagonSmartNet, showed a linear increase during the whole time period of the distance between Thiva and Chalkida, as well as between Thiva and Corinth, equal to 0.8 ± 0.1 mm/yr and 4.0 ± 0.1 mm/yr, respectively. These results are compatible with the extensional regime of the broad area that has been previously described [5,82]. The time series of the Uranus stations was similar to the one presented here. These data were acquired at a high rate (1 Hz). A close examination during the occurrence time of the three earthquakes with $M_w \geq 4$, with their epicenters located 2.8 km NE, 1.1 km NNE, and 0.6 km NW of the THIV GNSS station, respectively, yields that they caused no displacement waveform.

In an effort to define the source of the fluctuations in the vertical component, a joint interpretation of the local rainfall data and the geodetic data was performed. Rainfall data were recorded from a local meteorological station within Thiva. This station is operated by the METEO unit at the National Observatory of Athens and is part of the NOAAN network, consisting of 430 automatic surface weather stations in Greece, which monitors all basic meteorological variables, including rainfall, at 10 min intervals [83]. In Figure 9, the accumulated rainfall time series is shown upon the vertical THIV GNSS displacement time series, the latter being temporally shifted, after visual inspection, by 3 months towards the past, in order to minimize the difference between the rainfall variation and its impact to the ground observation. That is why the date of the alteration shown was previously mitigated from May 2019 to February 19 of the same year. In the diagram, a correlation of the peaks and valleys between the two time series within ±2 months is observed. Moreover, a similar correlation of the amplitudes of the two time series can be identified until October 2019. Then, there is no clear connection between the two datasets, except in October 2020. Towards the end of the examined period, we observe an intense decrease in the values of both curves. Thus, the vertical component of GNSS seems to be driven mainly by the water withdrawal to recovery equilibrium. From this, it can be inferred that the uplift observed after May 2019 can be attributed to the increase in the rainfall. The latter indicates that the observed vertical ground deformation is mainly controlled by the changes on the groundwater level horizon in the broad Thiva area and cannot directly be associated with the recorded seismic activity or be considered as a tectonic result.

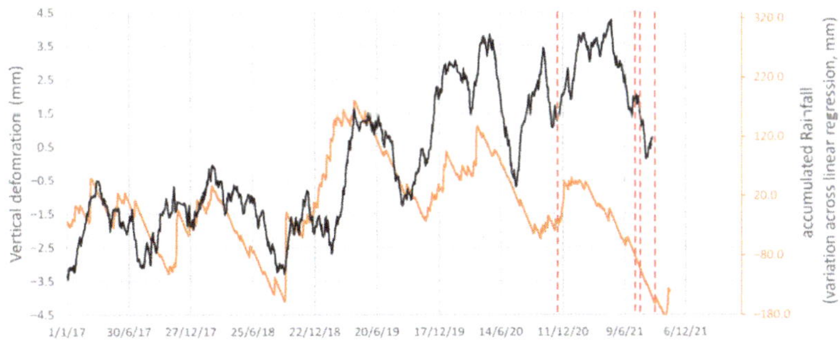

Figure 9. Two-month moving average time series of the vertical component of the THIV GNSS station (black) temporally shifted by 3 months towards the past and detrended accumulated rainfall time series (orange) after removal of its linear regression. The origin times of the $M_w > 4.0$ earthquakes are depicted with dashed red lines.

5.2. Imaging Geodesy

For the InSAR analysis, 177 acquisitions of Copernicus European Sentinel-1 SAR SLC over Thiva were used, available on the Copernicus Open Access Hub [78]. For the processing, a 90 m/pixel digital elevation model (DEM) was used [84]. The Sentinel-1 products were acquired in interferometric wide (IW) swath mode consisting of three sub-swaths with a series of bursts covering a swath of 250 km. Specifically, 89 acquisitions on ascending (track 102) and 88 images on descending (track 09) were used, covering the period from January 2018 to September 2021.

The Multi-temporal InSAR data processing was carried out following the Small BAseline Subset (SBAS) MT-InSAR approach [85] due to the characteristics of the study area (mostly covered by rural areas) using the ENVI SARscape® software (L3Harris Geospatial, Boulder, CO, USA). The preprocessing phase includes the orbit correction of every image, the burst selection over the study area, and the image coregistration using a master image in order to generate interferometric pairs. On the main processing, wrapped interferograms were generated with a multilook range and azimuth (5 × 1), and 90 m/pixel DEM was used to subtract the topographic phase. The wrapped interferograms were unwrapped using the minimum cost flow (MCF) method [86] and converted into line-of-sight (LOS) displacement. The reference for both tracks was set up in the location of the GNSS station THIV. Considering that InSAR displacement in LOS only measures the path length difference between the earth surface and the satellite, displacement decomposition was carried out, exploiting the ascending and descending sensing trajectories to recover the vertical (up–down) and horizontal (east–west) deformation.

The vertical displacement map over the study area (Figure 10) shows mainly subsiding sites with a maximum cumulative displacement of about −40 mm. Nevertheless, there are areas where localized uplift took place with maximum values of 20–30 mm. On the time series plot (Figure 10, inset panel), 12 scatterers were selected to observe the vertical deformation pattern during the period of satellite acquisition (June 2018 to September 2021). As observed in the plot, the scatterers that were located inside Thiva (9, 11, and 12) are relatively stable, compatible to the GNSS result. The same pattern is followed by those that are located south of Thiva (2, 3, and 4), while scatterers 1 and 5 show a subsidence trend, especially after the 2 December 2020 mainshock, up to −10 and −19 mm, respectively. The highest uplift deformation is observed on scatterer 4, up to 15 mm, not associated to the earthquake ($M_w > 4.0$) occurrences. A different pattern is recorded in the vicinity of Kallithea (scatterers 6, 7, 8, and 10) characterized by subsidence trends. The deformation pattern in these scatterers is altered significantly after the occurrence of the $M_w > 4.0$ earthquakes during the processing period, as can be seen in Figure 10 (red lines). Scatterers 6, 7, and 8 show an increased rate of subsidence caused by the 2 December 2020 event, while scatterer 10 is characterized by a smaller subsidence rate.

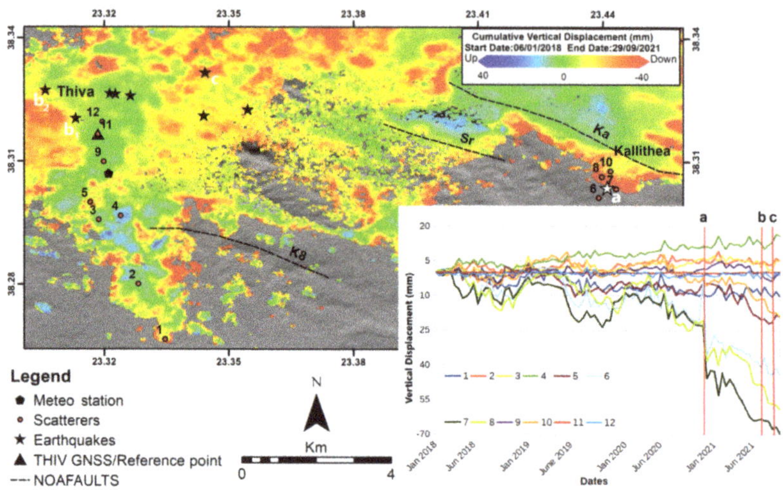

Figure 10. (Map) Vertical cumulative displacement inferred by the SBAS method. The relocated epicenter of the 2 December 2020 earthquake is presented with a white star, while black stars depict the relocated epicenters of major events ($M \geq 3.5$) of the 2021 seismic swarm. Faults (dashed black lines) after NOAFAULTS [15,16] as Sr: Soros; Ka: Kallithea; K8: Kallithea 8. (Lower-right inset panel) Diagram of the cumulated vertical SBAS time series for 12 points distributed over the study area (red dots with numbers on the map). Red vertical lines mark the origin time of the major events ($M_w > 4$) of the 2020–2021 sequence, also marked on the map as follows: a: 2 December 2020; b, including b_1: 11 July 2021 and b_2: 20 July 2021; and c: 2 September 2021. The reference point for both the displacement and the inset time series is the GNSS Station THIV (black triangle).

6. Discussion

The 2020–2021 seismic sequence at Thiva, Central Greece, is a peculiar case of an $M_w = 4.6$ earthquake (on 2 December 2020) occurring at a relatively small distance (6–12 km) from a quite productive seismic swarm, which started ~7 months later. The 2 December 2020 event was interesting on its own, given that it was an unusually shallow earthquake, with its rupture expanding downdip, which produced observable deformation of about two fringes, on Sentinel-1 interferograms, on the surface, permitting its association with the Kallithea fault [36], despite its relatively small magnitude. It is worth noting that earthquakes with similar shallow depths have been reported worldwide, as the 2014 $M_w = 5.1$ La Habra earthquake in California at a depth of 2 km [87], a $M_w = 4.9$ earthquake in France at a depth of 1 km [88], and a $M_w = 5.0$ earthquake in Ecuador at a depth of 1.5 km [89]. At the time, there was no indication that this earthquake would be followed by an intense swarm near Thiva, as previous seismicity in the area was generally low and no increase in the seismicity rate could be observed until July 2021 (Figure A4). In the seismic catalogue in [6] for the periods of 1900 to 2009, only few events with $M_w \geq 4.0$ are located in the broad area of Thiva; few sparse epicenters to the south are mainly related to aftershocks of the major earthquakes of 1981 in the eastern Gulf of Corinth owing to significant location uncertainties.

In this work, the parameters of four destructive events that have affected Thiva in the period of 1853–1917, all without any surface fault traces, were recalculated using macroseismic data (Table 1). These indicated seismic activation a few kilometers west of Thiva in 1853 and 1893, both of equivalent moment magnitude of $M_w = 6.2$, as well as an $M_w = 4.5$ event near Thiva in 1917. The epicenter of the $M_w = 6.0$ 1914 event is located between Thiva and Kallithea, possibly within the area covered by the recent 2021 spatial groups in Figure 4. Coseismic effects, including ground effects, were reported for previous earthquakes with $I > 8$ in Thiva and its vicinities, in accordance with the estimated macroseismic parameters presented in Table 1. A series of small shocks, of intensity $I < 5$, originating from the study area throughout 1915 (Figure 2b) were also recorded, likely attributed to a seismic swarm or to aftershocks triggered by the 1914 mainshock. Such activity, characterized by small events of intensity $I < 5$, has similarities to the recent seismicity. Since the beginning of the enhanced monitoring by the HUSN in 2007, only sparse seismic activity has been detected near Thiva with magnitudes mainly $M_L < 3$, with the largest event ($M_w = 4.0$) occurring near Yliki Lake on 25 June 2017 (see also the New Seismotectonic Atlas of Greece [13] for an overview of previous activity in the area).

The application of double-difference relocation in this study produced a high-resolution catalogue that permitted the distinction of several clusters within the seismicity cloud of the 2021 swarm (Figure 4). The spatiotemporal distribution of the sequence suggests that these spatial groups also exhibit strong temporal clustering, with a series of outbreaks related to the larger events of the sequence. The resolved focal mechanisms indicate mainly normal dip-slip and, to a lesser extent, oblique-normal faulting (Figure 6), compatible with the regional stress of the area [71]. Through detailed cross sections and a 3D model of the spatial distribution of hypocenters (Interactive Matlab Figure S1 in the Supplementary Material), it was possible to identify the likely geometries of the activated structures at depth. In contrast to the shallow 2 December 2020 mainshock, which is associated with the Kallithea fault by straightforward interpretation of the observed displacement pattern [36], the deformation due to the 2021 swarm did not reach the surface. This is explained by the larger focal depths of the major events of the swarm, combined with their low magnitudes. Furthermore, the epicentral area of the 2021 swarm is void of mapped faults (Figure 1). Specifically, the traces of known south-dipping faults in the east, including Kallithea fault, having a WNW–ESE trend, compatible with the focal mechanisms and general geometry of the structures involved in the 2021 seismicity, are terminated ~1–2 km before reaching the first epicenters of the swarm. The same is true for the mapped faults west of the swarm, which are less compatible in terms of their strike. To the south, a south-dipping fault is mapped near the edge of the southern cluster #4, which is associated with a north-dipping structure. To the north of the epicenters, there is a plain of agricultural land without mapped faults or any apparent anomaly in the topography. The only likely candidate fault that could be associated with some of the activated structures at depth is the south-dipping Yliki fault at the southern coast of Yliki Lake. Even so, the fault trace would have to be extended by a few km towards ESE (Figure 6) in order to match the updip extension of some of the activated structures at depth, particularly those related with the major ($M_w > 4.0$) events of the sequence. Other identified structures at depth seem to correspond to steeper faults that would outcrop south of Yliki fault. Two of the identified planes at the eastern part of the swarm suggest a north-dipping fault. The latter cannot be associated with any of the mapped faults on the surface.

The initiation of the 2021 seismic swarm took place near its western extreme, about 10 km WNW of the 2 December 2020 mainshock's relocated epicenter. In the constructed Coulomb stress transfer model for this earthquake, it became apparent that, even with a fault model derived by geodetic data [36], which produced a slightly overestimated moment magnitude ($M_w = 4.7$), stress loading for optimally oriented normal faults under the concurrent regional stress regime drops below +0.1 bar, which is regarded as the lower threshold capable for stress-trigger earthquakes [90], near the eastern margins of the swarm (Figure 8). Given that the nucleation of the 2021 sequence took place at its western end, with an $M_w = 4.2$ event on 11 July 2021, the Coulomb stress transfer due to the 2 December 2020 mainshock is not considered sufficient to trigger the 2021 swarm. This is further justified by the shallow depth of the 2 December 2020 event, but the result, concerning stress transfer at the seismogenic depths of the 2021 swarm, would be the same even if the 2020 mainshock's fault model was placed ~5 km deeper. Therefore, despite seeming counterintuitive, there appears to be no causative link between the 2 December 2020 mainshock and the occurrence of the 2021 swarm; that is, there is no evidence that the former accelerated the occurrence of the latter.

To further examine other possible triggering mechanisms, we employed instrumental and imaging geodesy to investigate pertinent local deformation transients that may have played a role in the stress loading of the area. As in reported cases in the literature, a slow slip event (e.g., [91]) or uplift/subsidence due to groundwater level changes (e.g., [92]) could have increased the applied stresses to the faults of the area, including Kallithea fault and the fault system that hosted the 2021 swarm. This would have the potential to induce seismicity at both places and indicate a common cause for their seismic excitation. However, no significant anomalies in the displacement components of the GNSS station THIV, located at Thiva, could be detected (Figure A6), with the station presenting the anticipated regional horizontal motion and a steady long-term uplift at a slightly increased rate of 1.18 ± 0.39 mm/yr since May 2019. It is noted that the variability of vertical displacement at THIV was found to partially correlate with the detrended variability of accumulated rainfall at a local meteorological station with a lag of ~3 months (Figure 9). This likely indicates a delayed response of the vertical ground displacement due to the groundwater equilibrium, but the uplift trend prevails. Similar cases of uplift incidences due to recharging of the aquifer after near-record rainfall were documented by [93] using both GNSS and InSAR, supported by well level signals. The seismic swarm was initiated at the time of a highly decreased rate of the detrended accumulated rainfall that started declining 1 year before. This observation alone cannot support the groundwater level changes as a triggering cause of the seismic activity.

By employing InSAR data and through the application of the SBAS method, downlift or uplift, cumulated through the period between 2018 and September 2021, with reference to the GNSS station THIV, were detected at several localities in the study area. The examination of the temporal variation of vertical displacement at selected scatterers (Figure 10) showed subsidence associated with the $M_w = 4.6$ 2 December 2020 earthquake in the vicinity of Kallithea fault. On the other hand, no significant anomalies in the vertical displacement were recognized at scatterers near or south of Thiva, that is, the epicentral area of the 2021 swarm. Although coseismic displacements on the surface were not expected to be detected for the major events of the 2021 swarm, due to the small magnitudes and larger focal depths compared with the 2 December 2020 mainshock, the lack of preseismic deformation indicates no detectable changes that could be associated with slow slip or groundwater level changes, affecting the local stress.

Although the initiation of the 2021 swarm could not be explained by Coulomb stress transfer due to the 2 December 2020 mainshock, its evolution can be partly attributed to stress triggering. The major earthquakes of the swarm in July 2021 occurred at its western end, and the swarm progressed eastward in a cascade of outbreaks related to major events. Following a small interval of low seismic activity in the second half of August 2021, a second stage of the swarm was initiated by an $M_w = 4.1$ event on 2 September. However, part of the continuing eastward migration of seismicity could possibly be attributed to pore-pressure diffusion, as seismicity is roughly bounded by a parabolic envelope of hydraulic diffusivity $D = 0.25$ m^2/s (Figure 5). On a larger temporal scale, seismicity tends to spread eastward at a rate of ~70 m/day since August 2021. During the continued monitoring of earthquake activity in the study area at SL-NKUA, after 16 October 2021, that is, the end of the period studied in this work, the seismic swarm gradually diminished. Fewer than 300 events of $M_L \leq 2.6$ occurred until the end of January 2022 east of Thiva, where seismic activity was already observed in the first half of October 2021.

Both the abovementioned hydraulic diffusivity and the migration rate are comparable with the respective values previously measured in fluid-associated earthquake triggering at the Western Corinth Gulf (e.g., [9,10,12,94,95]). Seismic swarms are commonly observed in areas related to volcanic activity, involving magmatic or hydrothermal fluids (e.g., the 2008–2009 swarm at Yellowstone Lake [96] or the 2008 swarm at Vogtland/NW Bohemia [97]), or induced by water injection in geothermal fields (e.g., the 1993 swarm in Soultz-sous-Foret, France [98]). In tectonic environments, where seismic swarms are less frequent, the fluids that drive seismicity can be of meteoric origin, as has been suggested for cases such as the 2003–2004 swarm in the Western Corinth Gulf [99] or the 2011 swarm in Oichalia, Southern Greece [100], as well as for the 2002 swarm in Hochstaufen Mountain, SE Germany [101], and the 2003–2004 swarm in the Ubaye-Argentera area in the southwestern French–Italian Alps [102]. Another possible source could be the subducting oceanic slab through dehydration, with fluids migrating upwards (e.g., [103]). The intrusion of fluids through fissures or the fracture network can induce seismicity either by localizing stress or by increasing pore pressure, reducing the effective normal stress, and facilitating aseismic creep. During the evolution of fluid-driven swarms, small asperities embedded within the faults are breaking, as creeping of the surrounding surface proceeds, generating multiplets, caused by repeated slip on the same fault patch. Larger asperities may generate the major events of a swarm, transferring static stress to neighboring fault patches where aftershocks are triggered, causing the seismicity rate to increase abruptly, then to gradually decay following Omori's law. Therefore, swarms evolve as a consequence of both fluid intrusion and stress transfer, as has been suggested for the case of the seismic swarm of 2000 in Vogtland/NW Bohemia [104], and even involve episodes of aseismic slip, as in the October 2015 seismic swarm in Malamata, in the Western Corinth Gulf [12]. Spatial leaps in the triggering of earthquake clusters (e.g., group #4 of the 2021 Thiva swarm) or delayed major events (e.g., the 2 September 2021 event in group #6) could also be explained by subcritical crack growth due to stress corrosion attributed to chemical action at the crack tips (e.g., the earthquake swarm at Hida Mountains, Central Japan [105]). Another possible mechanism for gradual fault weakening is the erosion of the fault walls by fluids, which has been suggested for the 2008 swarm in Vogtland/NW Bohemia [106]. The short time period of relative seismic quiescence between 15 and 31 August in the Thiva area suggests that stress-transfer-related triggering effectively ceased in the western part of the swarm, but stress-induced corrosion, fluid erosion, or another aseismic factor, such as creeping or intrusion of subsurface fluids, were undermining the faults of the eastern half. The signature pattern of fluid-driven pore-pressure diffusion is the observed parabolic envelope in the spatiotemporal projection of Figure 5, indicating an expanding triggering front, starting from an injection point that roughly coincides with the location of the 2 September 2021 event. It is likely that the latter earthquake created a new pathway for the propagation of pressurized fluids, which caused seismicity to migrate further eastwards.

As evidenced by historical data, Thiva has experienced destructive $M_w \geq 6.0$ earthquakes in the past. Even though there are no significant recent events in either Kallithea or Thiva, macroseismic data suggest a past activation of the area in between. This points out that the 2020–2021 earthquake sequence does not reflect the full seismic potential of the study area. The local densification of the seismological network has permitted the detailed mapping of the activated structures at depth and the enhancement of the detection threshold. The fault segmentation, leading to a cascade of small outbreaks, and the *b*-value of the Gutenberg–Richter law of the 2021 seismic swarm, which is near or slightly above unity ($b \approx 1.07$ by the least-squares method, or $b \approx 1.14$ by the maximum likelihood method for a magnitude of completeness of M_c = 1.6; Figure A5), indicates that this particular seismogenic volume is unlikely to produce a large event at this time. However, other neighboring fault zones could possibly be triggered due to the cumulative stress transfer caused by the major events of the sequence. This could have potential implications to the seismic hazard of the area, as several seismogenic sources have been recognized in the vicinity of Thiva (e.g., [107]).

It is also worth noting that the evolution of a seismic swarm is not always similar to the one of the present study. Recently, on 27 September 2021, while the Thiva swarm was still active, a destructive earthquake (M_w = 6.0) occurred in Arkalochori in central Crete [31,32]. In Arkalochori, the seismic activity was initiated about 4 months before the mainshock, in early June 2021, as an earthquake swarm. However, the evolution of the Arkalochori sequence was dramatically different, culminating in the M_w = 6.0 mainshock that caused one fatality and hundreds of damaged buildings. In that instance, the seismic swarm was characterized, retrospectively, as a foreshock sequence. It is, therefore, important to continue monitoring the microseismic activity of the area of Thiva, in case seismicity migrates to neighboring structures.

7. Conclusions

We performed a detailed seismotectonic analysis of the 2020–2021 earthquake sequence in Central Greece involving a mainshock–aftershock sequence in December 2020 at Kallithea fault and a seismic swarm in July–October 2021 near Thiva. Through high-resolution double-difference relocation of the earthquake catalogue, we identified a system of conjugate WNW–ESE-trending normal faults below Thiva with several fault segments being triggered at different times, producing a complex seismic swarm that evolved in a series of bursts. Seismicity mainly migrated from west to east, with triggering mainly attributed to stress redistribution due to the larger events of the sequence.

The swarm could be divided in two stages, the first concerning its western half, initiated by an M_w = 4.2 event on 11 July 2021, and the second triggered by an M_w = 4.1 event on 2 September 2021, activating its eastern half. In the former stage, the seismic activity can be partly related to the, inferred, eastward extension of the south-dipping Yliki fault at the southern margin of Yliki Lake, north of Thiva. However, activity at more steeply dipping structures could also be identified. The second stage, likely related to the eastern termination of Yliki fault, also generated seismic activity on some conjugate north-dipping structures with some indications of triggering by overpressurized fluids, with seismicity migrating eastwards following a pore-pressure diffusion front.

We applied a multidisciplinary approach to investigate possible triggering mechanisms of the swarm. Coulomb stress transfer due to the M_w = 4.6 2 December 2020 earthquake was considered insufficient to trigger the July–October 2021 swarm near Thiva, as stress loading much lower than +0.1 bar was measured at the nucleation site. Furthermore, surface deformation measurements through GNSS and InSAR did not show any significant anomalies, which could be related to phenomena such as subsidence due to groundwater withdrawal or slow slip, which could have triggered the 2021 seismic swarm. Despite the apparent association of the December 2020 event with the July–October 2021 swarm, due to their proximity in both space and time, no causal relation could be derived between the two sequences. Subsurface fluids under pressure seem to have played a significant role in the evolution of the 2021 seismic swarm, either causing local stress concentrations or facilitating slip by lowering the effective normal stresses or weakening the faults' walls through erosion. The rich seismic activity in the conjugate normal fault system below Thiva likely released the accumulated stress in this particular area. However, considering the previous historic records of major $M > 6$ events in the vicinity, the occurrence of a significant earthquake at the unruptured downdip extension of Kallithea fault or in the seismic gap between Kallithea and Yliki faults cannot be excluded.

Supplementary Materials: The following are available online at https://www.mdpi.com/article/10.3390/app12052630/s1, Excel File S1: Relocated catalogue of the 2020–2021 earthquake sequence in Thiva; Excel File S2: Focal mechanisms determined by first-motion polarities (FMPs); Interactive Matlab Figure S1: 3D model of the 2020–2021 earthquake sequence at Thiva.

Author Contributions: Conceptualization, G.K., P.E., V.S. and I.P.; methodology, G.K., V.K. (Vasilis Kapetanidis), I.S., V.K. (Vasiliki Kouskouna), P.E. and V.S.; software, V.K. (Vasilis Kapetanidis), N.S., P.E., A.K. (Andreas Karakonstantis), A.K. (Andreas Karavias), V.S. and T.G.; validation, G.K., I.S., V.K. (Vasiliki Kouskouna), P.E., I.K., J.D.A., P.P., N.V. and I.P.; investigation, V.K. (Vasilis Kapetanidis), I.S., N.S., A.K. (Andreas Karakonstantis), A.K. (Andreas Karavias) and T.G.; resources, G.K., V.K. (Vasiliki Kouskouna), P.E., V.S., I.K., J.D.A., P.P., N.V. and I.P.; data curation, G.K., V.K. (Vasilis Kapetanidis), I.S., N.S., V.K. (Vasiliki Kouskouna), P.E. and V.S.; writing—original draft preparation, G.K., V.K. (Vasilis Kapetanidis), I.S., N.S., A.K. (Andreas Karakonstantis), V.K. (Vasiliki Kouskouna), P.E., V.S. and I.P.; writing—review and editing, G.K., V.K. (Vasilis Kapetanidis), I.S., N.S., V.K. (Vasiliki Kouskouna), P.E. and V.S.; visualization, V.K. (Vasilis Kapetanidis), P.E., A.K. (Andreas Karavias) and V.S.; supervision, G.K.; project administration, P.P. and N.V.; funding acquisition, G.K. All authors have read and agreed to the published version of the manuscript.

Funding: This research received no external funding.

Institutional Review Board Statement: Not applicable.

Informed Consent Statement: Not applicable.

Data Availability Statement: The initial earthquake catalogues and arrival-time data are available at the database of SL-NKUA (http://www.geophysics.geol.uoa.gr/stations/gmapv3_db/index.php?lang=en; accessed on 31 January 2022) and GI-NOA (http://bbnet.gein.noa.gr/HL/databases/database; accessed on 31 January 2022). The relocated catalogue of this study is available in the Supplementary Material (Excel File S1). Seismological waveform data from permanent and temporary stations of the HUSN are available at the European Integrated Data Archive (EIDA) node hosted at GI-NOA (http://eida.gein.noa.gr/; accessed on 31 January 2022) [56]. Data of the meteorological station at Thiva are operated by the METEO unit at the National Observatory of Athens (https://penteli.meteo.gr/stations/thiva/; accessed on 28 January 2022). The Sentinel-1 SAR SLC images are available on the Copernicus Open Access Hub (https://scihub.copernicus.eu/dhus/#/home; accessed on 28 January 2022) [78]. The digital elevation model was extracted by the NASA's Shuttle Radar Topography Mission (SRTM) 3 arc-second (90 m/pixel) database (https://srtm.csi.cgiar.org/10; accessed on 10 October 2021) [84]. The GNSS data used in this study are provided by the private sector and are not available.

Acknowledgments: The authors would like to thank the three anonymous reviewers and the academic editor for the constructive and detailed comments that contributed to the improvement of the manuscript. We would like to thank the scientists and personnel who participated in the installation or maintenance of the permanent and temporary stations belonging to the HUSN. Sentinel-1 satellite imagery was acquired through the European Space Agency and Copernicus. We are grateful to Metrica S.A. and TreeComp for kindly sharing their GNSS data and M. Chanioti (sales manager of Inforest Research O.C.) for SARSCAPE s/w availability. We would also like to thank Charalambos Georgiou for the valuable discussions. Most maps and cross sections were drawn with the Generic Mapping Tools (GMT) software [108]. The Coulomb stress transfer model for the 2 December 2020 mainshock was constructed using the Coulomb 3.3 software [70].

Conflicts of Interest: The authors declare no conflict of interest.

Appendix A. Additional Macroseismic Data

Macroseismic intensity data points (IDPs) were re-evaluated to determine the macroseismic epicenter using the MEEP software [50].

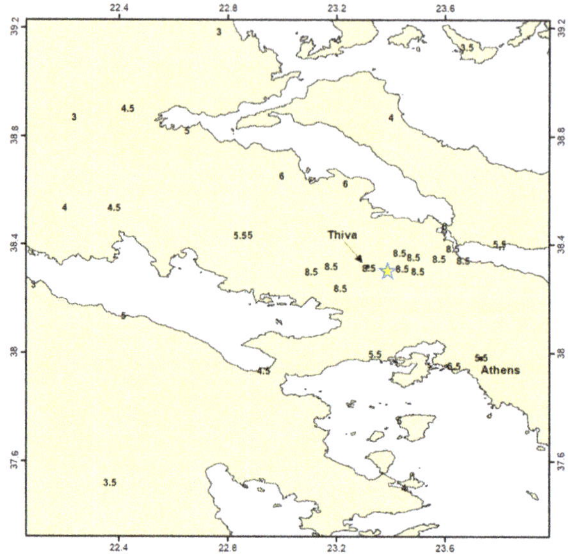

Figure A1. EMS98 intensity data points (IDPs) of the 17 October 1914 earthquake and the calculated macroseismic epicenter (star).

Appendix B. Seismological Methods

Statistics and additional information concerning the seismological data.

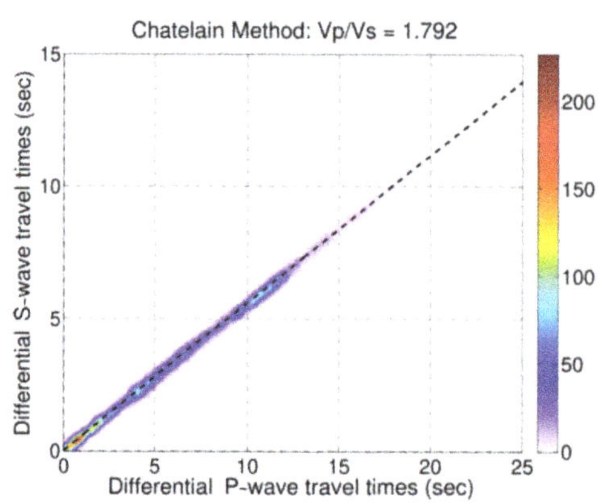

Figure A2. Chatelain [64] diagram for the determination of the average ratio $V_p/V_s = 1.79$ using P and S travel-time data from the 2020–2021 Thiva earthquake sequence.

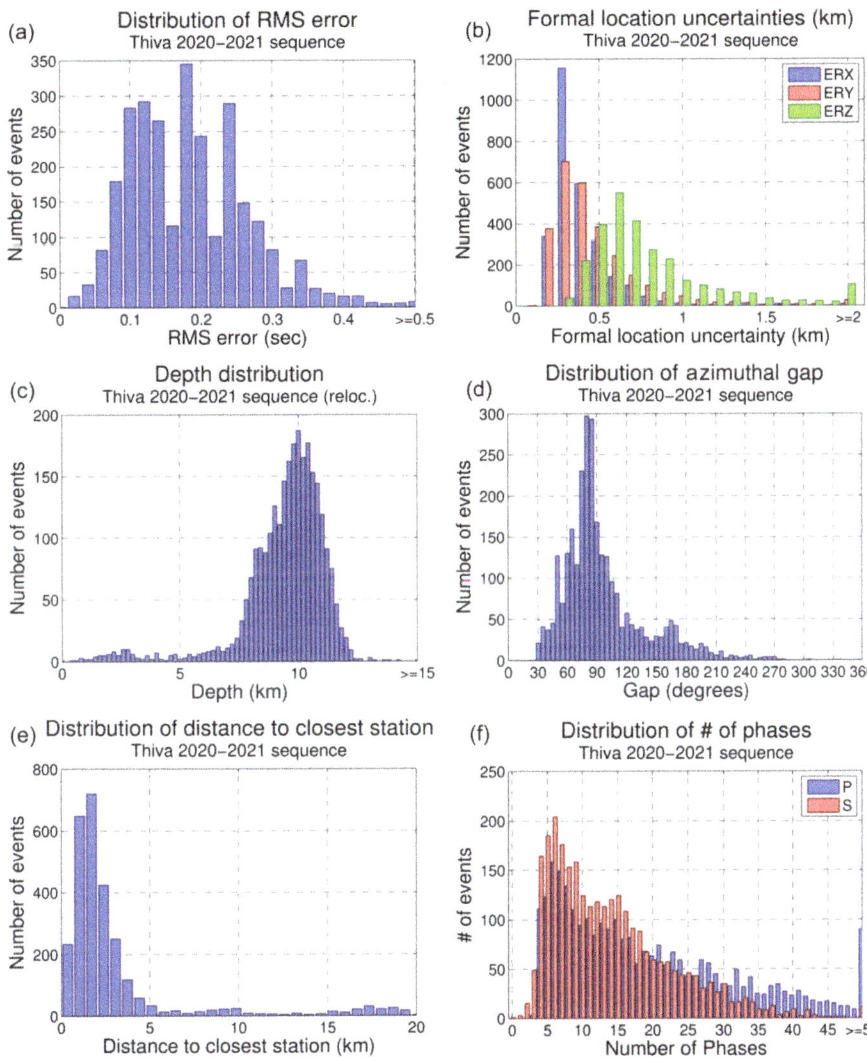

Figure A3. Histograms of location statistics and uncertainties for the 2020–2021 Thiva earthquake sequence: (**a**) root mean square (RMS) travel-time residuals, (**b**) horizontal (ERX, ERY) and vertical (ERZ) formal uncertainties reported by the location algorithm HypoInverse, (**c**) focal depths from the relocated catalogue with the HypoDD code, (**d**) azimuthal gap, (**e**) epicentral distance to the closest station, and (**f**) number of P- (blue) and S-wave (red) arrival times.

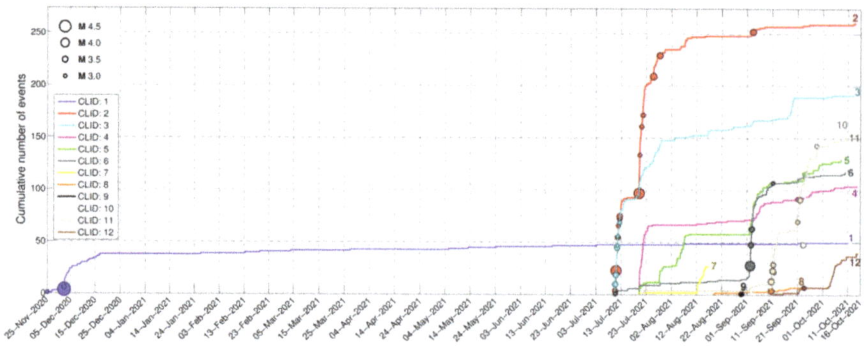

Figure A4. Cumulative number of events with $M \geq M_c = 1.6$ per spatial group, represented by different colors and numbers, for the period between 25 November 2020 and 16 October 2021. The occurrence of events with $M \geq 3.0$ is marked with circles with size proportional to magnitude. A close-up of the period between July and 16 October 2021 is presented in Figure 4b.

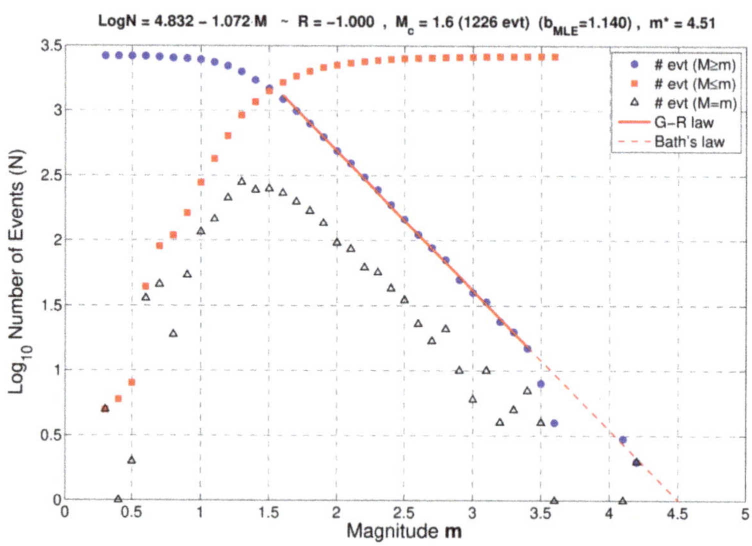

Figure A5. Gutenberg–Richter law of the 2021 earthquake swarm (groups #2–#12, excluding group #9; Figure 4) with blue dots showing the $\log_{10}$ number N of events with magnitude $M > m$, whereas red dots show the reverse relation. Open triangles show the number of events with magnitude M in the range $m - 0.05 \leq M \leq m + 0.05$. The straight red line shows the linear fit corresponding to a b-value of 1.072. The magnitude of completeness is $M_c = 1.6$. The b-value determined from the maximum likelihood method is $b_{MLE} = 1.14$. The red dashed line is the extension of the linear fit to $\log_{10}N = 0$, with $m^* = 4.51$ the largest expected magnitude according to the modified Bath's law [109].

Table A1. Focal mechanisms of major events determined by first-motion polarities (FMPs). CLID is the spatial group number. For more details, see Excel File S2 in the Supplementary Material.

Origin Time	Latitude (°N)	Longitude (°E)	Depth (km)	M_L	Strike$_1$ (°)	Dip$_1$ (°)	Rake$_1$ (°)	CLID
2 December 2020 12:37:37.7	38.3001	23.4327	3.4	3.3	240.2	46.6	−142.6	1
10 July 2021 12:54:17.7	38.3233	23.3276	9.8	3.2	293.9	37.1	−94.6	2
20 July 2021 07:36:20.4	38.3233	23.3254	10.0	2.9	298.1	31.5	−50.1	2
21 July 2021 03:54:53.9	38.3286	23.3064	9.9	3.1	238.3	38.3	−122.6	2
25 July 2021 19:12:44.1	38.3273	23.3055	10.9	3.6	261.8	41.2	−120.7	2
29 August 2021 18:15:23.0	38.2942	23.4932	11.8	3.2	317.1	44.3	−48.9	9
30 August 2021 06:08:12.1	38.2876	23.4836	13.0	2.9	325.7	67	−13.1	9
2 September 2021 09:03:54.2	38.3222	23.3335	10.9	3.0	248.7	39.7	−125.5	5
2 September 2021 14:46:15.8	38.3248	23.3370	9.5	3.3	286.9	34.7	−99.7	6
3 September 2021 05:13:51.6	38.3260	23.3263	8.2	3.5	290.8	32.9	−95.5	2
10 September 2021 13:51:33.0	38.3211	23.3439	9.6	3.5	287.1	36.6	−117.5	11
11 September 2021 05:28:16.0	38.3226	23.3545	9.7	3.5	279.3	35.6	−95.4	11
11 September 2021 09:42:38.8	38.3233	23.3503	10.7	3.3	293.8	32.5	−95.2	11
11 September 2021 17:45:48.8	38.3220	23.3427	10.6	3.2	278.7	38.6	−110.8	10
21 September 2021 01:50:54.8	38.3207	23.3564	11.1	3.2	248.5	42.4	−130.5	11
22 September 2021 01:03:07.7	38.3187	23.3574	9.3	3.4	295.3	28.9	−93	11
23 September 2021 19:28:56.2	38.3145	23.3376	11.3	3.1	279.6	38.6	−94.2	5
28 September 2021 11:41:55.0	38.3180	23.3442	10.3	3.1	98.4	36.6	−93.8	10
29 September 2021 00:20:39.0	38.3176	23.3433	11.3	2.9	238.1	48	−138.2	10
10 October 2021 05:37:41.1	38.3252	23.3513	10.6	2.8	242.5	48.7	−137.1	11

Appendix C. Geodetic Methods

Daily data from the continuous GNSS station THIV located in Thiva were processed over a period of 5 years, 2017 to 2021. Time series were produced, and the velocity vector was calculated.

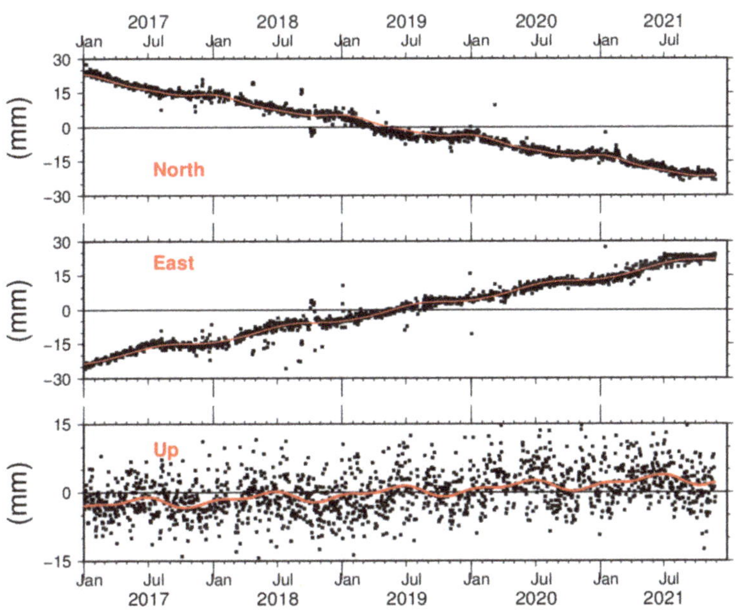

Figure A6. Time series of the three components for the THIV GNSS site. The fit (red line) produced by the software in [110].

References

1. Papazachos, B.C. Seismicity of the Aegean and surrounding area. *Tectonophysics* **1990**, *178*, 287–308. [CrossRef]
2. Armijo, R.; Lyon-Caen, H.; Papanastassiou, D. East-west extension and Holocene normal-fault scarps in the Hellenic arc. *Geology* **1992**, *20*, 491–494. [CrossRef]

3. McKenzie, D. Some remarks on the development of sedimentary basins. *Earth Planet. Sci. Lett.* **1978**, *40*, 25–32. [CrossRef]
4. Ganas, A.; Parsons, T. Three-dimensional model of hellenic arc deformation and origin of the cretan uplift. *J. Geophys. Res. Solid Earth* **2009**, *114*, B06404. [CrossRef]
5. Briole, P.; Ganas, A.; Elias, P.; Dimitrov, D. The GPS velocity field of the Aegean. New observations, contribution of the earthquakes, crustal blocks model. *Geophys. J. Int.* **2021**, *226*, 468–492. [CrossRef]
6. Makropoulos, K.; Kaviris, G.; Kouskouna, V. An updated and extended earthquake catalogue for Greece and adjacent areas since 1900. *Nat. Hazards Earth Syst. Sci.* **2012**, *12*, 1425–1430. [CrossRef]
7. Kaviris, G.; Elias, P.; Kapetanidis, V.; Serpetsidaki, A.; Karakonstantis, A.; Plicka, V.; De Barros, L.; Sokos, E.; Kassaras, I.; Sakkas, V.; et al. The Western Gulf of Corinth (Greece) 2020–2021 Seismic Crisis and Cascading Events: First Results from the Corinth Rift Laboratory Network. *Seism. Rec.* **2021**, *1*, 85–95. [CrossRef]
8. Pacchiani, F.; Lyon-Caen, H. Geometry and spatio-temporal evolution of the 2001 Agios Ioanis earthquake swarm (Corinth Rift, Greece). *Geophys. J. Int.* **2010**, *180*, 59–72. [CrossRef]
9. Kapetanidis, V.; Deschamps, A.; Papadimitriou, P.; Matrullo, E.; Karakonstantis, A.; Bozionelos, G.; Kaviris, G.; Serpetsidaki, A.; Lyon-Caen, H.; Voulgaris, N.; et al. The 2013 earthquake swarm in Helike, Greece: Seismic activity at the root of old normal faults. *Geophys. J. Int.* **2015**, *202*, 2044–2073. [CrossRef]
10. Kapetanidis, V.; Michas, G.; Kaviris, G.; Vallianatos, F. Spatiotemporal Properties of Seismicity and Variations of Shear-Wave Splitting Parameters in the Western Gulf of Corinth (Greece). *Appl. Sci.* **2021**, *11*, 6573. [CrossRef]
11. Kaviris, G.; Spingos, I.; Kapetanidis, V.; Papadimitriou, P.; Voulgaris, N.; Makropoulos, K. Upper crust seismic anisotropy study and temporal variations of shear-wave splitting parameters in the western Gulf of Corinth (Greece) during 2013. *Phys. Earth Planet. Inter.* **2017**, *269*, 148–164. [CrossRef]
12. De Barros, L.; Cappa, F.; Deschamps, A.; Dublanchet, P. Imbricated Aseismic Slip and Fluid Diffusion Drive a Seismic Swarm in the Corinth Gulf, Greece. *Geophys. Res. Lett.* **2020**, *47*, e2020GL087142. [CrossRef]
13. Kassaras, I.; Kapetanidis, V.; Ganas, A.; Tzanis, A.; Kosma, C.; Karakonstantis, A.; Valkaniotis, S.; Chailas, S.; Kouskouna, V.; Papadimitriou, P. The New Seismotectonic Atlas of Greece (v1.0) and Its Implementation. *Geosciences* **2020**, *10*, 447. [CrossRef]
14. Stucchi, M.; Rovida, A.; Gomez Capera, A.A.; Alexandre, P.; Camelbeeck, T.; Demircioglu, M.B.; Gasperini, P.; Kouskouna, V.; Musson, R.M.W.; Radulian, M.; et al. The SHARE European Earthquake Catalogue (SHEEC) 1000–1899. *J. Seismol.* **2013**, *17*, 523–544. [CrossRef]
15. Ganas, A.; Oikonomou, I.A.; Tsimi, C. NOAfaults: A digital database for active faults in Greece. *Bull. Geol. Soc. Greece* **2017**, *47*, 518. [CrossRef]
16. Ganas, A. NOAFAULTS KMZ layer Version 3.0 (2020 update) (V3.0) [Data set]. *Zenodo* **2020**. [CrossRef]
17. Kiratzi, A. The 12 June 2017 Mw 6.3 Lesvos Island (Aegean Sea) earthquake: Slip model and directivity estimated with finite-fault inversion. *Tectonophysics* **2018**, *724–725*, 1–10. [CrossRef]
18. Papadimitriou, P.; Kassaras, I.; Kaviris, G.; Tselentis, G.-A.; Voulgaris, N.; Lekkas, E.; Chouliaras, G.; Evangelidis, C.; Pavlou, K.; Kapetanidis, V.; et al. The 12th June 2017 Mw = 6.3 Lesvos earthquake from detailed seismological observations. *J. Geodyn.* **2018**, *115*, 23–42. [CrossRef]
19. Ganas, A.; Elias, P.; Kapetanidis, V.; Valkaniotis, S.; Briole, P.; Kassaras, I.; Argyrakis, P.; Barberopoulou, A.; Moshou, A. The July 20, 2017 M6.6 Kos Earthquake: Seismic and Geodetic Evidence for an Active North-Dipping Normal Fault at the Western End of the Gulf of Gökova (SE Aegean Sea). *Pure Appl. Geophys.* **2019**, *176*, 4177–4211. [CrossRef]
20. Triantafyllou, I.; Papadopoulos, G.A.; Lekkas, E. Impact on built and natural environment of the strong earthquakes of 23 April 1933, and 20 July 2017, in the southeast Aegean Sea, eastern Mediterranean. *Nat. Hazards* **2020**, *100*, 671–695. [CrossRef]
21. Papadimitriou, P.; Kapetanidis, V.; Karakonstantis, A.; Spingos, I.; Kassaras, I.; Sakkas, V.; Kouskouna, V.; Karatzetzou, A.; Pavlou, K.; Kaviris, G.; et al. First Results on the Mw = 6.9 Samos Earthquake of 30 October 2020. *Bull. Geol. Soc. Greece* **2020**, *56*, 251–279. [CrossRef]
22. Karakostas, V.; Tan, O.; Kostoglou, A.; Papadimitriou, E.; Bonatis, P. Seismotectonic implications of the 2020 Samos, Greece, M w 7.0 mainshock based on high-resolution aftershock relocation and source slip model. *Acta Geophys.* **2021**, *69*, 979–996. [CrossRef]
23. Kaviris, G.; Spingos, I.; Kapetanidis, V.; Papadimitriou, P.; Voulgaris, N. On the origin of upper crustal shear-wave anisotropy at Samos Island, Greece. *Acta Geophys.* **2021**, *69*, 1051–1064. [CrossRef]
24. Ganas, A.; Elias, P.; Briole, P.; Valkaniotis, S.; Escartin, J.; Tsironi, V.; Karasante, I.; Kosma, C. Co-seismic and post-seismic deformation, field observations and fault model of the 30 October 2020 Mw = 7.0 Samos earthquake, Aegean Sea. *Acta Geophys.* **2021**, *69*, 999–1024. [CrossRef]
25. Ganas, A.; Valkaniotis, S.; Briole, P.; Serpetsidaki, A.; Kapetanidis, V.; Karasante, I.; Kassaras, I.; Papathanassiou, G.; Karamitros, I.; Tsironi, V.; et al. Domino-style earthquakes along blind normal faults in Northern Thessaly (Greece): Kinematic evidence from field observations, seismology, SAR interferometry and GNSS. *Bull. Geol. Soc. Greece* **2021**, *58*, 37. [CrossRef]
26. Karakostas, V.; Papazachos, C.; Papadimitriou, E.; Foumelis, M.; Kiratzi, A.; Pikridas, C.; Kostoglou, A.; Kkallas, C.; Chatzis, N.; Bitharis, S.; et al. The March 2021 Tyrnavos, central Greece, doublet (Mw6.3 and Mw6.0): Aftershock relocation, faulting details, coseismic slip and deformation. *Bull. Geol. Soc. Greece* **2021**, *58*, 131. [CrossRef]
27. Lekkas, E.; Agorastos, K.; Mavroulis, S.; Kranis, C.; Skourtsos, E.; Carydis, P.; Gogou, M.; Katsetsiadou, K.-N.; Papadopoulos, G.; Triantafyllou, I.; et al. The Early March 2021 THESSALY Earthquake Sequence. 2021, pp. 1–195. Available online: https://edcm.edu.gr/images/docs/newsletters/Newsletter_22_2021-Thessaly-EQs.pdf (accessed on 28 January 2022).

28. Papadopoulos, G.A.; Agalos, A.; Karavias, A.; Triantafyllou, I.; Parcharidis, I.; Lekkas, E. Seismic and Geodetic Imaging (DInSAR) Investigation of the March 2021 Strong Earthquake Sequence in Thessaly, Central Greece. *Geosciences* **2021**, *11*, 311. [CrossRef]
29. Vallianatos, F.; Michas, G.; Hloupis, G. Seismicity Patterns Prior to the Thessaly (Mw6.3) Strong Earthquake on 3 March 2021 in Terms of Multiresolution Wavelets and Natural Time Analysis. *Geosciences* **2021**, *11*, 379. [CrossRef]
30. Kassaras, I.; Kapetanidis, V.; Ganas, A.; Karakonstantis, A.; Papadimitriou, P.; Kaviris, G.; Kouskouna, V.; Voulgaris, N. Seismotectonic analysis of the 2021 Damasi-Tyrnavos (Thessaly, Central Greece) earthquake sequence and implications on the stress field rotations. *J. Geodyn.* **2022**, *150*, 101898. [CrossRef]
31. Vallianatos, F.; Michas, G.; Hloupis, G.; Chatzopoulos, G. The Evolution of Preseismic Patterns Related to the Central Crete (Mw6.0) Strong Earthquake on 27 September 2021 Revealed by Multiresolution Wavelets and Natural Time Analysis. *Geosciences* **2022**, *12*, 33. [CrossRef]
32. Vassilakis, E.; Kaviris, G.; Kapetanidis, V.; Papageorgiou, E.; Foumelis, M.; Konsolaki, A.; Petrakis, S.; Evangelidis, C.; Alexopoulos, J.; Karastathis, V.; et al. The 27 September 2021, earthquake in central Crete (Greece). Detailed analysis of the earthquake sequence and indications for contemporary arc-parallel extension at the Hellenic arc. *Appl. Sci.* **2022**. *submitted*.
33. Sboras, S.; Ganas, A.; Pavlides, S. Morphotectonic Analysis of the Neotectonic and Active Faults of Beotia (Central Greece), Using G.I.S. Techniques. *Bull. Geol. Soc. Greece* **2017**, *43*, 1607–1618. [CrossRef]
34. Goldsworthy, M.; Jackson, J.; Haines, J. The continuity of active fault systems in Greece. *Geophys. J. Int.* **2002**, *148*, 596–618. [CrossRef]
35. Tsodoulos, I.M.; Koukouvelas, I.K.; Pavlides, S. Tectonic geomorphology of the easternmost extension of the Gulf of Corinth (Beotia, Central Greece). *Tectonophysics* **2008**, *453*, 211–232. [CrossRef]
36. Elias, P.; Spingos, I.; Kaviris, G.; Karavias, A.; Gatsios, T.; Sakkas, V.; Parcharidis, I. Combined Geodetic and Seismological Study of the December 2020 Mw = 4.6 Thiva (Central Greece) Shallow Earthquake. *Appl. Sci.* **2021**, *11*, 5947. [CrossRef]
37. Stewart, I.S.; Piccardi, L. Seismic faults and sacred sanctuaries in Aegean antiquity. *Proc. Geol. Assoc.* **2017**, *128*, 711–721. [CrossRef]
38. Nur, A.; Cline, E.H. Poseidon's Horses: Plate Tectonics and Earthquake Storms in the Late Bronze Age Aegean and Eastern Mediterranean. *J. Archaeol. Sci.* **2000**, *27*, 43–63. [CrossRef]
39. Ambraseys, N. *Earthquakes in the Mediterranean and Middle East*; Cambridge University Press: New York, NY, USA, 2009.
40. Eginitis, D. Le tremblement de terre de Thèbes du 17 Octobre 1914. *Ann. L'observatoire Natl. D'athènes* **1916**, *7*, 23–26.
41. Schmidt, J.F. *Studien über Erdbeben*; Alwin Georgi: Leipzig, Germany, 1879; 360p.
42. Ambraseys, N.N.; Jackson, J.A. Seismicity and associated strain of central Greece between 1890 and 1988. *Geophys. J. Int.* **1990**, *101*, 663–708. [CrossRef]
43. Papazachos, B.C.; Papazachou, C. *The Earthquakes of Greece*; Ziti Publications: Thessaloniki, Greek, 2003.
44. Goulandris, E. Sur le tremblement de terre de Thèbes du 17 Octobre 1914. *Ann. L'observatoire Natl. D'athènes* **1916**, *7*, 47–50.
45. Galanopoulos, A.G. *A catalogue of Shocks with I0 ≥ VI or M ≥ 5 for the Years 1801–1958*; Seismological Laboratory, Athens University: Athens, Greece, 1960; p. 119.
46. Kárnik, V. *Seismicity of the European Area, Part 1*; Reidel, D., Ed.; Springer Science & Business Media: Berlin, Germany, 1969.
47. Shebalin, N.V.; Kárnik, V.; Hadzievski, D. Catalogue of earthquakes of the Balkan region. In *Part I, UNDP-UNESCO Survey of the Seismicity of the Balkan Region*; UNESCO: Skopje, North Macedonia, 1974.
48. Kouskouna, V. Updating the macroseismic intensity database of 19th century damaging earthquakes in Greece: A case study in Samos Island. *Acta Geophys.* **2021**, *69*, 1101–1111. [CrossRef]
49. Grunthal, G. *European MakroseismicScale 1998, EMS-98*; European Seismological Commission: Luxembourg, 1998.
50. Musson, R.M.W.; Jimenéz, M.J. Macroseismic Estimation of Earthquake Parameters. NA4 Deliverable D3, NERIES Project. 2008. Available online: http://emidius.mi.ingv.it/neries_NA4/docs/NA4_D3.pdf (accessed on 28 January 2022).
51. Kouskouna, V.; Kaperdas, V.; Sakellariou, N. Comparing calibration coefficients constrained from early to recent macroseismic and instrumental earthquake data in Greece and applied to eighteenth century earthquakes. *J. Seismol.* **2020**, *24*, 293–317. [CrossRef]
52. Gasperini, P.; Bernardini, F.; Valensise, G.; Boschi, E. Defining seismogenic sources from historical earthquake reports. *Bull. Seismol. Soc. Am.* **1999**, *89*, 94–110.
53. Bakun, W.H.; Wentworth, C.M. Estimating earthquake location and magnitude from seismic intensity data. *Bull. Seismol. Soc. Am.* **1997**, *87*, 1502–1521.
54. Shumila, V.I. Algorithms of macroseismic intensity field interpretation with application to Vrancea region. In Proceedings of the 24th ESC General Assembly, Athina, Greece, 19–24 September 1994; p. 133.
55. Bakun, W.H.; Scotti, O. Regional intensity attenuation models for France and the estimation of magnitude and location of historical earthquakes. *Geophys. J. Int.* **2006**, *164*, 596–610. [CrossRef]
56. Evangelidis, C.P.; Triantafyllis, N.; Samios, M.; Boukouras, K.; Kontakos, K.; Ktenidou, O.-J.; Fountoulakis, I.; Kalogeras, I.; Melis, N.S.; Galanis, O.; et al. Seismic Waveform Data from Greece and Cyprus: Integration, Archival, and Open Access. *Seismol. Res. Lett.* **2021**, *92*, 1672–1684. [CrossRef]
57. Klein, F.W. *User's Guide to HYPOINVERSE-2000: A Fortran Program to Solve for Earthquake Locations and Magnitudes, Open-File Rep. 2002-171*; U.S. Geological Survey: Reston, VA, USA, 2002. [CrossRef]
58. Kaviris, G.; Papadimitriou, P.; Makropoulos, K. Magnitude Scales in Central Greece. *Bull. Geol. Soc. Greece* **2007**, *40*, 1114–1124. [CrossRef]

59. Papadimitriou, P.; Kaviris, G.; Makropoulos, K. Evidence of shear-wave splitting in the eastern Corinthian Gulf (Greece). *Phys. Earth Planet. Inter.* **1999**, *114*, 3–13. [CrossRef]
60. Papadimitriou, P.; Kaviris, G.; Karakonstantis, A.; Makropoulos, K. The Cornet seismological network: 10 years of operation, recorded seismicity and significant applications. *Ann. Geol Des Pays Hell.* **2010**, *45*, 193–208.
61. Kaviris, G. Study of Seismic Source Properties of the Eastern Gulf of Corinth. Ph.D. Thesis, Geophysics-Geothermics Department, Faculty of Geology, University of Athens, Athens, Greece, 2003. (In Greek).
62. Karakonstantis, A. 3-D Simulation of Crust and Upper Mantle Structure in the Broader Hellenic Area through Seismic Tomography. Ph.D. Thesis, Department of Geophysics-Geothermics, Faculty of Geology, University of Athens, Athens, Greece, 2017. (In Greek).
63. D'Ajello Caracciolo, F.; Console, R. Earthquake location in tectonic structures of the Alpine Chain: The case of the Constance Lake (Central Europe) seismic sequence. *Acta Geophys.* **2021**, *69*, 1163–1175. [CrossRef]
64. Chatelain, J.L. *Etude fine de la Sismicité en Zone de Collision Continentale au Moyen d'un Réseau de Stations Portables: La région Hindu-Kush Pamir, Thèse de 3eme Cycle*; Université de Grenoble: Saint-Martin-d'Hères, France, 1978; p. 219.
65. Woessner, J.; Wiemer, S. Assessing the quality of earthquake catalogues: Estimating the magnitude of completeness and its uncertainty. *Bull. Seismol. Soc. Am.* **2005**, *95*, 684–698. [CrossRef]
66. Waldhauser, F. hypoDD—A Program to Compute Double-Difference Hypocenter Locations. U.S. Geological Survey Open-File Report 01-113. 2001. Available online: https://pubs.usgs.gov/of/2001/0113/ (accessed on 28 January 2022). [CrossRef]
67. Kapetanidis, V. Spatiotemporal Patterns of Microseismicity for the Identification of Active Fault Structures Using Seismic Waveform Cross-Correlation and Double-Difference Relocation. Ph.D. Thesis, Department of Geophysics-Geothermics, Faculty of Geology and Geoenvironment, University of Athens, Athens, Greece, 2017.
68. Shapiro, S.A.; Huenges, E.; Borm, G. Estimating the crust permeability from fluid-injection-induced seismic emission at the KTB site. *Geophys. J. Int.* **1997**, *131*, F15–F18. [CrossRef]
69. Kagan, Y.Y. 3-D rotation of double-couple earthquake sources. *Geophys. J. Int.* **1991**, *106*, 709–716. [CrossRef]
70. Toda, S.; Stein, R.S.; Sevilgen, V.; Lin, J. Coulomb 3.3 Graphic-Rich Deformation and Stress-Change Software for Earthquake, Tectonic, and Volcano Research and Teaching-User Guide. U.S. Geological Survey Open-File Report 2011–1060. 2011. Available online: http://pubs.usgs.gov/of/2011/1060/ (accessed on 28 January 2022). [CrossRef]
71. Kapetanidis, V.; Kassaras, I. Contemporary crustal stress of the Greek region deduced from earthquake focal mechanisms. *J. Geodyn.* **2019**, *123*, 55–82. [CrossRef]
72. Briole, P.; Elias, P.; Parcharidis, I.; Bignami, C.; Benekos, G.; Samsonov, G.; Kyriakopoulos, C.; Stramondo, S.; Chamot-Rooke, N.; Drakatou, M.L.; et al. The seismic sequence of January-February 2014 at Cephalonia Island (Greece): Constraints from SAR interferometry and GPS. *Geophys. J. Int.* **2015**, *203*, 1528–1540. [CrossRef]
73. Ganas, A.; Elias, P.; Bozionelos, G.; Papathanassiou, G.; Avallone, A.; Papastergios, A.; Valkaniotis, S.; Parcharidis, I.; Briole, P. Coseismic deformation, field observations and seismic fault of the 17 November 2015 M = 6.5, Lefkada Island, Greece earthquake. *Tectonophysics* **2016**, *687*, 210–222. [CrossRef]
74. Elias, P.; Briole, P. Ground Deformations in the Corinth Rift, Greece, Investigated Through the Means of SAR Multitemporal Interferometry. Geochemistry, Geophys. *Geosystems* **2018**, *19*, 4836–4857. [CrossRef]
75. Gatsios, T.; Cigna, F.; Tapete, D.; Sakkas, V.; Pavlou, K.; Parcharidis, I. Copernicus sentinel-1 MT-InSAR, GNSS and seismic monitoring of deformation patterns and trends at the methana volcano, Greece. *Appl. Sci.* **2020**, *10*, 6445. [CrossRef]
76. Chousianitis, K.; Papanikolaou, X.; Drakatos, G.; Tselentis, G.-A. NOANET: A Continuously Operating GNSS Network for Solid-Earth Sciences in Greece. *Seismol. Res. Lett.* **2021**, *92*, 2050–2064. [CrossRef]
77. Sakkas, V. Ground Deformation Modelling of the 2020 Mw6.9 Samos Earthquake (Greece) Based on InSAR and GNSS Data. *Remote Sens.* **2021**, *13*, 1665. [CrossRef]
78. Copernicus Open Access Hub. Available online: https://scihub.copernicus.eu/dhus/#/home (accessed on 28 January 2022).
79. METRICA. Available online: https://metrica.gr/ (accessed on 28 January 2022).
80. URANUS Powered by TREE COMPANY. Available online: https://uranus.gr/ (accessed on 28 January 2022).
81. Dach, R.; Lutz, S.; Walser, P.; Fridez, P. *Bernese GNSS Software Version 5.2*; User manual; Astronomical Institute, University of Bern, Bern Open Publishing: Bern, Switzerland, 2015.
82. Chousianitis, K.; Ganas, A.; Evangelidis, C.P. Strain and rotation rate patterns of mainland Greece from continuous GPS data and comparison between seismic and geodetic moment release. *J. Geophys. Res. Solid Earth* **2015**, *120*, 3909–3931. [CrossRef]
83. Lagouvardos, K.; Kotroni, V.; Bezes, A.; Koletsis, I.; Kopania, T.; Lykoudis, S.; Mazarakis, N.; Papagiannaki, K.; Vougioukas, S. The automatic weather stations NOANN network of the National Observatory of Athens: Operation and database. *Geosci. Data J.* **2017**, *4*, 4–16. [CrossRef]
84. SRTM Data. Available online: https://srtm.csi.cgiar.org/10 (accessed on 28 January 2022).
85. Berardino, P.; Fornaro, G.; Lanari, R.; Sansosti, E. A new algorithm for surface deformation monitoring based on small baseline differential SAR interferograms. *IEEE Trans. Geosci. Remote Sens.* **2002**, *40*, 2375–2383. [CrossRef]
86. Costantini, M. A novel phase unwrapping method based on network programming. *IEEE Trans. Geosci. Remote Sens.* **1998**, *36*, 813–821. [CrossRef]
87. Fielding, E.J.; Simons, M.; Owen, S.; Lundgren, P.; Hua, H.; Agram, P.; Liu, Z.; Moore, A.; Milillo, P.; Polet, J.; et al. Rapid Imaging of Earthquake Ruptures with Combined Geodetic and Seismic Analysis. *Procedia Technol.* **2014**, *16*, 876–885. [CrossRef]

88. Causse, M.; Cornou, C.; Maufroy, E.; Grasso, J.-R.; Baillet, L.; El Haber, E. Exceptional ground motion during the shallow Mw 4.9 2019 Le Teil earthquake, France. *Commun. Earth Environ.* **2021**, *2*, 14. [CrossRef]
89. Champenois, J.; Baize, S.; Vallee, M.; Jomard, H.; Alvarado, A.; Espin, P.; Ekström, G.; Audin, L. Evidences of Surface Rupture Associated With a Low-Magnitude (M w 5.0) Shallow Earthquake in the Ecuadorian Andes. *J. Geophys. Res. Solid Earth* **2017**, *122*, 8446–8458. [CrossRef]
90. Stein, R.S. The role of stress transfer in earthquake occurrence. *Nature* **1999**, *402*, 605–609. [CrossRef]
91. Cheloni, D.; D'Agostino, N.; Selvaggi, G.; Avallone, A.; Fornaro, G.; Giuliani, R.; Reale, D.; Sansosti, E.; Tizzani, P. Aseismic transient during the 2010-2014 seismic swarm: Evidence for longer recurrence of M $\geq$ 6.5 earthquakes in the Pollino gap (Southern Italy)? *Sci. Rep.* **2017**, *7*, 576. [CrossRef]
92. Tiwari, D.K.; Jha, B.; Kundu, B.; Gahalaut, V.K.; Vissa, N.K. Groundwater extraction-induced seismicity around Delhi region, India. *Sci. Rep.* **2021**, *11*, 10097. [CrossRef]
93. King, N.E.; Argus, D.; Langbein, J.; Agnew, D.C.; Bawden, G.; Dollar, R.S.; Liu, Z.; Galloway, D.; Reichard, E.; Yong, A.; et al. Space geodetic observation of expansion of the San Gabriel Valley, California, aquifer system, during heavy rainfall in winter 2004–2005. *J. Geophys. Res. Solid Earth* **2007**, *112*, B03409. [CrossRef]
94. Duverger, C.; Lambotte, S.; Bernard, P.; Lyon-Caen, H.; Deschamps, A.; Nercessian, A. Dynamics of microseismicity and its relationship with the active structures in the western Corinth Rift (Greece). *Geophys. J. Int.* **2018**, *215*, 196–221. [CrossRef]
95. Mesimeri, M.; Karakostas, V. Repeating earthquakes in western Corinth Gulf (Greece): Implications for aseismic slip near locked faults. *Geophys. J. Int.* **2018**, *215*, 659–676. [CrossRef]
96. Massin, F.; Farrell, J.; Smith, R.B. Repeating earthquakes in the Yellowstone volcanic field: Implications for rupture dynamics, ground deformation, and migration in earthquake swarms. *J. Volcanol. Geotherm. Res.* **2013**, *257*, 159–173. [CrossRef]
97. Fischer, T.; Horálek, J.; Hrubcová, P.; Vavryčuk, V.; Bräuer, K.; Kämpf, H. Intra-continental earthquake swarms in West-Bohemia and Vogtland: A review. *Tectonophysics* **2014**, *611*, 1–27. [CrossRef]
98. Bourouis, S.; Bernard, P. Evidence for coupled seismic and aseismic fault slip during water injection in the geothermal site of Soultz (France), and implications for seismogenic transients. *Geophys. J. Int.* **2007**, *169*, 723–732. [CrossRef]
99. Bourouis, S.; Cornet, F.H. Microseismic activity and fluid fault interactions: Some results from the Corinth Rift Laboratory (CRL), Greece. *Geophys. J. Int.* **2009**, *178*, 561–580. [CrossRef]
100. Kassaras, I.; Kapetanidis, V.; Karakonstantis, A.; Kouskouna, V.; Ganas, A.; Chouliaras, G.; Drakatos, G.; Moshou, A.; Mitropoulou, V.; Argyrakis, P.; et al. Constraints on the dynamics and spatio-temporal evolution of the 2011 Oichalia seismic swarm (SW Peloponnesus, Greece). *Tectonophysics* **2014**, *614*, 100–127. [CrossRef]
101. Kraft, T.; Wassermann, J.; Schmedes, E.; Igel, H. Meteorological triggering of earthquake swarms at Mt. Hochstaufen, SE-Germany. *Tectonophysics* **2006**, *424*, 245–258. [CrossRef]
102. Leclère, H.; Fabbri, O.; Daniel, G.; Cappa, F. Reactivation of a strike-slip fault by fluid overpressuring in the southwestern French-Italian Alps. *Geophys. J. Int.* **2012**, *189*, 29–37. [CrossRef]
103. Halpaap, F.; Rondenay, S.; Perrin, A.; Goes, S.; Ottemöller, L.; Austrheim, H.; Shaw, R.; Eeken, T. Earthquakes track subduction fluids from slab source to mantle wedge sink. *Sci. Adv.* **2019**, *5*, eaav7369. [CrossRef] [PubMed]
104. Hainzl, S. Seismicity patterns of earthquake swarms due to fluid intrusion and stress triggering. *Geophys. J. Int.* **2004**, *159*, 1090–1096. [CrossRef]
105. Aoyama, H.; Takeo, M.; Ide, S. Evolution mechanisms of an earthquake swarm under the Hida Mountains, central Japan, in 1998. *J. Geophys. Res.* **2002**, *107*, 2174. [CrossRef]
106. Vavryčuk, V.; Hrubcová, P. Seismological evidence of fault weakening due to erosion by fluids from observations of intraplate earthquake swarms. *J. Geophys. Res. Solid Earth* **2017**, *122*, 3701–3718. [CrossRef]
107. Caputo, R.; Chatzipetros, A.; Pavlides, S.; Sboras, S. The greek database of seismogenic sources (GreDaSS): State-of-the-art for northern greece. *Ann. Geophys.* **2012**, *55*, 859–894. [CrossRef]
108. Wessel, P.; Smith, W.H.F. New, improved version of generic mapping tools released. *Eos. Trans. Am. Geophys. Union* **1998**, *79*, 579. [CrossRef]
109. Shcherbakov, R.; Turcotte, D.L. A modified form of Båth's law. *Bull. Seismol. Soc. Am.* **2004**, *94*, 1968–1975. [CrossRef]
110. Wu, D.; Yan, H.; Shen, Y. TSAnalyzer, a GNSS Time Series Analysis Software. *GPS Solut.* **2017**, *21*, 1389–1394. [CrossRef]

applied sciences

MDPI

Article

The 27 September 2021 Earthquake in Central Crete (Greece)—Detailed Analysis of the Earthquake Sequence and Indications for Contemporary Arc-Parallel Extension to the Hellenic Arc

Emmanuel Vassilakis [1,*], George Kaviris [2], Vasilis Kapetanidis [2], Elena Papageorgiou [3], Michael Foumelis [3], Aliki Konsolaki [1], Stelios Petrakis [1], Christos P. Evangelidis [4], John Alexopoulos [2], Vassilios Karastathis [4], Nicholas Voulgaris [2] and Gerassimos-Akis Tselentis [1,4]

1 Section of Geography and Climatology, Department of Geology and Geoenvironment, School of Sciences, National and Kapodistrian University of Athens, Panepistimiopolis Zografou, 15784 Athens, Greece; alikikons@geol.uoa.gr (A.K.); spetrakis@geol.uoa.gr (S.P.); tselenti@noa.gr (G.-A.T.)
2 Section of Geophysics-Geothermics, Department of Geology and Geoenvironment, School of Sciences, National and Kapodistrian University of Athens, Panepistimiopolis Zografou, 15784 Athens, Greece; gkaviris@geol.uoa.gr (G.K.); vkapetan@geol.uoa.gr (V.K.); jalexopoulos@geol.uoa.gr (J.A.); voulgaris@geol.uoa.gr (N.V.)
3 Department of Physical and Environmental Geography, Aristotle University of Thessaloniki (AUTH), 54124 Thessaloniki, Greece; elenpapageo@geo.auth.gr (E.P.); mfoumelis@geo.auth.gr (M.F.)
4 Institute of Geodynamics, National Observatory of Athens, Lofos Nymfon, 11810 Athens, Greece; cevan@noa.gr (C.P.E.); karastathis@noa.gr (V.K.)
* Correspondence: evasilak@geol.uoa.gr

Citation: Vassilakis, E.; Kaviris, G.; Kapetanidis, V.; Papageorgiou, E.; Foumelis, M.; Konsolaki, A.; Petrakis, S.; Evangelidis, C.P.; Alexopoulos, J.; Karastathis, V.; et al. The 27 September 2021 Earthquake in Central Crete (Greece)—Detailed Analysis of the Earthquake Sequence and Indications for Contemporary Arc-Parallel Extension to the Hellenic Arc. *Appl. Sci.* **2022**, *12*, 2815. https://doi.org/10.3390/app12062815

Academic Editor: Valerio Comerci

Received: 26 January 2022
Accepted: 4 March 2022
Published: 9 March 2022

Featured Application: Field validation of combined remote sensing and seismological data after a large earthquake.

Abstract: The Arkalochori village in central Crete was hit by a large earthquake (M_w = 6.0) on 27 September 2021, causing casualties, injuries, and severe damage to the infrastructure. Due to the absence of apparent surface rupture and the initial focal mechanism solution of the seismic event, we initiated complementary, multi-disciplinary research by combining seismological and remote sensing data processing, followed by extensive field validation. Detailed geological mapping, fault surface measuring accompanied with tectonic analysis, fault photorealistic model creation by unmanned aerial system data processing, post-seismic surface deformation analysis by DInSAR image interpretation coupled with accurately relocated epicenters recorded by locally established seismographs have been carried out. The combination of the results obtained from these techniques led to the determination of the contemporary tectonic stress regime that caused the earthquake in central Crete, which was found compatible with extensional processes parallel to the Hellenic arc.

Keywords: Arkalochori; Messara Basin; Heraklion Basin; Kastelli fault zone; supra detachment basin; fault segmentation; double-difference relocation; DInSAR

1. Introduction

The broader area of central Crete represents a neotectonic structure in the vicinity of the Hellenic Trench, which comprises two post-orogenic basins with trending orientations normal to each other, forming a uniform basin complex [1]. The northernmost part of the latter, the Heraklion Basin (HB), trends approximately N-S, whereas adjacent to its southern margin, the Messara Basin (MB) developed trending E-W [2,3]. We refer to this area as the Heraklion–Messara Basin complex, as it is covered by the same Miocene formations, implying that it shares a common paleo-environmental history, even though there are significant structural differences between the two basins.

In particular, MB (Figure 1) acts as a supra-detachment basin [2,4] which lies on top of the hanging wall of the south Cretan extensional detachment [3]. The latter is located at the active margin of the rapidly southwestwards-moving Aegean micro-plate [5–7]. It is an extensional, south-dipping structure, similar to the north-dipping Cretan Detachment [8–10].

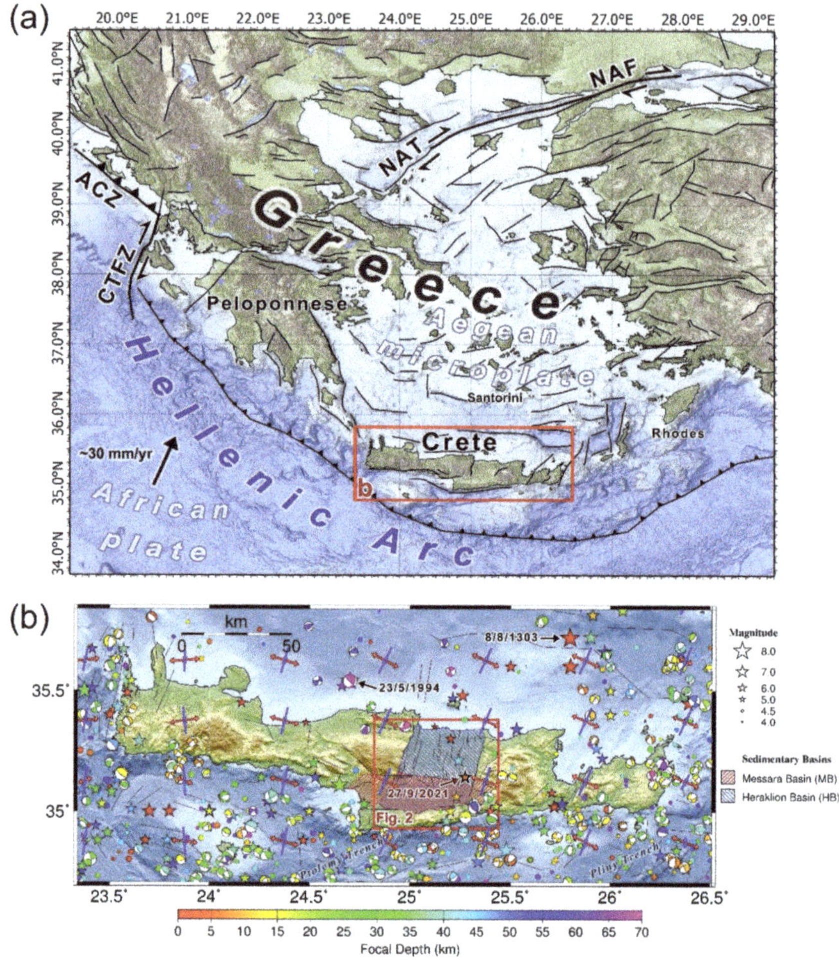

Figure 1. (**a**) Location of Crete (red rectangle) in southern Greece. Bold lines marked with dents or arrows indicate major tectonic structures in the vicinity of the Hellenic Arc; ACZ: Apulian Collision Zone, CTFZ: Cephalonia Transform Fault Zone, NAT: North Aegean Trough, NAF: North Anatolian Fault. Lines indicate the traces of seismogenic sources from the Euro-Mediterranean Seismic Hazard Model 2013 (ESHM13; [11]). The NNE-pointing vector at the lower-left indicates the direction and convergence rate of the Nubian plate [12]. (**b**) Seismotectonic map of Crete (red rectangle in panel (**a**). Black lines are active faults from the NOAfaults v3.0 database [13]. Beachballs indicate focal mechanisms of earthquakes ($M_w \geq 4.0$) from the compilation of [14] and sources referenced therein (see also Data Availability). Circles ($M < 5.0$) and stars ($M \geq 5.0$) represent past seismicity of years 1000–1899 (SHEEC; [15]) and 1900–2009 [16] for events with focal depths down to 70 km. The maximum (S_1) and minimum (S_3) principal stress axes of the regional crustal stress field [14] are

depicted by blue and red arrows, respectively, with arrow length proportional to the cosine of the plunge angle. The epicenter of the Arkalochori earthquake is presented by a star within MB and HB areas.

MB has been considered an E-W-trending, typical graben structure [17]; however, it is characterized by the absence of classic continuous bounding faults at the north and south margins of the basin. Its northern margin is probably covered with Miocene and/or Pliocene sediments. Therefore, the exact boundaries between MB and HB can hardly be observed and, consequently, several studies, with different arguments [18–20], implicate arc-parallel extension dynamics, trending along the local Hellenic trench axis (Figure 1).

Intense historical and recent seismicity reveal the continuous contemporary tectonic activity of the entire Cretan territory (e.g., [21]). Destructive earthquakes have occurred in the broader Crete region since antiquity and during the pre-instrumental era, until 1900 AD. The earthquake of 8 August 1303 [15,22], with an estimated magnitude of 8.3 [23], is one of the largest that has ever been reported in the entire Mediterranean region. This earthquake also caused a large tsunami that was observed at Crete, Rhodes, Peloponnese, Egypt, Syria, the Adriatic Sea, and along the eastern Mediterranean coast. In the vicinity of the study area, two major historical earthquakes have been reported (Figure 1). The first is the 1 July 1494 ($M = 5.4$) earthquake [15] that occurred close to Heraklion and caused damage to church towers and private buildings, while large waves were observed in the harbor [22]. One century later, on 26 November 1595, another earthquake ($M = 6.4$) took place a few kilometers off the Heraklion city, causing severe damage and destruction in the entire Crete island [15,22]. It is worth noticing that no large events ($M \geq 6.0$) occurred in the study area during the instrumental period (after 1900), with the nearest one being recorded on 23 May 1994 ($M_w = 6.1$) and that was an intermediate depth (60–70 km) event (Figure 1b) located approximately 40 km WNW of Heraklion [16]. Tsapanos [24] assessed the seismic hazard for the three largest cities of Crete and obtained maximum peak ground acceleration (PGA) values for Heraklion between 0.130 g and 0.165 g for rock and soft soil conditions, respectively.

One of the major structures that seem to control the HB-MB complex is the NNE-SSW fault zone that controls its eastern margin, located only a few kilometers east of Arkalochori village (Figure 2). It bounds the easternmost post-Alpine sediment outcrops of the HB-MB in a morphologically clear way, even though several active structures seem to intersect. It was only recently (27 September 2021) that the lethal "Arkalochori earthquake" occurred in this area, causing one fatality, and resulting in hundreds of damaged buildings, mainly of old stone-built masonries. As the authorities reported, more than 6150 houses have been declared uninhabitable due to excessive damage in the mainly affected municipality of Minoa Pediadas (Figure 1), as a result of the mainshock and the major aftershocks.

The earthquake of 27 September 2021, 06:17:21 UTC, occurred near Arkalochori, Central Crete, ~25 km SSE of Heraklion city [25,26]. Its focal mechanism is characterized by an SSW-NNE to SW-NE-trending, nearly dip-slip normal faulting. Considering the solutions proposed by different agencies (Table S1), except for a few outliers, its strike generally ranges N200° E–N230° E and its dip angle varies between 40° and 60°. The magnitude of the main event was initially determined as $M_L = 5.8$ by the Seismological Laboratory of the National and Kapodistrian University of Athens (SL-NKUA) and the Geodynamic Institute of the National Observatory of Athens (GI-NOA), but later measurements of seismic moment upgraded its value to $M_w = 5.9$–6.0. The largest aftershock ($M_w = 5.3$) occurred on 28 September 04:48:08 UTC, one day after the mainshock, with a similar focal mechanism. According to the report of ITSAK [27], the recorded PGA at station ARK1 in the epicentral area (Arkalochori) was 609 cm s^{-2} in the horizontal component (N-S) and 806 cm s^{-2} (~0.82 g) in the vertical one, with the duration of strong ground motion (>0.1 g) being 6 s. The horizontal to vertical spectral ratio reached 8 at the dominant frequencies of 0.65Hz and 1.4 Hz. The accelerometric recordings at station ARK1 far exceeded the provisions of the National Building Code [28] for effective PGA = 0.24 g at the epicentral region in the seismic hazard zone II and for soil classes A and B.

Herein, we present a multi-disciplinary work, based on extensive fieldwork and detailed geological mapping of the affected area at central Crete, combined with results from seismological and geodetic imaging techniques. Our aim is to understand the contemporary seismotectonic regime of a densely populated area close to the active Hellenic arc and provide useful data for a realistic model concerning the neotectonic evolution of Crete. We intend to examine whether arc-parallel extension is a temporal phenomenon that is related to the Arkalochori earthquake or a dominant contemporary stress component.

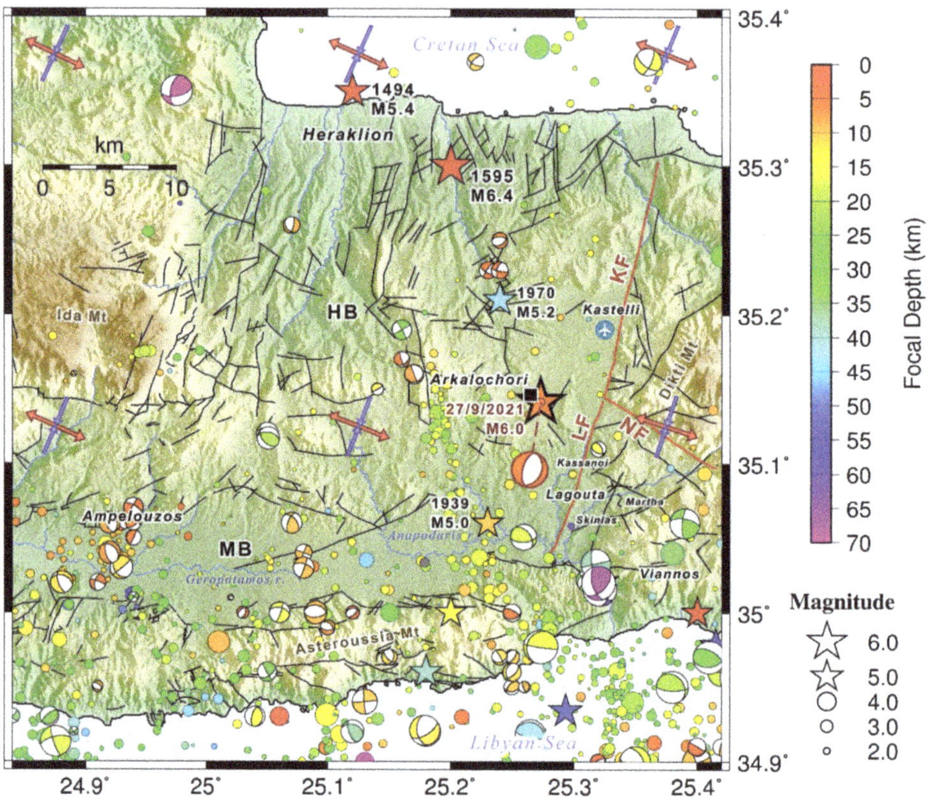

Figure 2. Close up of the study area, marked with a red rectangle in Figure 1b. Black lines show the mapped tectonic structures. Seismicity of the period from 2008 to May 2021 is from GI-NOA. The Arkalochori earthquake is represented by a star with bold outline. Note the offset beachball linked with a dashed callout line representing its focal mechanism solution from GI-NOA (see Table S1 at Supplementary Materials) and the faults, described in this work, are drawn in red (KF: Kastelli Fault, NF: Nipiditos Fault, LF: Lagouta Fault). Additional focal mechanisms are from [29].

2. Geological Setting

The study area is located at central Crete (Figure 1b) and, specifically, at the easternmost part of the supra-detachment MB, which is bounded by active fault systems producing high relief landscape [3]. It is the area where HB shares significant structures with MB and, consequently, it is quite complex to distinguish the margin between these Neogene basins with a single discipline method. In the following sections, we present interpretations on the geological setting of the study area, based on geomorphology, lithology, and seismotectonics.

2.1. Geomorphology

The general morphology of the wider area is rather mountainous, with steep slopes which very often exceed the order of 45°, due to active tectonic structures that divide it into fault blocks that move either independently or in groups, depending on the local dominant tectonic regime [3]. The latter seems to have been changing since Miocene and is either expressed with the change of sediment facies or imprinted on the drainage network patterns, which are developed within a hilly area, with tilted monocline blocks, between higher relief mountains, implying a listric geometry of the involved faults (e.g., [30–32]). The highest mountain is Dikti Mt which exceeds an elevation of 2100 m and bounds both HB to the east and MB to the N-NE (Figure 2). The much lower elevation E-W-trending Asteroussia mountain chain (~1200 m) is the geomorphological structure that acts as the southern margin of MB (Figure 2).

2.2. Lithology

The geology of the area consists of Alpine (mainly Triassic to Late Jurassic limestones/dolomites, flysch, marbles, and ophiolites) and post-Alpine formations (silts, clays, conglomerates, and sandstones of Middle-Upper Miocene). In many places, they are covered by Quaternary alluvial sands and conglomerates of small consistency, as a result of the erosion of the aforementioned formations. More specifically:

- The Alpine basement formations can be distinguished into two major groups, regarding their tectonic emplacement on the Cretan nappe pile [1]. They are members of the tectono-stratigraphic succession which builds the island of Crete as a stacked pile of 10 geotectonic units with a total thickness of about 10–12 km [33]. The upper part of the pile consists of five units (Pindos, Arvi, Miamou, Vatos, and Asteroussia) which are tectonically emplaced beneath the ophiolites, that is the uppermost nappe, whilst the lower four units (Mani, W. Crete, Arna, and Tripolis) comprise the basement of Crete [1,33]. At the study area of easternmost MB, the Alpine basement outcrops are limited to the following rocks:
    - Limestones and Dolomites (Mani Unit). Carbonate rocks consisting of thin-bedded crystalline marbles with layers of siltstones (L. Jurassic—L. Eocene). They appear heavily folded with isoclinal asymmetric folds.
    - Limestones and Dolomites (Tripolis Unit). Intercalations of unbedded to thick-bedded neritic dark-colored and partially bituminous Limestones and Dolomites, with sparse appearances of evaporites.
    - Flysch (Pindos Unit). It consists of heavily folded layers of Sandstones, Schists, and Conglomerates, with the characteristic appearance of large carbonate olistoliths that originated from inner units.
    - Ophiolitic nappe. Limited appearance at the top of E. Asteroussia Mt, emplaced between the Flysch and the Viannos formation sediments as its basement rock.

  The entire nappe pile behaves homogenously till after the Middle Miocene, when the compressional, almost N-S-trending, orogenic stress field is converted into an extensional one due to the African plate slab roll-back [9,34–36], which happened at the edge of the Aegean microplate [37]. Within this context, Crete is a large part of the Mid-Miocene Southern Aegean core complex exhumation that took place between 24 and 15 Ma [9] which hosts several supra-detachment depositional areas [3,4], with MB being one of them.
- The Cretan nappe pile is segmented by many faults and covered by sedimentary formations representing different paleo-environments, as follows (from deeper to shallower):
    - Viannos Formation (Mid. Miocene). Sandy/silty deposits of lacustrine origin, coexisting with alluvial conglomerates. The average thickness of the formation is about 400 m. Its outcrops may be found at most of the eastern MB margin,

either covering unconformably the basement rocks or in tectonic contact with them through ruptured fault zones.

- Skinias Formation. It is a transitional succession consisting of Middle Miocene silts, with a thickness of ~200 m, yielding the sea intrusion within the MB area. At the area between Skinias and Martha villages, the Skinias silts were deposited in a deep (over 200 m) marine environment. The base of the deposits consists of Pelites, whereas layers of conglomerates and sand have been observed in the area close to Martha village.
- Ampelouzos Formation. It is deposited unconformably above both aforementioned formations. The lower part of the formation includes a variety of sedimentary, well-layered deposits, transitioning gradually from shallow to deeper (fan conglomerates to homogenous continental shelf sandstones). According to fossils found in the continental and shallow-marine deposits, the formation is dated as Middle to Upper Miocene [38].
- Alluvial deposits. Unconsolidated sands and conglomerates with pebbles originated from all the above formations, mainly along riverbeds.

2.3. Seismotectonic Setting

The vast majority of the large events affecting Crete, before the 2021 Arkalochori sequence, have occurred offshore, either to the north or to the south, with the latter probably related to the intra-crustal graben system, i.e., the Ptolemy and Pliny trenches [39] (Figure 1b). The only knowledge regarding the local seismic activity in the study area comes from a temporary seismological network, which consisted of 10 analog stations, installed in the area of Heraklion during the period of September to December 1995 [29]. Data recorded by this network revealed the existence of shallow onshore activity. During that period, the seismicity was mainly concentrated along the eastern margin of the Heraklion Basin, decreasing from east to west, with epicenters also located in the vicinity of Arkalochori. Furthermore, a W-E-trending decrease in the sedimentary cover has been identified offshore, north of Crete [40].

In a broader view frame, the location of Crete Island, north of the Hellenic Arc (Figure 1a), comprises an area of complex geodynamics, with transpressional tectonics dominating the south of Crete, in contrast with normal and oblique-slip faulting onshore, yielding a heterogeneous stress field [14,41]. On the other hand, an extension has been observed to the north, between Santorini and Crete [42–45]. The latter reveals extension in a general E-W direction, with NNE-SSW shortening observed in western Crete [46]. According to [14], the sub-horizontal S_1 axis is in an SW-NE (in the east) to the N-S direction (in the west), likely affected by the proximity of the region to the subduction interface and compression parallel to the direction of convergence, which becomes more oblique to the arc towards the east. However, the sub-horizontal S_3 axis also reveals E-W (in the west) to WNW-ESE (in the east) extension onshore, even in the region where the 2021 Arkalochori earthquake occurred, in accordance with the observed normal fault scarps, trending in a general N-S direction [19]. The resulting regional shear stress favors SW-NE left-lateral faults, mainly offshore, in agreement with [41,47]. On the other hand, fault sources, determined by [11], indicate mainly E-W normal faulting to the north and south of Crete, incompatible with the resolved stress. It should be taken into account that the aforementioned stress field was determined using offshore earthquakes at crustal depths, both to the north and the south of Crete, as no large onshore events were located onshore until recently. The stress field in the region of Crete was characterized as transpressional by [14] due to the involvement of shallow events related to strike-slip (primary stress state) or reverse faulting (secondary state). In central Crete, due to the absence of focal mechanism data for shallow onshore events, the stress field is largely a result of interpolation from resolved nodes of neighboring regions. The stress shape value of 0.5 indicates instability in both S_1 and S_3 axes. However, the direction of S_3 seems to be generally stable in a WNW-ESE direction in the region from central Crete to the volcanic arc in the north.

This complex stress regime is also revealed by the focal mechanisms, as all types (normal, strike-slip, and reverse) can be observed offshore, either to the north or to the south (Figure 1). Focal depths increase towards the north, due to the African slab rollback phenomenon. Onshore focal mechanisms at shallow depths are few. De Chebalier et al. [48] worked on the western edge of Crete and obtained fault-plane solutions of shallow events, revealing either approximately N-S normal or strike-slip faulting. Delibasis et al. [29] constrained 29 fault-plane solutions close to the 2021 Arkalochori earthquake sequence, which also indicate normal or reverse motion with, in some cases, a significant strike-slip component.

3. Materials and Methods

We combined data from different sources, performing a multidisciplinary study, to further understand the geotectonic regime that caused the 2021 earthquake activity and possibly reveal its association with the mapped fault pattern. Therefore, our methodology included field mapping accompanied by the necessary laboratory work (e.g., GIS techniques for tectonic geomorphology, photogrammetric processing of unmanned aerial system images), seismological data analysis and interpretation, as well as SAR interferometry processing.

3.1. Field Mapping

The fieldwork included detailed mapping of the outcrops, especially the post-Alpine sediments, as well as of the fault surfaces that crosscut the stratigraphic contacts between them. At several places, a drone was used for image acquisition, followed by photogrammetric processing and the generation of high-resolution digital surface models and ortho-photographs.

A quite clear marginal structure, trending NNE-SSW, which delineates the eastern side of the MB-HB, can be identified, even on a medium resolution shaded relief of the study area in central Crete (Figure 2). This represents the major Kastelli fault zone, which constitutes the western boundary of Dikti Mt that lies to the east [3]. One of its segments, just uphill the new international Kastelli Airport (Figure 2), was captured by the UAS and modeled through structure-from-motion techniques, aiming to map its trace with the highest possible detail, especially at areas where the morphology is relatively smooth. About 320 images were processed and a point cloud of more than 100 million points was generated for the construction of a photorealistic 3D model, with a spatial resolution of 2.5 cm (Figure 3).

Figure 3. Photorealistic 3D model of a Kastelli fault segment, which, at some places, can be clearly identified by a morphological discontinuity (white arrows) or is quite difficult to distinguish where the morphology has been smoothened by the erosion.

The mapped faults were measured along with the kinematic indicators that were found on their surfaces and interpreted within the frame of kinematic analysis. The classification and clustering of the field measurements were performed using algorithms and methods already published and incorporated in software packages, such as M.I.M. v.4.0.1 [49].

3.2. Seismological Data and Methods

The 2021 seismic crisis in the wider area of Arkalochori began in the form of an earthquake swarm in early June 2021. The situation, however, changed dramatically after the occurrence of the M_w = 6.0 mainshock of 27 September 2021. Although several permanent stations of the Hellenic Unified Seismological Network operate on the island of Crete, at the time of the mainshock, the closest station to the epicentral area was KNSS of the Hellenic Mediterranean University (HC network at formerly Technological Educational Institute of Crete [50]), at a distance of about 12–22 km (Figure S1). As a result, the seismicity of the first stage of the sequence (foreshock swarm) could not be determined with high precision, due to the lack of data from stations at local distances, which would help constrain the focal depths. Resolving hypocentral locations with data from more distant stations is further problematic due to the inhomogeneity of the velocity structure, because of the proximity of the area to the Hellenic subduction zone. The day after the mainshock, four temporary seismological stations (CRE1-4) were deployed by GI-NOA (HL Network [51]) around the aftershock zone (Figure S1b). The inclusion of data from these stations drastically improved the hypocentral locations, even with the use of a generic velocity model. We collected the available catalogues, including manually analyzed phase data from SL-NKUA and GI-NOA at all operational stations, and compiled a dataset of ~2500 events for the period between 1 June and 18 October 2021. We divided the data to be processed in two periods, one before (Period A) and one after (Period B) 28 September, when the temporary local network was deployed.

We used the HypoInverse code [52] to locate the sequence, initially with a generic model used by routine analysis at SL-NKUA, as well as with the local velocity model of [29] for Central Crete. The latter model was preferred, as it yielded lower root mean square (RMS) travel-time residuals and locations errors, while providing a better general image regarding the distribution of epicenters, in terms of scattering, especially between events that form spatial clusters. Another issue concerns the distance weighting parameters for the two periods, i.e., the maximum epicentral distance of stations to be used. After several tests, the maximum distance was set to ~280 km for the first period and ~120 km for the second one. A V_P/V_S ratio of 1.78 was measured using the Chatelain [53] diagram for the first period, whereas a V_P/V_S = 1.76 was preferred for the second period with the shorter distance limit. The application of station corrections, especially for the first period, plays an important role to reduce artifacts of the initial epicentral distribution, where spurious epicenters of small magnitude events seemed to spread northwards of the main seismicity cloud.

To further improve the hypocentral distribution, with emphasis on the second period, we applied a relocation procedure using the HypoDD code [54]. The algorithm works by minimizing the double difference between calculated and observed travel times in combinations of event pairs with neighboring hypocenters, at inter-event distances much smaller than their respective hypocentral distance from a station. This method enhances the hypocentral distribution by minimizing relative location uncertainties between correlated events, caused by discrepancies between the velocity model and the real structure. Furthermore, waveform cross-correlation data are used to minimize uncertainties due to arrival-time reading inconsistencies in groups of events with similar waveforms, called multiplets.

We used station KNSS (Figure S1b) as the main reference station for the identification of multiplets, as this is the closest station that was operational during both periods, with the exception of its N-S component that was damaged after the occurrence of the 27 September mainshock. Recordings of the temporary station CRE1 (Figure S1b) were also used to enhance the reference cross-correlation measurements, particularly those of smaller events, during the second period. The waveforms were band-pass filtered in the range 2–8 Hz, for KNSS, and 2–15 Hz, for CRE1, before cross-correlation was calculated for the full P and S waveform signals in all combinations of event pairs with available data. The cross-correlation maximum (XC_{max}) of every pair was registered in a matrix for every component. The matrices produced for different reference stations and components were then combined, retaining the RMS value of the XC_{max} determined for each pair. The nearest neighborhood linkage was applied and a C_{th} = 0.81 optimal threshold was selected, maximizing the difference between the size of the largest multiplet and the sum of clustered events [55]. This procedure created 234 multiplets containing a total of 1725 events. The P- and S-wave windows for each pair of events within the same multiplet were then cross-correlated for all stations with available picks, measuring the XC_{max} and its respective time lag to be used as input for HypoDD.

The relocation procedure was applied separately for the two periods, acknowledging issues due to the different network geometry, so as not to degrade the data quality of the second period. On the other hand, we chose to relocate the mainshock and its early aftershocks (roughly a hundred), which occurred before the deployment of the temporary local stations, together with events that were located with data from the local network. This manages to improve the relative locations of the first aftershocks by exploiting their links with the rest of the events of the second period. Each relocation procedure was mainly divided into two types of sets, one with stronger a priori weights to the catalogue data and the other with stronger weights to the cross-correlation data.

3.3. SAR Data and Interferometric Processing

Space-born synthetic aperture radar (SAR) imagery is routinely utilized for measuring co-seismic surface displacements based on the DInSAR technique. DInSAR allows mapping the co-seismic motion by using two satellite images taken before and after an earthquake [56–58]. The systematic availability of SAR data from the Copernicus Sentinel-1 mission, with a revisiting time of 6 days over Europe (12 days globally), as well as the wide swath coverage of 250 km, enables the near-real-time response to seismic events [58–63].

The contribution of platform-based solutions in the early response phase of an earthquake has been well demonstrated [64,65]. The Geohazards Exploitation Platform (GEP; https://geohazards-tep.eu, accessed on 23 January 2022) focuses on mapping hazard-prone land surfaces and monitoring terrain motion, providing access to several Earth Observation (EO) missions and a broad range of relevant online services and development on cloud processing resources [66].

The present work utilized Sentinel-1 data acquired during the period from 23 September to 1 October 2021, with different viewing geometries to optimize the displacement map generation. For the interferometric processing, the GEP DIAPASON service was exploited, considering precise orbit state vectors and SRTM Digital Elevation Model 3-arcsecond ($\approx$90 m) data to remove the topographic phase. Interferometric measurements correspond to movements detected along the Line-of-Sight (LoS) of the satellite, i.e., the oblique direction between the satellite and the Earth's surface. Positive LoS values indicate uplift or motion towards the satellite, whereas negative ones correspond to subsidence or motion away from the sensor. The DInSAR processing scheme implemented herein is described in detail in [67,68]. Four differential interferograms were generated for the descending 036 (D036) and 109 (D109) orbital tracks, as well as for the ascending 102 (A102) and 029 (A029) ones, with temporal separation of 6 to 12 days and spatial resolution of approx. 45 m. Figure 4 and Figure S2 in the Supplementary Materials show the obtained displacement fields, while Table 1 details the interferometric pairs considered for the various geometries. Based on their temporal coverage, it is important to note that each co-seismic displacement map is affected by a different contribution of post-seismic motion. The interferograms for tracks A102 and D036 represent those with the smallest and largest post-seismic contribution, respectively (Figure 4).

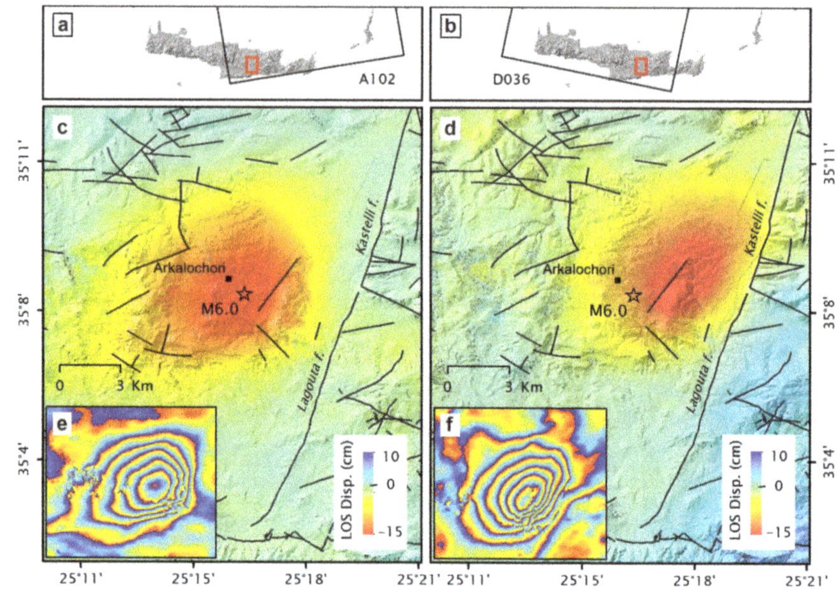

Figure 4. Co-seismic LoS displacement maps, as derived using GEP DIAPASON service, for ascending track A102 (23 September 2021–29 September 2021) and descending D036 (25 September 2021–1 October 2021) (**c**,**d**) and the corresponding wrapped differential interferograms (**e**,**f**). The spatial extend of Sentinel-1 acquisition frames for both geometries are also indicated (**a**,**b**). The difference in the temporal span following the mainshock (27 September 2021) for the two opposite geometries is worth noting. While A102 spans approx. two days, D036 extends over five days after the earthquake, denoting larger contribution of post-seismic motion. Black lines correspond to fault zones (after [3]). The relocated epicenter of the 27 September 2021 mainshock is marked with a star.

Table 1. Overview of the interferometric pairs considered in DInSAR processing, including the LoS incidence angles of the measurements. The time span of each DInSAR pair following the mainshock is also indicated.

Mission	Orbit	Track	Incidence Angle (Degrees)	Observation Period	Span after Mainshock
Sentinel 1B-Sentinel 1A	Ascending	102	43.8	23 September 2021–29 September 2021	2 days
Sentinel 1A-Sentinel 1B	Descending	036	43.8	25 September 2021–1 October 2021	5 days
Sentinel 1B-Sentinel 1B	Ascending	029	33.8	18 September 2021–30 September 2021	3 days
Sentinel 1B-Sentinel 1A	Descending	109	33.8	24 September 2021–30 September 2021	3 days

Furthermore, due to the opposite geometries, independent LoS measurements were decomposed into vertical (Up) and E-W-motion components, to facilitate the optimal interpretation of motion patterns. The horizontal and vertical components were calculated using both Sentinel-1 geometries in a comprehensive decomposition scheme [69], considering the variability of the incidence angles within the scenes. In order to simplify the geometrical assumptions, only measurements of similar incidence angles were considered in the decomposition (i.e., orbital tracks A102—D036 and A029—D109) (see Table 1).

4. Data Analysis and Results

The combination of data from various and heterogeneous sources is a complex task that needs to be evaluated and prioritized to examine their compatibility grade. The field data analysis provides us with strong arguments for establishing plausible planes of movement, which host the hypocenters of an earthquake. Therefore, it offers a tool for filtering them with higher relevance to the contemporary tectonic regime. In addition, the field data were used to validate the surface deformation detected through DInSAR data and attribute it to specific geological structures (e.g., active faulting).

4.1. Field Data Analysis

The length of the Kastelli fault zone exceeds 30 km since it can be traced from the northern coast of Crete to Asteroussia Mt to the south, within the recent sediments of MB. It was mapped and measured at several places, especially where the Alpine basement outcrops hosted the fault plane. It was quite frequent that striations along with other kinematic indicators were overprinted on the calcite coating of the fault plane (Figure 5).

Figure 5. Fault plane of a Kastelli fault segment measured at 74/263, on which striations measured at 61/220 were found (parallel to the pen), confirming the oblique normal movement with left-lateral component. Measurements at several places are plotted in the inset diagram.

The fault slip data recorded in the field were used as inputs within the stress inversion method TRM [70] and a stress tensor with S_1: 90/079, S_2: 00/349, and S_3: 00/079 (R = 0.25) was defined. The resolved stress tensor, although calculated with fault planes that do not deviate strongly from each other, is rather compatible with the E-W-trending extension, which in turn agrees with the Arkalochori earthquake focal mechanism.

During the fieldwork, it was rather obvious that the morphological discontinuity, which the Kastelli fault zone has generated through its activity, is gradually eliminated and finally disappeared within the early Miocene lacustrine sediments of the Viannos, Skinias, and Ampelouzos units [71] and, hence, the fault trace which comprises its southern segment (Lagouta fault, according to [3]) is not very clearly delineated.

Despite this, a significant number of smaller fault traces that were mapped around Lagouta and Kassani villages (Figure 6a) yield that the Kastelli fault zone continues to the south and ends up to an E-W-trending fault, almost parallel and antithetic to the south Cretan detachment [3]. In addition, a pair of conjugate syn-sedimentary faults was located, measured, and analyzed (Figure 6b) from a kinematic and dynamic perspective. The measurements of the two faults (p1 88/114, p2 72/150) and the sliding directions recorded on them (s_1 27/025, s_2 19/066), respectively, were found to be compatible—after tectonic analysis—with left-lateral strike-slip faulting and, specifically, with a stress field which is defined by the axes S_1: 24/046, S_2: 63/200 and S_3: 11/311 [3].

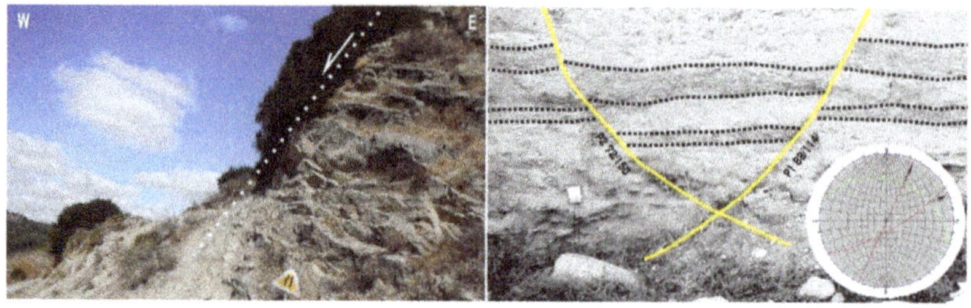

(a) (b)

Figure 6. Field photographs of (**a**) the Lagouta normal fault, which comprises the southern segment of the Kastelli fault zone, and (**b**) a pair of conjugate syn-sedimentary faults, within Tortonian deposits, compatible with left-lateral strike-slip faulting.

The left-lateral component of this movement is in agreement with the en-echelon formation of numerous faults found along the previously activated E-W-trending, MB marginal fault zone, at the southern edge of this segment at Asteroussia Mt [3]. The monoclinal stratigraphy of the area yields that the same west-dipping fault zone has also a strong normal component. This conclusion is based on the fact that the outcrops of the older post-alpine sediments (Viannos formation) are found beneath more recent successions that have been mapped in detail on the western fault block and simultaneously at higher altitudes at the eastern one, yielding that the latter has been relatively uplifted (Figure 7).

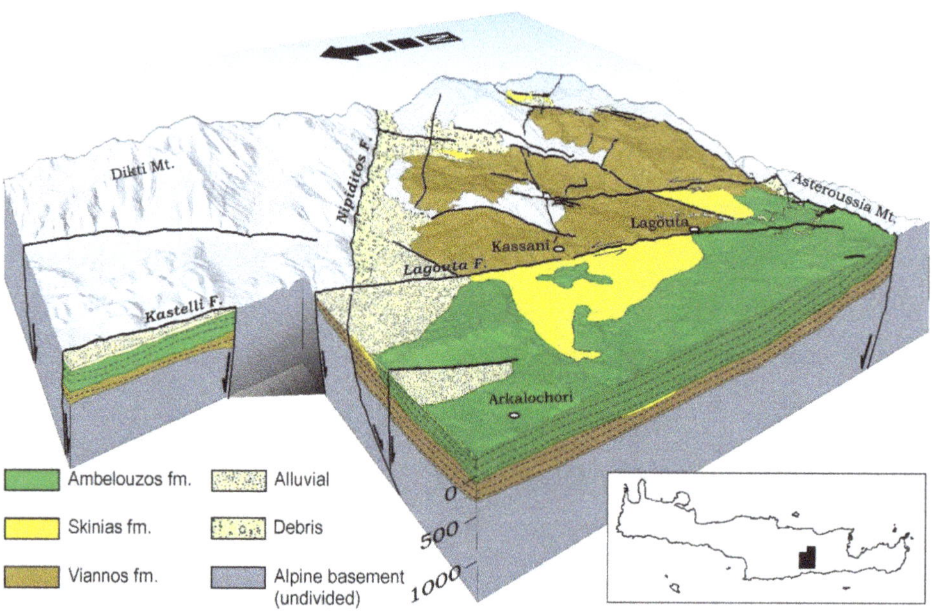

Figure 7. Simplified 3D block diagram of the eastern marginal area of MB. Arkalochori village is located at the hanging wall of Lagouta fault but also at the hanging wall of Nipiditos fault, which is segmenting the Kastelli fault zone.

4.2. Seismological Results

Regarding the initial locations of earthquake hypocenters (Figure 8a) of the first period, before 28 September 2021, focal depths remain uncertain, due to the lack of local stations, mainly ranging between 3 and 20 km (Figure S3c), with the majority of events being concentrated near a discontinuity of the velocity model at ~6 km [29]. On the other hand, more constrained focal depth values, between 5 and 13 km (Figure S3d), are determined for the second period. Location statistics for the two periods are presented in Table 2 and Figures S3 and S4. Average vertical location errors are nearly halved during Period B and horizontal location errors are reduced by ~26%, as a result of the deployment of stations at local distances, which also reduced the azimuthal gap (Figure S4).

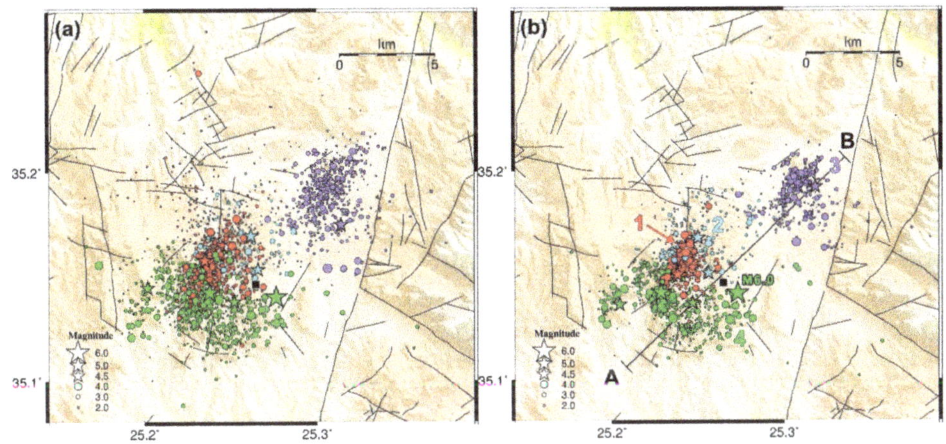

Figure 8. Seismicity of the study region for the period between 1 June and 18 October 2021. (**a**) Initial locations with HypoInverse, (**b**) results of double-difference relocation with HypoDD. Colors and numerical labels in panel (**b**) represent the foreshock swarm activity (group #1, red) and the three spatial aftershock groups: #2 (cyan) middle, sparse group, #3 (blue) northern cluster near Kastelli airport, #4 (green) southern group, including the mainshock. Major events ($M \geq 4.5$) are depicted as stars. Profile line A-B in panel (**b**) is used for the spatiotemporal projection of Figure 9. The location of Arkalochori village is marked with a black square.

Table 2. Initial location statistics for Period A (1 June–27 September 2021, $V_P/V_S = 1.78$, long-distance weighting) and Period B (28 September–18 October 2021, $V_P/V_S = 1.76$, short-distance weighting). ERH, ERZ are horizontal and vertical location errors, "min.stat.dist" denotes the epicentral distance of the closest station with available arrival-time data (see also Figures S3 and S4 in the Supplementary Materials).

	Period A	**Period B**
Number of events	843	1657
Mean RMS error (s)	0.219	0.128
Median ERH (km)	0.530	0.390
Mean ERZ (km)	3.471	1.064
Median ERZ (km)	2.180	0.900
Median Depth (km)	7.120	8.370
Median min.stat.dist. (km)	17	6.3
Mean azim. gap (°)	138	110

In order to assess the 2021 Arkalochori earthquake sequence in more detail and to identify the activated tectonic structures, a relocation procedure was applied, as previously described in Section 3.2, to both periods (Figure 8b). The events distribution was divided into four groups. Group #1 (red) is a temporal cluster that concerns the first period of the foreshock swarm activity, whereas groups #2, #3, and #4 (cyan, blue, and green, respectively) were determined by spatial clustering of the events of the second period. This was achieved by applying Ward's linkage to the matrix of inter-event epicentral distances of the relocated hypocenters [72]. This procedure of agglomerative hierarchical clustering, also known as the "minimum variance method", fuses two clusters into a new one, provided that the resulting increase in the sum of squares (i.e., its objective function) between the objects of the new cluster is minimized, compared to the outcome of any other potential combinations of cluster fusions. After creating the clustering hierarchy with Ward's linkage by setting an appropriate threshold to the fusion level, we divided the spatial distribution into three clusters (i.e., groups #2, #3, and #4) to aid the description.

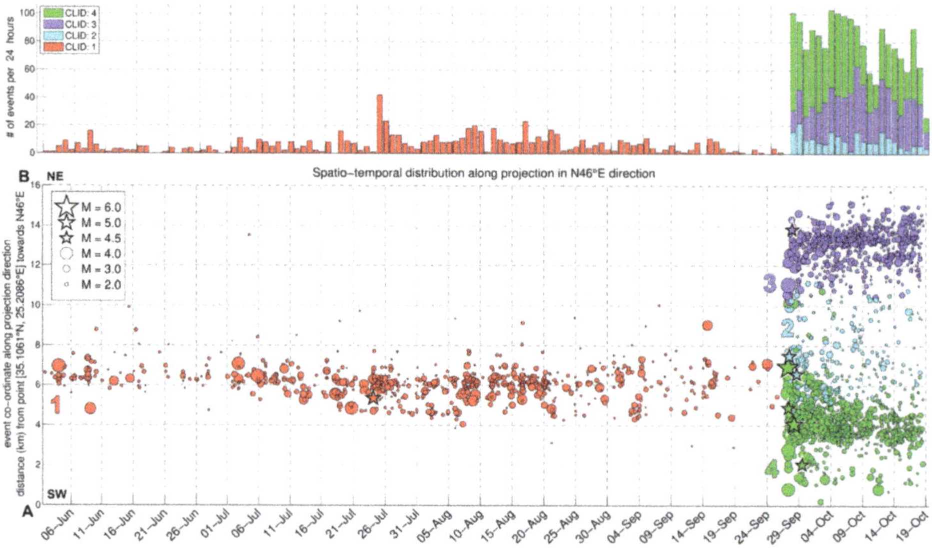

Figure 9. (**Top**) Stacked histogram of the number of seismic events per day in the study area between 1 June and 18 October 2021. (**Bottom**) Spatiotemporal projection along the SW-NE-oriented profile A-B of Figure 8b. Colors and numbers (CLID) represent groups 1–4 of Figure 8b. Events with $M \geq 4.5$ are depicted as stars.

The evolution of the earthquake sequence is examined through the spatiotemporal projection of Figure 9, along the SW-NE-oriented, 16 km-long profile A-B of Figure 8b. The foreshock swarm began in June 2021 with a few small events ($M_L \leq 3.0$) and an $M_L = 4.2$ event on 4 June which did not, however, cause any significant outbreak. Minor bursts occurred on 8 June and then on 18 July including three $M_L = 4.1$–4.2 events. During that period, a slow migration of seismicity towards the southwest can be observed in Figure 9. Then on 24 July 02:07:37 UTC, an $M_L = 4.8$ occurred, triggering a subsequence. Afterwards, apart from a few more minor bursts, the seismicity rate gradually diminished, notably to the point where only 17 events were located in the period between 16 and 25 September, with a complete absence of detectable events for about 41 h preceding the $M_w = 6.0$ mainshock of 27 September 2021, 06:17:21 UTC. The whole foreshock group #1 is roughly distributed in an area of 3.0 km × 1.7 km, elongated in an N-S direction, with its centroid ~2.5 km WNW of Arkalochori. The main event's epicenter is located adjacent to the foreshock swarm, near its eastern edge.

The occurrence of the mainshock, near the middle of profile A-B, caused seismicity to spread rapidly nearly through the whole length of the aftershock zone, both towards the southwest (group #4) and the northeast (group #3). The aftershocks appear to have little overlap with the foreshock zone, specifically with the sparser middle group #2 (cyan) and the northern margin of the southern group #4 (green). The two major groups, #4 south-west of the foreshocks and group #3 (blue) towards the NE, the latter with epicenters near the Kastelli airport, are separated by the less populated group #2 (cyan). This deficit of aftershocks between groups #3 and #4 is likely associated with an area of maximum coseismic slip during the mainshock. The rupture of a large asperity in that region apparently caused redistribution of stress towards the northern and southern margins of the main fault, while the central part was relaxed due to stress drop. The largest aftershock was an $M_w = 5.3$ event that occurred on 28 September 04:48:08 UTC, about 4 km west of the mainshock. The epicenter of the main event is located at the NE edge of group #4, whereas the largest aftershock occurred towards its western part, near the southwestern tip of the foreshock group. The major aftershocks include 11 more events with $M_L \geq 4.0$ between 27 and 29 September, with only smaller events being detected through the rest of the sequence during the period of study (up to 18 October 2021). Regarding the period after 18 October, routine monitoring of seismicity at SL-NKUA detected five more events with $4.0 \leq M_L \leq 4.5$ between 20 and 22 of October, two associated with cluster #3 and three with cluster #4, while another $M_L = 4.2$ event occurred on 29 December associated with the latter cluster. However, no significant changes in the overall spreading of the aftershocks' distribution were observed, besides a small, isolated cluster that was detected ~10 km NNW of Kastelli, activated between 16 and 18 January 2022, with the largest event having $M_L = 3.6$.

The three groups of the relocated aftershocks, if considered as a whole, appear to be well distributed on a plane, striking ~N216° E and dipping ~45° (Figure 10, red-dashed line), as determined by the least-squares fit on the hypocenters. This is quite compatible with focal mechanism solutions for the mainshock reported by most agencies, taking into account the WNW-dipping nodal plane (Table S1). Regarding specific spatial groups of the aftershock sequence, the best-fit plane for the northern cluster #3 strikes N181° E and dips 49° westwards (Figure 10, blue dashed line in profile c_1-c_2), whereas the mid-southern part (groups #2 and #4) has a least-squares plane that trends N216° E and dips 53° WNW (Figure 10, green-dashed line in profiles a_1-a_2 and b_1-b_2). The more north-southwards orientation of the northern cluster's plane makes it more compatible with the Kastelli fault. The fault is steeply dipping at the surface (74°), but the aftershocks' distribution indicates a lower dip angle at the hypocentral depths of 6–9 km, implying listricity of the fault. Cluster #3 is fairly more concentrated than group #4 in the south, which appears to include several smaller sub-clusters, some of which may be related to the activation of antithetic structures (e.g., Figure 10, black-dashed lines with a question mark in profile a_1–a_2). Normal faulting is indicated for the mainshock, the major aftershock, and several other aftershocks and foreshocks. However, oblique and strike-slip faulting is also observed for several events, including some foreshocks (Figure 10). Although the mainshock's focal mechanism is in line with the rest of the aftershocks, it is noted that its hypocenter, along with those of aftershocks that occurred before the deployment of the local network, is not well constrained at depth. Hence, it is reasonable to assume that its true focal depth should be deeper, i.e., between 6 and 14 km, which is an estimated range for its centroid (Table S1).

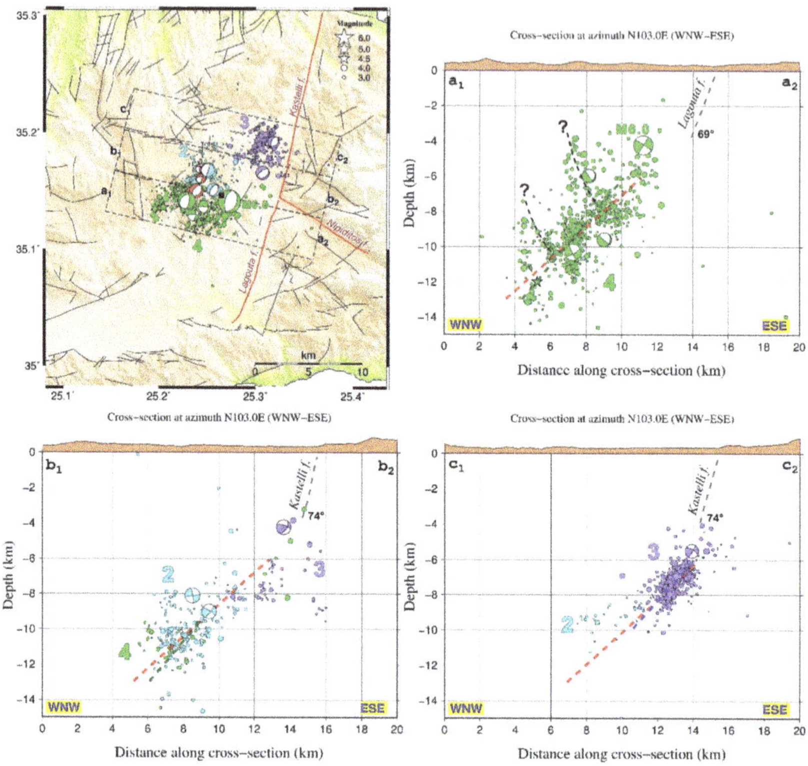

Figure 10. Map and cross-sections of the relocated 2021 Arkalochori aftershock sequence (period 27 September–18 October 2021). Colors and numbers correspond to the aftershock spatial groups 2–4 presented in Figure 7b. The cross-section profiles are drawn in an N103° E direction, roughly perpendicular to the strike of the mapped Kastelli and Lagouta faults, whose downdip extension is presented in the cross-sections with gray-dashed lines. The red dashed line is the section of the least-squares plane fitted to the relocated hypocenters of period B, whereas the green (**a_1–a_2**, **b_1–b_2**) and blue (**c_1–c_2**) dashed lines are the planes fitted to groups #2 and #4 (green), and group #3 (blue), respectively. The possible antithetic structures are marked with black-dashed lines (**a_1–a_2**). Focal mechanisms of the mainshocks and other major events of the sequence, including foreshocks (red beachballs, on the map only) are from the database of GI-NOA. The location of Arkalochori village is marked with a black square on the map.

4.3. DInSAR Analysis

In terms of satellite observations, the systematic acquisition strategy of the Copernicus Sentinel-1 mission ensured the availability of SAR data to investigate the earthquake sequence. Interferometric coherence levels for 6- to 12-days pairs are maintained, allowing for the extraction of reliable DInSAR ground motion measurements.

Independent results from four different satellite tracks (Figures 4 and S2) are consistent, indicating co-seismic downthrow of the epicentral area, extending over an area of about 38 km^2 to the east of the Lagouta and Kastelli faults. The ground displacement patterns are comparable, retaining an elliptical shape of the co-seismic fringes within the epicentral area. The increased ellipticity of the fringes for pairs covering a larger time span after the mainshock (tracks D036, A029, and D109) implies the contribution of post-seismic motion, compared to the more concentric fringes of the A102 track, containing only 2 days of post-seismic activity. In addition, the direction of the major axis of the elliptical fringe pattern is following the aftershock distribution, retaining roughly an NE-SW trend. The spatial continuity of the interferometric fringes yields that the rupture did not reach the surface. The higher density of fringes at the eastern part of the epicentral area implies higher ground displacement gradients, and thus, proximity to the upper part of the activated rupture zone (i.e., NNE striking, west-dipping fault).

However, examining in more detail the DInSAR displacement fields for the various independent measurements, variability arises both in terms of magnitude and location of ground motion maxima (Figure 4). Specifically, for the pair with the shortest temporal extent (2 days) after the mainshock (A102, 23 September 2021–29 September 2021), LoS ground displacement is located in the vicinity of the relocated mainshock epicenter, with values reaching −18 cm. On the contrary, for the pair with the largest temporal extent (track D036, 5 days), the location maximum has been shifted approx. 2.5 km towards ENE and the observed ground displacement increased to −20 cm (Figure 4). The other two displacement pairs (A029 and D109), spanning 3 days within the post-seismic period, show comparable motion (downlift of about −20 cm) with the 5-days D036 pair (Figure S2). The shift of the displacement maxima can be attributed to the amount of post-seismic motion contained in the DInSAR measurements (see Table 1), and it is consistent with the seismological observations indicating migration of aftershocks towards the northern margins (cluster #3) of the mainshock epicentral area. It can also be assumed that no surface deformation occurred, or at least with a magnitude detectable by DInSAR, after three days of post-seismic activity.

In fact, a few centimeters of variation of LoS displacement measurements (still within the error budget of the conventional DInSAR technique) might also be related to the difference in viewing the angle between the ascending and descending satellite observations, and the contribution of the horizontal (mainly E-W) motion component. Therefore, the interpretation of the observed LoS displacements is in advance favored by the decomposition of the ascending and descending interferograms (Figures 11 and S5). The estimated vertical motion indicates down throw reaching −22 cm (Figure 11a), concentrated within the mainshock epicentral region, whereas a more complex pattern, with a maximum of +8 cm, is presented by the E-W horizontal component (Figure 11b).

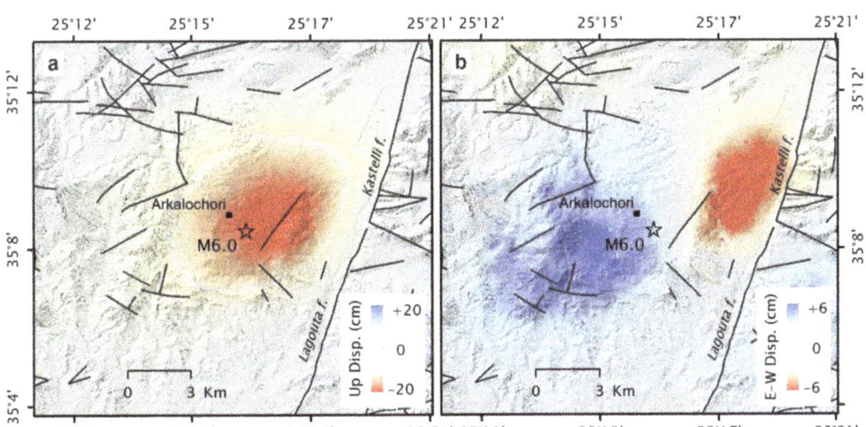

Figure 11. Vertical (**a**) and E-W (**b**) ground displacement maps as decomposed using Sentinel-1 DInSAR LoS observations from ascending A102 and descending D036 tracks (see Table 1). Black lines correspond to fault zones (after [3]). The relocated epicenter of the 27 September 2021 mainshock is marked with a star.

5. Discussion and Conclusions

The recent Arkalochori earthquake sequence, which was recorded before and after the mainshock (27 September 2021), seems to be a quite prominent opportunity to discuss the contemporary tectonic regime of MB. The latter has a very significant geotectonic placement in the context of the African plate subduction that takes place under the Aegean microplate, as it is an onshore post-orogenic basin in the vicinity of the active Hellenic Arc [4,73]. Crete is at a crucial location right above the subducting slab and it is more than certain that the impact of the plate convergence is imprinted on the island, at all scales [74].

The fault pattern that was mapped in detail revealed the different generations of fault-block movements, as a result of the stress field changes since the integration of the Alpine nappe pile buildup [1]. Several direction changes of the main stress axes have been imprinted on the fault structures and the surface morphology [3]. It seems that the easternmost area of MB had been under N-S-trending compression during the orogen buildup, followed by an N-S-trending extension during the slab rollback and the orogen collapse through the detachments' operation (either north or south dipping) [1,8,10,75].

From an evolutionary point of view, we argue that after the generation of the nappe pile on Crete and the south-dipping detachment fault, several almost E-W-trending fault zones were activated (as segments of the south-dipping detachment) since they crosscut rocks that belong to different Alpine units. These fault zones remained active until after the end of late Serravallian, when the deposition of the lacustrine Viannos sediments was completed, as large fault blocks comprising Serravallian age successions are found tilted towards NNE.

A transtensional geotectonic regime, which seems to be dominating the entire Aegean microplate [76], is expressed on the Cretan territory by nearly NNE-SSW-trending strike-slip fault zones [20,77] and one of them crosscuts the central area of the island and simultaneously was acting as the marginal structure of the HB-MB system [2]. It should be mentioned that the main fault zones that were found along the eastern margin of MB are striking NNE-SSW, NW-SE, and E-W. The kinematic analysis of these data showed that most of the faults that were activated after the Early Miocene were either normal or strike-slip faults with left-lateral movement components [3]. According to this analysis, the NNE-SSW-trending shear zone was also affecting the previously generated E-W-trending fault zones, which were inherited by the Early Miocene N-S extensional period, when the south Cretan detachment was still active [1]. The impact on these E-W-trending structures is crucial as they seem to have been segmented and some of the segments have been reactivated during or shortly after the Pliocene (e.g., Nipiditos fault).

Specifically, during Tortonian, while the sediments of the Ampelouzos formation were still depositing, the dominating stress field had a strong left-lateral component, which is evidenced by the syn-sedimentary structure analysis (Figure 6b). The stress field seems to be active during the Tortonian since the syn-sedimentary faults were deforming the Ampelouzos unit strata (see Section 2.2 for lithology description), which were deposited before 10 Ma in a coastal to shallow depth marine environment [78]. A strong case scenario, considering both tectonic and stratigraphic data, is that the same fault zone seems to have a larger left-lateral component during Tortonian and the stress field gradually changed during Messinian, when the normal component dominates, increasing the compatibility of the stress field with arc-parallel extension. The latter seems to be active up to the contemporary era and is tectonically responsible for the Arkalochori earthquake.

The tectonic analysis of the collected measurements during the fieldwork provided strong arguments of stress field changes along the NNE-SSW-trending fault zones, as two groups of structures were found on fault surfaces revealing lateral movement and extension along the NW-SE direction. As explained above, the lateral movement on the same fault surfaces is older than the normal one, which in turn coincides with arc-parallel extension.

Finally, we refer to the Kastelli fault zone and in particular its southernmost segment (Lagouta fault), which is a west-dipping structure hosting a normal displacement movement with a strong left-lateral oblique component inherited by the dominating stress field during Tortonian. It was exactly this structure that was activated during the Arkalochori earthquake, as the hypocenters of the sequence clearly delineate the above-mentioned fault surface. It is obvious that the two hypocentral clusters that have been described, form two discrete segments that belong to the same fault zone. It is the WNW-ESE-trending Nipiditos fault that seems to be the structural reason for this segmentation. The northern cluster is formed on the surface of the Kastelli fault and the southern cluster on the Lagouta fault (Figure 10). Even though all events belong to the same sequence, it is rather clear that the northern cluster is located at more shallow depths compared to the southern cluster.

Concerning the seismological data, the 27 September $M_w = 6.0$ mainshock was a reminder that strong earthquakes do occur onshore Crete. It was preceded by a rich sequence of over 700 foreshocks that was considered a swarm, due to its largest event occurring halfway through 24 July. However, it did not cause much concern at the time, as similar swarms have been known to occur elsewhere in Greece, without leading to a strong event of $M_w \approx 6.0$, unlike the case of the Arkalochori earthquake. The herein analysis highlights the importance of local seismological networks to the reduction in location uncertainties. The applied double-difference relocation has managed to improve the relative locations of the foreshock epicenters, which are concentrated in the vicinity of the 27 September mainshock, although their hypocenters could not be adequately constrained. The mainshock apparently broke a large asperity of a west-dipping normal fault and distributed stresses towards its northern and southern edges, triggering aftershocks mainly at two large groups, separated by a spatial gap, where the asperity was located. Similar cases have been previously reported in other significant earthquakes on normal faults in Greece, including the 1999 Athens [79,80], the 2017 Kos [81,82], and the 2020 Samos [83–86] earthquakes. The herein presented relocation analysis of the sequence during its second period, i.e., with data from the temporary local network that was deployed by GI-NOA, permitted the detailed delineation of the main activated structures which could be related to mapped faults on the surface, allowing for a small degree of listricity, as well as contingent smaller antithetic, conjugate faults at depth. The spatiotemporal evolution of the sequence indicated triggering of seismicity throughout most of the aftershock zone soon after the mainshock, attributed to coseismic stress transfer, followed by slower migration towards its outer edges, indicating possible after slip.

Based on the post-earthquake fieldwork, both activated fault segments (Kastelli and Lagouta) did not show any signs of surface raptures along their traces, which agrees with the interferometric findings. A straightforward interpretation of horizontal motion patterns derived from the DInSAR analysis points out that the observed migration of motion during the post-seismic period (interferograms of different temporal spans) is mainly in agreement with the aftershock distribution and clustering, and to a lesser extent linked to actual co-seismic motion patterns. This is more evident by the collocation of the horizontal motion lobes to the aftershock clusters #3 and #4, as well as the absence of motion in the central part of the epicentral area where the mainshock nucleated. The vertical down-throw at the Arkalochori village area is also in accordance with the local stress field, being attributed to the hanging wall of the normal westward dipping Lagouta and Kastelli faults (Figure 6).

The activation of the specific fault zone is also implied by the increased displacement gradients at the easternmost part of the epicentral area. However, the −22 cm of vertical motion, based on the decomposition of opposite DInSAR geometries (Figure 10), appears to be contaminated by post-seismic deformation. A more conservative estimate would attribute −18 cm of LoS motion to the main earthquake event, as calculated by the interferometric pair of the shortest temporal span after the mainshock. Additional displacement detected by other DInSAR pairs, also present in the decomposition, is attributed to the migration of motion towards NE, in accordance with the clustering and the shallower aftershocks' depth nearby the northern Kastelli fault segment.

In conclusion, the 2021 Arkalochori earthquake is a characteristic implication of the contemporary existence of a local arc-parallel extensional regime adjacent to major arc-normal strike-slip zones on the forearc region of the Aegean microplate. The consistency between seismological, geodetic, and field observations has been well demonstrated, highlighting the complementarity of multi-disciplinary approaches. The availability of dense seismological recordings and the contribution of Earth Observation platform-based solutions, validated, when possible, with field observations ensure proper mapping and tectonic interpretation of seismic events.

Supplementary Materials: The following supporting information can be downloaded at: https://www.mdpi.com/article/10.3390/app12062815/s1, Table S1: Fault plane solutions for the mainshock of the 2021 Arkalochori sequence, as determined by different agencies. Figure S1: Seismological and accelerometric stations of the Hellenic Unified Seismological Network (HUSN) with available data during the study period (1 June–18 October 2021). (a) Stations in the broader area of Southern Greece, (b) zoom in the area marked with a red rectangle in panel (a), showing stations at distances less than 60 km from the epicenter of the 27 September 2021, mainshock at Arkalochori, Central Crete, marked with a star. Figure S2: Co-seismic LoS displacement maps, as derived using GEP DIAPASON service, for ascending track A029 (18 September 2021–30 September 2021) and descending D109 (24 September 2021–30 October 2021) (c,d). The spatial extend of Sentinel-1 acquisition frames for both geometries are also indicated (a,b). Black lines correspond to fault zones (after [3]). The relocated epicenter of the 27 September 2021, mainshock is marked with a star. Figure S3: Initial location statistics (as reported by HypoInverse). (a,b) RMS travel-time residuals, (c,d) focal depth, (e,f) horizontal (ERX, ERY) and vertical (ERZ) formal location uncertainties, (a,c,e) for Period A, between 1 June and 27 September 2021, and (b,d,f) for Period B, between 28 September and 18 October 2021. Figure S4: Initial location statistics. (a,b) Distribution of distance from the closest station with available data, (c,d) azimuthal gap, (a,c) for Period A, between 1 June and 27 September 2021, and (b,d) for Period B, between 28 September and 18 October 2021. Figure S5: Vertical (a) and E-W (b) ground displacement maps as decomposed using Sentinel-1 DInSAR LoS observations from ascending A029 and descending D109 tracks (see Table 1 in manuscript). Black lines correspond to fault zones (after [3]). The relocated epicenter of the 27 September 2021, mainshock is marked with a star. File E1: Relocated earthquake catalogue (1 June–18 October 2021). References [50,51,87–89] are cited in the Supplementary Materials.

Author Contributions: Conceptualization, E.V., J.A. and G.K.; methodology, E.V., G.K., V.K. (Vasilis Kapetanidis), J.A., N.V. and M.F.; software, E.V., V.K. (Vasilis Kapetanidis) and E.P.; validation, E.V., A.K., C.P.E. and S.P.; resources, E.V., G.K., V.K. (Vasilis Kapetanidis), E.P., M.F., C.P.E., V.K. (Vassilios Karastathis), G.-A.T. and J.A.; data curation, A.K., V.K. (Vasilis Kapetanidis), C.P.E. and E.P.; writing—original draft preparation, E.V., G.K., V.K. (Vasilis Kapetanidis), E.P., M.F. and J.A.; writing—review and editing, E.V., G.K., V.K. (Vasilis Kapetanidis), M.F., J.A. and N.V.; visualization, E.V., V.K. (Vasilis Kapetanidis) and M.F.; supervision, E.V.; project administration, J.A. and N.V.; funding acquisition, E.V., M.F., J.A., G.-A.T. and N.V. All authors have read and agreed to the published version of the manuscript.

Funding: This research received no external funding.

Institutional Review Board Statement: Not applicable.

Informed Consent Statement: Not applicable.

Data Availability Statement: Initial earthquake catalogues and arrival-time data are available at the database of SL-NKUA (http://www.geophysics.geol.uoa.gr/stations/gmapv3_db/index.php?lang=en; accessed on 7 November 2021) and GI-NOA (http://bbnet.gein.noa.gr/HL/databases/database; accessed on 7 November 2021). Waveform data from permanent and temporary stations of the HUSN are available at the National EIDA Node hosted at GI-NOA (http://eida.gein.noa.gr/; accessed on 7 November 2021; [90]. Focal mechanisms of the mainshock from various agencies (Table S1) were acquired from https://www.seismicportal.eu/mtws/ (accessed on 22 January 2022). Focal mechanisms for the aftershocks are from GI-NOA (http://bbnet.gein.noa.gr/HL/seismicity/mts/automatic-moment-tensor-gisola; accessed on 7 November 2021). The relocated catalogue of this study is available in the Supplementary Materials.

Acknowledgments: The authors would like to express their appreciation to the three anonymous reviewers for their constructive comments that helped improving the initial version of the manuscript, as well as the editors for giving us the opportunity to publish our work. We thank the scientists and personnel who have participated in the installation or maintenance of the permanent and temporary stations belonging to the HUSN and routine analysis of earthquake data.

Conflicts of Interest: The authors declare no conflict of interest.

References

1. Papanikolaou, D.; Vassilakis, E. Thrust faults and extensional detachment faults in Cretan tectono-stratigraphy: Implications for Middle Miocene extension. *Tectonophysics* **2010**, *488*, 233–247. [CrossRef]
2. Vassilakis, E.; Alexopoulos, J. Recognition of strike-slip faulting on the supra-detachment basin of Messara (central Crete Island) with remote sensing image interpretation techniques. In Proceedings of the 4th EARSeL Workshop on Remote Sensing and Geology, Mykonos, Greece, 24–25 May 2012; pp. 108–115.
3. Vassilakis, E. Study of the Tectonic Structure of Messara Basin, Central Crete, with the Aid of Remote Sensing Techniques and G.I.S. Ph.D. Thesis, National & Kapodestrian University of Athens, Athens, Greece, 2006.
4. Van Hinsbergen, D.J.J.; Meulenkamp, J. Neogene supradetachment basin development on Crete (Greece) during exhumation of the South Aegean core complex. *Basin Res.* **2006**, *18*, 103–124. [CrossRef]
5. Kahle, H.G.; Cocard, M.; Peter, Y.; Geiger, A.; Reilenger, R.; Barka, A.; Veis, G. GPS-derived strain rate field within the boundary zones of the Eurasian, African, and Arabian Plates. *J. Geophys. Res.* **2000**, *105*, 23353–23370. [CrossRef]
6. McClusky, S.; Balassanian, S.; Barka, A.; Demir, C.; Ergintav, S.; Georgiev, I.; Gurkan, O.; Hambourger, M.; Hurst, K.; Kahle, H.-G.; et al. Global Positioning System constraints on plate kinematics and dynamic in the eastern Mediterranean and Caucasus. *J. Geophys. Res.* **2000**, *105*, 5695–5719. [CrossRef]
7. Nyst, M.; Thatcher, W. New constraints on the active tectonic deformation of the Aegean. *J. Geophys. Res.* **2004**, *109*, B11406. [CrossRef]
8. Fassoulas, C.; Kilias, A.; Mountrakis, D. Postnappe stacking extension and exhumation of HP/LT rocks in the island of Crete, Greece. *Tectonics* **1994**, *13*, 127–138. [CrossRef]
9. Jolivet, L.; Goffe, B.; Monie, P.; Truffert, C.; Patriat, M.; Bonneau, M. Miocene detachment in Crete and exhumation P-T-t paths of high pressure metamorphic rocks. *Tectonics* **1996**, *15*, 1129–1153. [CrossRef]
10. Ring, U.; Brachert, T.; Fassoulas, C. Middle Miocene graben development in Crete and its possible relation to large-scale detachment faults in the southern Aegean. *Terra Nova* **2001**, *13*, 297–304. [CrossRef]
11. Woessner, J.; Laurentiu, D.; Giardini, D.; Crowley, H.; Cotton, F.; Grünthal, G.; Valensise, G.; Arvidsson, R.; Basili, R.; Demircioglu, M.B.; et al. The 2013 European Seismic Hazard Model: Key components and results. *Bull. Earthq. Eng.* **2015**, *13*, 3553–3596. [CrossRef]
12. Reilinger, R.; McClusky, S.; Paradissis, D.; Ergintav, S.; Vernant, P. Geodetic constraints on the tectonic evolution of the Aegean region and strain accumulation along the Hellenic subduction zone. *Tectonophysics* **2010**, *488*, 22–30. [CrossRef]
13. Ganas, A.; Oikonomou, I.A.; Tsimi, C. NOAfaults: A digital database for active faults in Greece. *Bull. Geol. Soc. Greece* **2017**, *47*, 518–530. [CrossRef]
14. Kapetanidis, V.; Kassaras, I. Contemporary crustal stress of the Greek region deduced from earthquake focal mechanisms. *J. Geodyn.* **2019**, *123*, 55–82. [CrossRef]
15. Stucchi, M.; Rovida, A.; Gomez Capera, A.A.; Alexandre, P.; Camelbeeck, T.; Demircioglu, M.B.; Gasperini, P.; Kouskouna, V.; Musson, R.M.W.; Radulian, M.; et al. The SHARE European Earthquake Catalogue (SHEEC) 1000–1899. *J. Seismol.* **2013**, *17*, 523–544. [CrossRef]
16. Makropoulos, K.; Kaviris, G.; Kouskouna, V. An updated and extended earthquake catalogue for Greece and adjacent areas since 1900. *Nat. Hazards Earth Syst. Sci.* **2012**, *12*, 1425–1430. [CrossRef]
17. Zachariasse, W.J.; van Hinsbergen, D.J.J.; Fortuin, A.R. Formation and fragmentation of a late Miocene supradetachment basin in central Crete: Implications for exhumation mechanisms of high-pressure rocks in the Aegean forearc. *Basin Res.* **2011**, *23*, 678–701. [CrossRef]
18. Angelier, J.; Brebion, P.; Lauriat-Rage, A.; Muller, C. Late Cenozoic biostratigraphy and neotectonic evolution of Crete. *Ann. Geol. Pays Hell.* **1979**, *1*, 9–17.
19. Armijo, R.; Lyon-Caen, H.; Papanastassiou, D. East-west extension and Holocene normal-fault scarps in the Hellenic arc. *Geology* **1992**, *20*, 491–494. [CrossRef]
20. Caputo, R.; Catalano, S.; Monaco, C.; Romagnoli, G.; Tortorici, G.; Tortorici, L. Active faulting on the island of Crete (Greece). *Geophys. J. Int.* **2010**, *183*, 111–126. [CrossRef]
21. Werner, V.; Baika, K.; Fischer, P.; Hadler, H.; Obrocki, L.; Willershäuser, T.; Tzigounaki, A.; Tsigkou, A.; Reicherter, K.; Papanikolaou, I.; et al. The sedimentary and geomorphological imprint of the AD 365 tsunami on the coasts of southwestern Crete (Greece)—Examples from Sougia and Palaiochora. *Quat. Int.* **2018**, *473*, 66–90. [CrossRef]
22. Papazachos, B.C.; Papazachou, C.B. *The Earthquakes of Greece*; Ziti: Thessaloniki, Greece, 2003; p. 304.

23. Guidoboni, E.; Comastri, A. *Catalogue of Earthquakes and Tsunamis in the Mediterranean Area from the 11th to the 15th Century*; INGV-SGA: Roma, Italy, 2005; p. 1037.
24. Tsapanos, T.M. A seismic hazard scenario for the main cities of Crete island, Greece. *Geophys. J. Int.* **2003**, *153*, 403–408. [CrossRef]
25. Vallianatos, F.; Michas, G.; Hloupis, G.; Chatzopoulos, G. The Evolution of Preseismic Patterns Related to the Central Crete (Mw6.0) Strong Earthquake on 27 September 2021 Revealed by Multiresolution Wavelets and Natural Time Analysis. *Geosciences* **2022**, *12*, 33. [CrossRef]
26. Triantafyllou, I.; Karavias, A.; Koukouvelas, I.; Papadopoulos, G.A.; Parcharidis, I. The Crete Isl. (Greece) Mw6.0 Earthquake of 27 September 2021: Expecting the Unexpected. *GeoHazards* **2022**, *3*, 106–124. [CrossRef]
27. ITSAK. *Arkalochori Earthquakes, M 6.0 on 27/09/2021 & M 5.3 on 28/09/2021: Preliminary Report—Recordings of the ITSAK Accelerometric Network and Damage on the Natural and Built Environment*; ITSAK Research Unit: Thessaloniki, Greece, 2021; p. 44. (In Greek)
28. EPPO. *New National Building Code*; EPPO: Athens, Greece, 2003; p. 72.
29. Delibasis, N.D.; Ziazia, M.; Voulgaris, N.; Papadopoulos, T.; Stavrakakis, G.N.; Papanastassiou, D.; Drakatos, G. Microseismic activity and seismotectonics of Heraklion Area (central Crete Island, Greece). *Tectonophysics* **1999**, *308*, 237–248. [CrossRef]
30. Dula, W.F., Jr. Geometric Models of Listric Normal Faults and Rollover Folds1. *AAPG Bull.* **1991**, *75*, 1609–1625. [CrossRef]
31. Williams, G.; Vann, I. The geometry of listric normal faults and deformation in their hangingwalls. *J. Struct. Geol.* **1987**, *9*, 789–795. [CrossRef]
32. Wernicke, B.; Burchfiel, B.C. Modes of extensional tectonics. *J. Struct. Geol.* **1982**, *4*, 105–115. [CrossRef]
33. Papanikolaou, D.; Vassilakis, E. Middle Miocene E-W tectonic horst structure of Crete through extensional detachment faults. *Earth Environ. Sci.* **2008**, *2*, 012003. [CrossRef]
34. Faccenna, C.; Jolivet, L.; Piromallo, C.; Morelli, A. Subduction and the depth of convection in the Mediterranean mantle. *J. Geophys. Res.* **2003**, *108*, 2099. [CrossRef]
35. Brun, J.-P.; Faccenna, C. Exhumation of high-pressure rocks driven by slab rollback. *Earth Planet. Sci. Lett.* **2008**, *272*, 1–7. [CrossRef]
36. Kaviris, G.; Fountoulakis, I.; Spingos, I.; Millas, C.; Papadimitriou, P.; Drakatos, G. Mantle dynamics beneath Greece from SKS and PKS seismic anisotropy study. *Acta Geophys.* **2018**, *66*, 1341–1357. [CrossRef]
37. Royden, L.H.; Papanikolaou, D.J. Slab segmentation and late Cenozoic disruption of the Hellenic arc. *Geochem. Geophys. Geosyst.* **2011**, *12*, Q03010. [CrossRef]
38. Peterek, A.; Schwarze, J. Architecture and Late Pliocene to recent evolution of outer-arc basins of the Hellenic subduction zone (south-central Crete, Greece). *J. Geodyn.* **2004**, *38*, 19–55. [CrossRef]
39. Becker, D.; Meier, T.; Bohnhoff, M.; Harjes, H.P. Seismicity at the convergent plate boundary offshore Crete, Greece, observed by an amphibian network. *J. Seismol.* **2010**, *14*, 369–392. [CrossRef]
40. Kokinou, E.; Vallianatos, F. Seismic velocity Structure and Waveform Modelling in the southern Hellenic Arc (offshore Crete). In Proceedings of the 2nd IASME/WSEAS International Conference on Geology and Seismology (GES'08), Cambridge, UK, 23–25 February 2008.
41. Vassilakis, E.; Alexopoulos, J.; Farangitakis, G.P. Combination of Earth Observation and Seismic Reflection Data Analysis for the Definition of Strike Slip Fault Zones in Central Crete. In Proceedings of the EGU2020, Vienna, Austria, 4–8 May 2020.
42. Kaviris, G.; Papadimitriou, P.; Kravvariti, P.; Kapetanidis, V.; Karakonstantis, A.; Voulgaris, N.; Makropoulos, K. A detailed seismic anisotropy study during the 2011–2012 unrest period in the Santorini Volcanic Complex. *Phys. Earth Planet. Inter.* **2015**, *238*, 51–88. [CrossRef]
43. Papadimitriou, P.; Kapetanidis, V.; Karakonstantis, A.; Kaviris, G.; Voulgaris, N.; Makropoulos, K. The Santorini Volcanic Complex: A detailed multi-parameter seismological approach with emphasis on the 2011–2012 unrest period. *J. Geodyn.* **2015**, *85*, 32–57. [CrossRef]
44. Kokinou, E.; Moisidi, M.; Tsanaki, I.; Tsakalaki, E.; Tsiskaki, E.; Sarris, A.; Vallianatos, F. A seismotectonic study for the Heraklion basin in Crete (Southern Hellenic arc, Greece). *Int. J. Geol.* **2008**, *2*, 9–16.
45. Floyd, M.A.; Billiris, H.; Paradissis, D.; Veis, G.; Avallone, A.; Briole, P.; McClusky, S.; Nocquet, J.M.; Palamartchouk, K.; Parsons, B.; et al. A new velocity field for Greece: Implications for the kinematics and dynamics of the Aegean. *J. Geophys. Res.* **2010**, *115*, B10403. [CrossRef]
46. Hatzfeld, D.; Besnard, M.; Makropoulos, K.; Hatzidimitriou, P. Microearthquake seismicity and fault-plane solutions in the southern Aegean and its geodynamic implications. *Geophys. J. Int.* **1993**, *115*, 799–818. [CrossRef]
47. Sakellariou, D.; Mascle, J.; Lykousis, V. Strike slip tectonics and transtensional deformation in the Aegean region and the Hellenic arc: Preliminary results. *Bull. Geol. Soc. Greece* **2017**, *47*, 647. [CrossRef]
48. De Chabalier, J.B.; Lyon-Caen, H.; Zollo, A.; Deschamps, A.; Bernard, P.; Hatzfeld, D. A detailed analysis of microearthquakes in western Crete from digital three-component seismograms. *Geophys. J. Int.* **1992**, *110*, 347–360. [CrossRef]
49. Yamaji, A.; Sato, K. *Multiple Inverse Method Software Package User's Guide*, v. 4.0; Kyoto University: Kyoto, Japan, 2004.
50. Technological Educational Institute of Crete. *Seismological Network of Crete*; 10.7914/SN/HC; International Federation of Digital Seismograph Networks: Crete, Greece, 2006.
51. National Observatory of Athens, Institute of Geodynamics. *National Observatory of Athens Seismic Network*; 10.7914/SN/HL; International Federation of Digital Seismograph Networks: Athens, Greece, 1975.

52. Klein, F.W. *User's Guide to HYPOINVERSE-2000, a Fortran Program to Solve for Earthquake Locations and Magnitudes*, 2002-171; United States Department Of The Interior Geological Survey: Menlo Park, CA, USA, 2002; 123.
53. Chatelain, J.L. Etude Fine de la Sismicité en Zone de Collision Continentale au Moyen d'un Réseau de Stations Portables: La Région Hindu-Kush Pamir. Doctoral Dissertation, Université Scientifique et Médicale de Grenoble, Grenoble, France, 1978.
54. Waldhauser, F. *hypoDD-A Program to Compute Double-Difference Hypocenter Locations*, open-file report, 01-113; U.S. Geological Survey: Menlo Park, CA, USA, 2001.
55. Kapetanidis, V. Spatiotemporal Patterns of Microseismicity for the Identification of Active Fault Structures Using Seismic Waveform Cross-Correlation and Double-Difference Relocation. Doctoral Dissertation, University of Athens, Athens, Greece, 2017.
56. Massonnet, D.; Rossi, M.; Carmona, C.; Adragna, F.; Peltzer, G.; Feigl, K.; Rabaute, T. The displacement field of the Landers earthquake mapped by radar interferometry. *Nature* **1993**, *364*, 138–142. [CrossRef]
57. Papadaki, E.S. Monitoring subsidence at Messara basin using radar interferometry. *Environ. Earth Sci.* **2014**, *72*, 1965–1977. [CrossRef]
58. Svigkas, N.; Atzori, S.; Kiratzi, A.; Tolomei, C.; Salvi, S. Isolation of swarm sources using InSAR: The case of the February 2017 seismic swarm in western Anatolia (Turkey). *Geophys. J. Int.* **2019**, *217*, 1479–1495. [CrossRef]
59. Elias, P.; Spingos, I.; Kaviris, G.; Karavias, A.; Gatsios, T.; Sakkas, V.; Parcharidis, I. Combined Geodetic and Seismological Study of the December 2020 Mw = 4.6 Thiva (Central Greece) Shallow Earthquake. *Appl. Sci.* **2021**, *11*, 5947. [CrossRef]
60. Cornou, C.; Ampuero, J.-P.; Aubert, C.; Audin, L.; Baize, S.; Billant, J.; Brenguier, F.; Causse, M.; Chlieh, M.; Combey, A.; et al. Rapid response to the Mw 4.9 earthquake of November 11, 2019 in Le Teil, Lower Rhône Valley, France. *Comptes Rendus. Géosci.* **2021**, *353*, 441–463. [CrossRef]
61. Karakostas, V.; Papazachos, C.; Papadimitriou, E.; Foumelis, M.; Kiratzi, A.; Pikridas, C.; Kostoglou, A.; Kkallas, C.; Chatzis, N.; Bitharis, S.; et al. The March 2021 Tyrnavos, central Greece, doublet (Mw6.3 and Mw6.0): Aftershock relocation, faulting details, coseismic slip and deformation. *Bull. Geol. Soc. Greece* **2021**, *58*, 131–178. [CrossRef]
62. Kaviris, G.; Elias, P.; Kapetanidis, V.; Serpetsidaki, A.; Karakonstantis, A.; Plicka, V.; De Barros, L.; Sokos, E.; Kassaras, I.; Sakkas, V.; et al. The Western Gulf of Corinth (Greece) 2020–2021 Seismic Crisis and Cascading Events: First Results from the Corinth Rift Laboratory Network. *Seism. Rec.* **2021**, *1*, 85–95. [CrossRef]
63. Tolomei, C.; Caputo, R.; Polcari, M.; Famiglietti, N.A.; Maggini, M.; Stramondo, S. The Use of Interferometric Synthetic Aperture Radar for Isolating the Contribution of Major Shocks: The Case of the March 2021 Thessaly, Greece, Seismic Sequence. *Geosciences* **2021**, *11*, 191. [CrossRef]
64. Foumelis, M.; Papazachos, C.; Papadimitriou, E.; Karakostas, V.; Ampatzidis, D.; Moschopoulos, G.; Kostoglou, A.; Ilieva, M.; Minos-Minopoulos, D.; Mouratidis, A.; et al. On rapid multidisciplinary response aspects for Samos 2020 M7.0 earthquake. *Acta Geophys.* **2021**, *69*, 1025–1048. [CrossRef]
65. Le Cozannet, G.; Kervyn, M.; Russo, S.; Ifejika Speranza, C.; Ferrier, P.; Foumelis, M.; Lopez, T.; Modaressi, H. Space-Based Earth Observations for Disaster Risk Management. *Surv. Geophys.* **2020**, *41*, 1209–1235. [CrossRef]
66. Foumelis, M.; Papadopoulou, T.; Bally, P.; Pacini, F.; Provost, F.; Patruno, J. Monitoring Geohazards Using On-Demand And Systematic Services On Esa's Geohazards Exploitation Platform. In Proceedings of the IGARSS 2019—2019 IEEE International Geoscience and Remote Sensing Symposium, Yokohama, Japan, 28 July–2 August 2019; pp. 5457–5460.
67. Foumelis, M.; Parcharidis, I.; Lagios, E.; Voulgaris, N. Evolution of post-seismic ground deformation of the Athens 1999 earthquake observed by SAR interferometry. *J. Appl. Geophys.* **2009**, *69*, 16–23. [CrossRef]
68. Papageorgiou, E.; Foumelis, M.; Parcharidis, I. Long-and Short-Term Deformation Monitoring of Santorini Volcano: Unrest Evidence by DInSAR Analysis. *IEEE J. Sel. Top. Appl. Earth Obs. Remote Sens.* **2012**, *5*, 1531–1537. [CrossRef]
69. Samieie-Esfahany, S.; Hanssen, R.F.; van Thienen-Visser, K.; Muntendam-Bos, A. On the Effect of Horizontal Deformation on InSAR Subsidence Estimates. In Proceedings of the Fringe 2009 Workshop, Frascati, Italy, 30 November–4 December 2009.
70. Tranos, M.D. TR method (TRM): A separation and stress inversion method for heterogeneous fault-slip data driven by Andersonian extensional and compressional stress regimes. *J. Struct. Geol.* **2015**, *79*, 57–74. [CrossRef]
71. Meulenkamp, J.E.; van der Zwaan, G.L.; van Wamel, W.A. On late Miocene to recent vertical motions in the Cretan segment of the Hellenic Arc. *Tectonophysics* **1994**, *234*, 53–72. [CrossRef]
72. Ward, J.H. Hierarchical Grouping to Optimize an Objective Function. *J. Am. Stat. Assoc.* **1963**, *58*, 236–244. [CrossRef]
73. Papanikolaou, D.; Royden, L. Disruption of the Hellenic arc: Late Miocene extensional detachment faults and steep Pliocene-Quaternary normal faults—Or what happened at Corinth? *Tectonics* **2007**, *26*, TC5003. [CrossRef]
74. Van Hinsbergen, D.J.J.; Zachariasse, W.J.; Wortel, M.J.R.; Meulenkamp, J.E. Underthrusting and exhumation: A comparison between the External Hellenides and the hot Cycladic and cold South Aegean core complexes (Greece). *Tectonics* **2005**, *24*, TC2011. [CrossRef]
75. Van Hinsbergen, D.J.J.; Hafkenscheid, E.; Spakman, W.; Meulenkamp, J.; Wortel, R. Nappe stacking resulting from subduction of oceanic and continental lithosphere below Greece. *Geology* **2005**, *33*, 325–328. [CrossRef]
76. Sakellariou, D.; Tsampouraki-Kraounaki, K. Plio-Quaternary Extension and Strike-Slip Tectonics in the Aegean. In *Transform Plate Boundaries and Fracture Zones*; Duarte, J.C., Ed.; Elsevier: Amsterdam, The Netherlands, 2019; pp. 339–374. [CrossRef]
77. Mason, J.; Schneiderwind, S.; Pallikarakis, A.; Wiatr, T.; Mechernich, S.; Papanikolaou, I.; Reicherter, K. Fault structure and deformation rates at the Lastros-Sfaka Graben, Crete. *Tectonophysics* **2016**, *683*, 216–232. [CrossRef]

78. Krijgsman, W.; Hilgen, F.J. Langereis, C.G.; Zachariasse, W.J. The age of the Tortonian/Messinian boundary. *Earth Planet. Sci. Lett.* **1994**, *121*, 533–547. [CrossRef]
79. Papadimitriou, P.; Voulgaris, N.; Kassaras, I.; Kaviris, G.; Delibasis, N.; Makropoulos, K. The Mw = 6.0, 7 September 1999 Athens Earthquake. *Nat. Hazards* **2002**, *27*, 15–33. [CrossRef]
80. Kapetanidis, V.; Karakonstantis, A.; Papadimitriou, P.; Pavlou, K.; Spingos, I.; Kaviris, G.; Voulgaris, N. The 19 July 2019 earthquake in Athens, Greece: A delayed major aftershock of the 1999 Mw = 6.0 event, or the activation of a different structure? *J. Geodyn.* **2020**, *139*, 101766. [CrossRef]
81. Ganas, A.; Elias, P.; Kapetanidis, V.; Valkaniotis, S.; Briole, P.; Kassaras, I.; Argyrakis, P.; Barberopoulou, A.; Moshou, A. The July 20, 2017 M6.6 Kos Earthquake: Seismic and Geodetic Evidence for an Active North-Dipping Normal Fault at the Western End of the Gulf of Gökova (SE Aegean Sea). *Pure Appl. Geophys.* **2019**, *176*, 4177–4211. [CrossRef]
82. Triantafyllou, I.; Papadopoulos, G.A.; Lekkas, E. Impact on built and natural environment of the strong earthquakes of April 23, 1933, and July 20, 2017, in the southeast Aegean Sea, eastern Mediterranean. *Nat. Hazards* **2020**, *100*, 671–695. [CrossRef]
83. Triantafyllou, I.; Gogou, M.; Mavroulis, S.; Lekkas, E.; Papadopoulos, G.A.; Thravalos, M. The Tsunami Caused by the 30 October 2020 Samos (Aegean Sea) Mw7.0 Earthquake: Hydrodynamic Features, Source Properties and Impact Assessment from Post-Event Field Survey and Video Records. *J. Mar. Sci. Eng.* **2021**, *9*, 68. [CrossRef]
84. Papadimitriou, P.; Kapetanidis, V.; Karakonstantis, A.; Spingos, I.; Kassaras, I.; Sakkas, V.; Kouskouna, V.; Karatzetzou, A.; Pavlou, K.; Kaviris, G.; et al. First Results on the Mw=6.9 Samos Earthquake of 30 October 2020. *Bull. Geol. Soc. Greece* **2020**, *56*, 251–279. [CrossRef]
85. Karakostas, V.; Tan, O.; Kostoglou, A.; Papadimitriou, E.; Bonatis, P. Seismotectonic implications of the 2020 Samos, Greece, Mw 7.0 mainshock based on high-resolution aftershock relocation and source slip model. *Acta Geophys.* **2021**, *69*, 979–996. [CrossRef]
86. Kaviris, G.; Spingos, I.; Kapetanidis, V.; Papadimitriou, P.; Voulgaris, N. On the origin of upper crustal shear-wave anisotropy at Samos Island, Greece. *Acta Geophys.* **2021**, *69*, 1051–1064. [CrossRef]
87. Aristotle University of Thessaloniki. Aristotle University of Thessaloniki Seismological Network [Data Set]. International Federation of Digital Seismograph Networks. 1981. Available online: https://www.fdsn.org/networks/detail/HT/ (accessed on 26 January 2022).
88. (ITSAK) Institute of Engineering Seimology Earthquake Engineering. ITSAK Strong Motion Network [Data Set]. International Federation of Digital Seismograph Networks. 1981. Available online: https://www.fdsn.org/networks/detail/HI/ (accessed on 26 January 2022).
89. University of Athens. Hellenic Seismological Network, University of Athens, Seismological Laboratory [Data Set]. International Federation of Digital Seismograph Networks. 2008. Available online: https://www.fdsn.org/networks/detail/HA/ (accessed on 26 January 2022).
90. Evangelidis, C.P.; Triantafyllis, N.; Samios, M.; Boukouras, K.; Kontakos, K.; Ktenidou, O.J.; Fountoulakis, I.; Kalogeras, I.; Melis, N.S.; Galanis, O.; et al. Seismic Waveform Data from Greece and Cyprus: Integration, Archival, and Open Access. *Seismol. Res. Lett.* **2021**, *92*, 1672–1684. [CrossRef]

applied sciences

MDPI

Article

Inventory of Historical and Recent Earthquake-Triggered Landslides and Assessment of Related Susceptibility by GIS-Based Analytic Hierarchy Process: The Case of Cephalonia (Ionian Islands, Western Greece)

Spyridon Mavroulis [1,*], Michalis Diakakis [1], Haralambos Kranis [1], Emmanuel Vassilakis [2], Vasilis Kapetanidis [3], Ioannis Spingos [3], George Kaviris [3], Emmanuel Skourtsos [1], Nicholas Voulgaris [3] and Efthymis Lekkas [1]

1 Department of Dynamic Tectonic Applied Geology, Faculty of Geology and Geoenvironment, School of Sciences, National and Kapodistrian University of Athens, Panepistimiopolis Zografou, 15784 Athens, Greece; diakakism@geol.uoa.gr (M.D.); hkranis@geol.uoa.gr (H.K.); eskourt@geol.uoa.gr (E.S.); elekkas@geol.uoa.gr (E.L.)

2 Department of Geography and Climatology, Faculty of Geology and Geoenvironment, School of Sciences, National and Kapodistrian University of Athens, Panepistimiopolis Zografou, 15784 Athens, Greece; evasilak@geol.uoa.gr

3 Department of Geophysics and Geothermy, Faculty of Geology and Geoenvironment, School of Sciences, National and Kapodistrian University of Athens, Panepistimiopolis Zografou, 15784 Athens, Greece; vkapetan@geol.uoa.gr (V.K.); ispingos@geol.uoa.gr (I.S.); gkaviris@geol.uoa.gr (G.K.); voulgaris@geol.uoa.gr (N.V.)

* Correspondence: smavroulis@geol.uoa.gr

Citation: Mavroulis, S.; Diakakis, M.; Kranis, H.; Vassilakis, E.; Kapetanidis, V.; Spingos, I.; Kaviris, G.; Skourtsos, E.; Voulgaris, N.; Lekkas, E. Inventory of Historical and Recent Earthquake-Triggered Landslides and Assessment of Related Susceptibility by GIS-Based Analytic Hierarchy Process: The Case of Cephalonia (Ionian Islands, Western Greece). *Appl. Sci.* **2022**, *12*, 2895. https://doi.org/10.3390/app12062895

Academic Editor: Saro Lee

Received: 18 February 2022
Accepted: 9 March 2022
Published: 11 March 2022

Abstract: Cephalonia, located in the middle of the central Ionian Islands, has been affected by destructive earthquakes during both the instrumental and the historical period. Despite the fact that it is widely studied from several scientific viewpoints, limited research has been conducted so far regarding the earthquake-triggered landslides (ETL) and the related susceptibility. In the context of the present study, an inventory with 67 ETL from 11 earthquakes that occurred from 1636 to 2014 is presented. Given this record, the study further examines the ETL susceptibility exploiting 10 landslide causal factors in the frame of a GIS-based Analytic Hierarchy Process (AHP). Four factors (i.e., slope, PGA, tectonic structures and lithology) were associated in a higher degree to the locations where ETL occurred on the island. Based on the comparison of the ETL inventory and the landslide susceptibility index (LSI) map, the distribution of ETL in Cephalonia is not random, as their majority (82%) were generated within high to critically high susceptible zones. This fact, along with the AUC values of 80.3%, reveals a fair-to-good accuracy of the landslide susceptibility assessment and indicate that the contribution of the studied variables to the generation of ETL was effectively determined.

Keywords: earthquake-induced landslides; landslide inventory; event inventories; rockfalls; contemporary sources; landslide susceptibility; earthquake-induced landslide susceptibility; Analytic Hierarchy Process; Ionian Sea; Cephalonia

1. Introduction

Among the earthquake environmental effects (EEE) as defined by Michetti et al. [1], the earthquake-triggered landslides (ETL) prevail in a variety of environments from coastal to mountainous areas. They are characterized by high potential to cause not only economic but also human losses [2–5]. Casualties due to ETL are attributed to the collapse of buildings and other infrastructure due to slope failures and mobilization of unstable geological formations [6], as well as to the transport of material with high velocities over terrains with gentle slopes [7]. ETL are responsible for the 70% of all earthquake-induced human losses, which are not directly attributed to the earthquake ground motion [2], but to secondary phenomena. At least 90% of the ETL fatalities are attributed to rockfalls, rock avalanches and rapid soil flows, despite the fact that rock avalanches and rapid soil flows are neither common nor frequent events [7].

The consequences of ETL on society are not only direct, i.e., destruction of buildings and infrastructure, but also indirect, being associated with failures and damage to lifelines including road network, dams and utilities (e.g., [8–10]). This damage has high potential to cause disruption of various socio-economic activities including transportation as well as communication breakdown. This, in turn, can result in a delay in the emergency response and recovery actions and an increase in the number of casualties, due to delayed first aid and transport of the injured to health facilities.

ETL have been reported since 1789 BC in China [11]. The most used ETL classification has been provided by Keefer [7], who has analyzed data from a set of 40 historical earthquakes worldwide having a magnitude ranging from 5.2 to 9.5 and generating in a variety of geographic, geological and seismotectonic environments. This classification is based on the nature of the movement, the degree of the internal disruption of the mobilized material and the morphological, lithological and geological properties of the affected areas. Fourteen individual types of ETL are classified into three main categories: (i) disrupted slides and falls, (ii) coherent slides and (iii) lateral spreads and flows [7]. The first category comprises falls, slides and avalanches in rocks and soils, the second one slumps and block slides in rocks and soils as well as slow earth flows and the third one soil lateral spreads, rapid soil flows and subaqueous landslides [7]. The most common type of ETL comprises highly disrupted landslides, which includes falls, slides and avalanches involving either rock or soil, generated along steep source slopes and extending further on relatively gentle slopes [7,12].

The total number of ETL generally increases with earthquake magnitude [7] ranging from a few tens for the majority of $M < 5.5$ earthquakes to several thousand in $M > 8.0$ earthquakes.

Since the extensive research on ETL worldwide conducted by Keefer [7] and Rodriguez et al. [13], several ETL studies have been performed both at national and local level. Among others, Hancox et al. [14,15] and Rosser et al. [16] have published relevant research and databases for New Zealand, Papadopoulos and Plessa [17] for Greece, and Prestinizi and Romeo [18] and Martino et al. [19] for Italy.

Taking into account the existing literature on seismicity and EEE in the Mediterranean, we see that the Lefkada, Cephalonia, Ithaki and Zakynthos islands have been repeatedly hit by strong and destructive earthquakes with many triggered EEE (e.g., Rondoyanni et al. [20] for Lefkada Island and Mavroulis et al. [21] for Zakynthos Island). This strong seismicity is attributed to the proximity of the southern Ionian Islands to the: (i) Cephalonia Transform Fault Zone (CTFZ), which is a NE-SW striking and SE-dipping right-lateral strike-slip fault zone, the most seismic active structure in Greece and the Eastern Mediterranean and (ii) the northwesternmost part of the Hellenic Trench, along which the African plate is subducting beneath Eurasia (Figure 1).

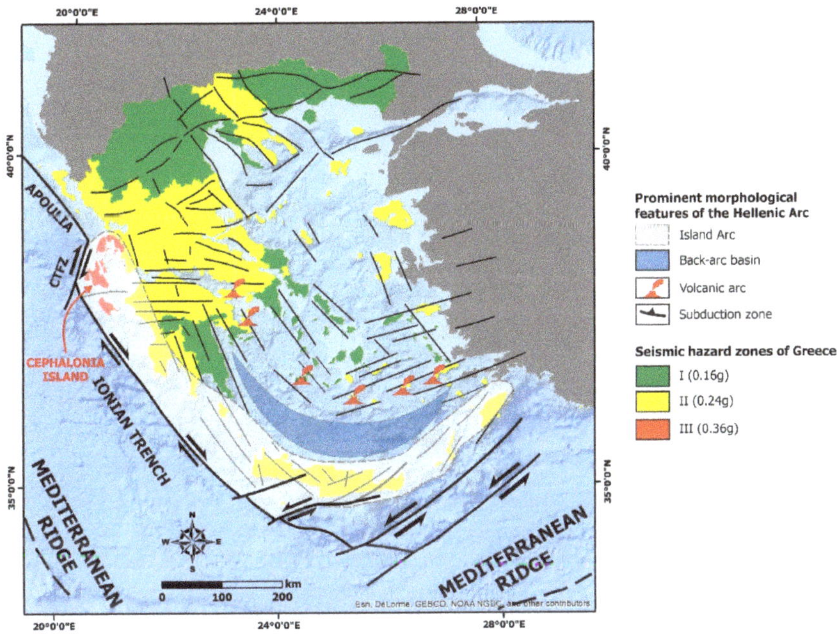

Figure 1. The Cephalonia Island is located at the northwestern part of the Hellenic Arc, east of the Cephalonia Transform Fault Zone (CTFZ). The island belongs to the zone III of the current Greek Building Code with a Peak Ground Acceleration (PGA) value of 0.36 g for a return period of 475 years [22].

However, there is a gap regarding the inventory of ETL in Cephalonia, especially the historical ones, which are poorly recorded and mapped. The filling of this gap is considered to be of high importance since this area has suffered from the strongest and most destructive earthquake in the Ionian Sea and from the most destructive earthquake sequence in recent history of Greece: the 1867 earthquake and the August 1953 seismic sequence, respectively, with adverse effects on the local population, the building stock and the natural environment [21,23]. Moreover, Cephalonia is a highly touristic island, with millions of visitors spending time in various landslide-prone locations, especially along the coast.

In this context, we used all available sources of information to create an inventory of landslides, which have been induced from destructive earthquakes generated not only during the instrumental period, but also during the historical pre-instrumental period in Cephalonia (Figure 2). The compilation of an ETL inventory constitutes a preliminary step toward the assessment of the landslide susceptibility in one of the most seismic active areas of the world, for which, however, no relevant information has been published so far.

Landslide susceptibility mapping is an important tool that contributes to the mitigation of the adverse effects of disasters induced by ETL not only in Cephalonia but also worldwide. In addition, it is another significant step towards more effective and responsible urban and land-use planning and related policies, procedures, guidelines [24] and improvements in seismic and landslide hazard and risk assessment and mitigation.

In this study, we assess the landslide susceptibility of the island in order to initially detect and then highlight the areas susceptible to ETL. This assessment is composed of several steps, including the development of landslide causal factors datasets, the implementation of the Analytic Hierarchy Process (AHP) adapted to the needs of the study, the calculation of the landslide susceptibility index (LSI) and finally the validation of the resulting LSI map (Figure 2).

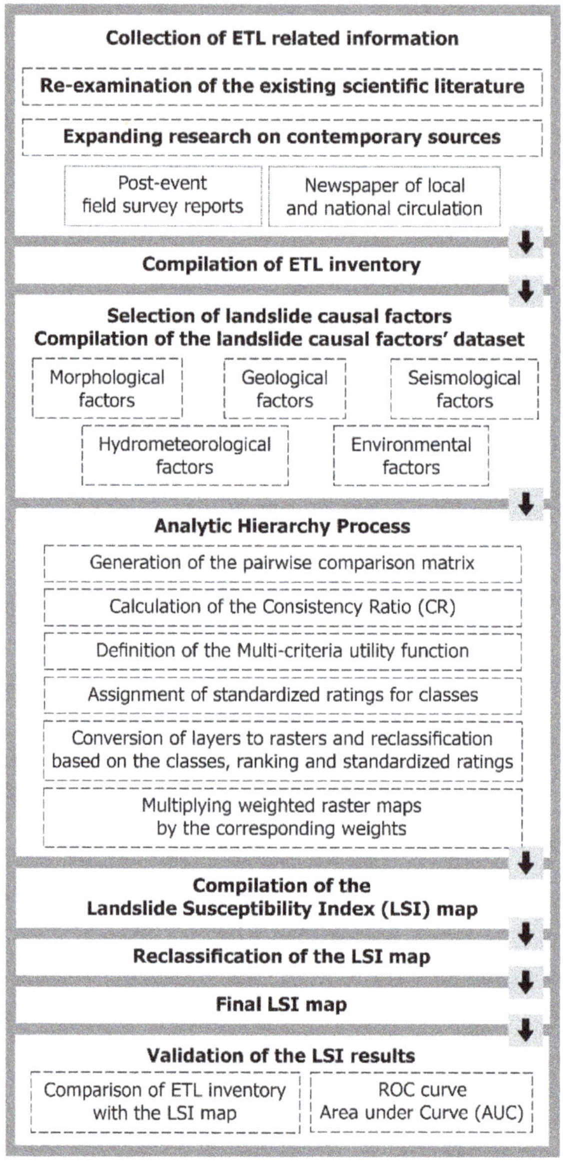

Figure 2. Flow chart showing the approach followed for the compilation of the ETL inventory and the respective landslide susceptibility assessment in Cephalonia.

2. Geological and Seismotectonic Setting

The geological structure of Cephalonia comprises alpine formations that belong to the Paxi and Ionian geotectonic units and post-alpine deposits of Pliocene to Quaternary lying uncomformably on the alpine basement [25–33] (Figure 3). The Paxi unit occurs in the largest part of Cephalonia (Figure 3) and includes mainly carbonate rocks from Triassic to Middle Miocene and a Middle Miocene–Early Pliocene clay-clastic sequence, which includes marls, clays and limestones [32,33] (Figure 3).

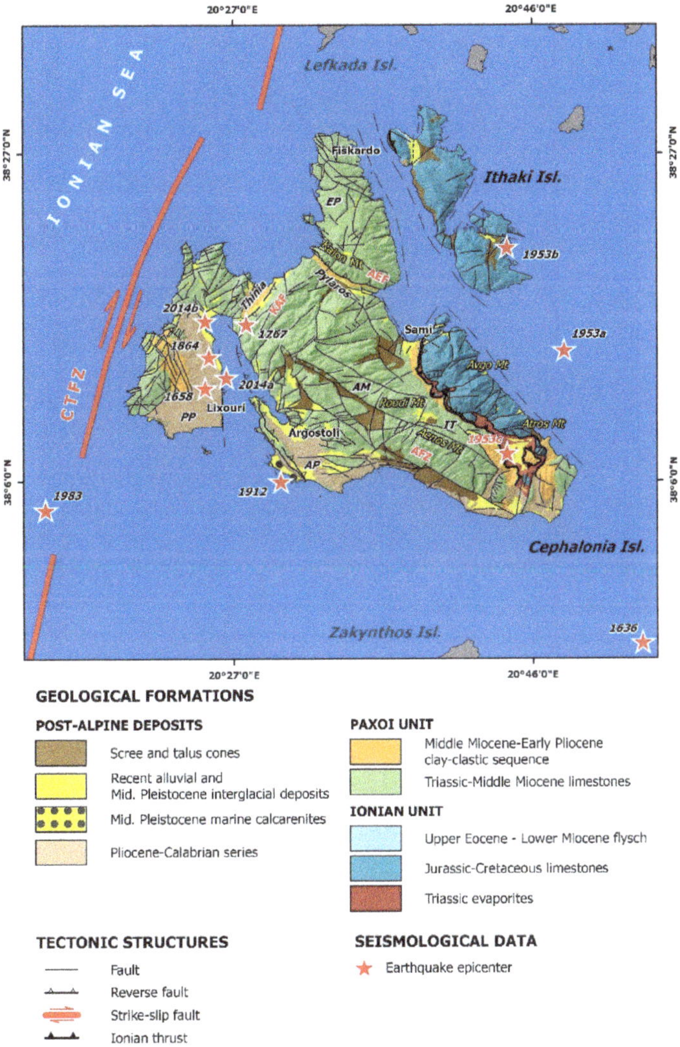

Figure 3. The neotectonic map of Cephalonia along with epicenters of historical and recent earthquakes that triggered landslides in the island. Fault blocks are also noted (EP: Erissos peninsula, AM: Aenos Mt fault block, AP: Argostoli peninsula, PP: Paliki Peninsula) along with their marginal faults (AEF: Ayia Efimia Fault, KAF: Kontogourata–Agonas Fault, CTFZ: Cephalonia Transform Fault Zone) and other major structures (IT: Ionian Thrust).

The Ionian unit is the allochthon tectonic nappe, which crops out in the eastern part of Cephalonia emplaced on top of the Paxi unit (Figure 3). It includes a sequence of Triassic evaporites and limestone breccia as well as Jurassic–Cretaceous fine-grained limestones, red nodular limestones, and shales with the Upper Eocene–Lower Miocene flysch at the top of the sequence [32,33]. The western boundary of the Ionian unit is defined by the Ionian thrust (IT, Figure 3), which is the most external tectonic structure of the Hellenides. The Paxi and Ionian geotectonic units are affected by major reverse faults [31] (Figure 3).

The Pliocene–Quaternary deposits include the Pliocene–Calabrian sequence and the Pleistocene–Holocene formations [32,33]. The Pliocene–Calabrian sequence is a marine sequence, observed in an elevated coastal area with width ranging from 2 to 10 km, especially in the largest part of the Argostoli peninsula (AP) and in the eastern and western part of the Paliki peninsula (PP) (Figure 3). It has developed over Paxoi formations, with thickness ranging from 200 to 500 m. The lower part of the sequence is composed of conglomerates, breccia, calcarenites and limestones of the Lower Pliocene. The intermediate part includes yellowish marls with siltstones, sandstone intercalations, sands and conglomerates of the Middle-Upper Pliocene. The upper part consists of marls with intercalations of sands and coarse-grained calcarenites [29,32,33]. The Middle-Upper Pleistocene–Holocene formations occur in several places, mainly in the Argostoli peninsula (Figure 3). These are continental deposits and are classified into scree, marine calcarenites of the Middle Pleistocene, Pleistocene interglacial deposits, Pleistocene scree and Holocene deposits [32,33]. Recent deposits also occur in many sites. Red terrestrial clay sands and coastal conglomerates as well as red calcareous crusts are found in the southwestern part of AP and in the area located WSW of Skala (Figure 3). Scree and alluvial fans cover the Pliocene formations along the foothills of Aenos Mt. These formations cover many limestone slopes. Loose fine-grained and coarse-grained alluvial deposits consist of sands, cobbles and clays on narrow beaches and wide valleys of Cephalonia.

Regarding its neotectonic structure, the island is composed of the following fault blocks with different kinematic evolution [33] (Figure 2).

The fault block of Aenos Mt (AM) is located in the central and eastern part of Cephalonia. It is bounded to the southwest by the eponymous fault (AFZ—Aenos fault in Figure 3), to the northwest by the Kontogourata-Agon fault (KAF—Kontogourata-Agon fault in Figure 3), to the northeast by the Ayia Efimia fault (AEF—Ayia Efimia Fault in Figure 3) and to the southeast by the Palaeokastro fault (PF—Palaeokastro fault in Figure 3). The Aenos fault block is characterized by intense uplift and intense incision since the Lower Pliocene [32,34]. It comprises an anticline with a NW–SE to N–S trending and westward plunging axis extending for a distance of about 20 km from Myrtos in the north to Kolaitis area in the southeast.

The fault block of Erissos peninsula (EP) is located in the northern part of Cephalonia. It is bounded to the south by the Ayia Efimia fault, and it is characterized by uplift and erosion. Since the Pleistocene, the Erissos peninsula presents the same evolution as the Aenos fault block.

The Paliki peninsula is located in the western part of Cephalonia (Figure 3). It is bounded to the west by the CTFZ and especially by its southern segment, located offshore western Cephalonia. The evolution of this fault block is affected by the CTFZ throughout Pliocene–Pleistocene, an impact which is expressed through intense uplift [34]. Complex movements attributed to evaporites of the Ionian unit control local vertical movements [35,36].

The Argostoli peninsula is extended southwest and south of Aenos Mt (Figure 3). It is bounded to the east and south by the Aenos fault (Figure 3). A possible increase in compression on the Argostoli peninsula during the Pleistocene resulted in the development of back-thrust faults, exclusively in this fault block.

The main seismogenic structure in the Cephalonia area is the Cephalonia Transform Fault Zone [37–39] (Figure 3). It is the most seismically active structure in the Ionian Sea and one of the most active in the Eastern Mediterranean region. A large proportion of the high seismicity in the studied part of the Ionian Sea has been attributed to this structure. Based on the historical and recent seismicity, Cephalonia has been frequently stricken by large and destructive earthquakes [40] with significant impact on the local population, buildings and infrastructures (e.g., [41,42]). A special characteristic of the large earthquakes in the southern Ionian Islands is that they are usually generated as twin or cluster events with occurrence period ranging from a few days to 5 years [43]. Typical examples from Cephalonia are the 1953 earthquakes on 9–12 August and the 2014 earthquakes on 26 January and 3 February (e.g., [44]).

3. ETL Inventory in Cephalonia Island

3.1. Recent Studies on Landslide Inventory in Greece

The research for the systematic inventory of landslides in Greece is not so widespread. It mainly includes studies presenting maps with landslides throughout Greece [45–49] and maps of landslide-affected settlements [50] as well as several studies with results from field reconnaissance surveys in several earthquake- and landslide-affected areas, where ETL are reported and recorded among other effects.

The first attempt for the development of a complete recording system of the landslides in Greece was made by Koukis and Ziourkas [45]. They conducted a statistical analysis on 800 case histories of landslides in Greece from 1949 to 1986. The related information was retrieved from 1500 coded engineering reports and studies conducted by different public corporations including the Hellenic Authority of Geological and Mining Research (former Institute of Geology and Mineral Exploration of Greece) and the Central Union of Greek Municipalities. Their study includes cases in villages and their surroundings as well as along the road network and refers to the landslide frequency distribution and their consequences on different geological formations, altitudes, rainfall, slope angle, etc., while certain interrelations were studied and an engineering geological map as well as landslide distribution and frequency zone maps were compiled.

A similar approach was followed by Koukis et al. [46], who further evaluated data from 1116 case histories of landslides from 1949 to 1991, expanding the dataset of Koukis and Ziourkas [45]. They investigated the frequency distribution of landslide control parameters, and their results were presented in tables and graphs. Analogous extensions of the number of landslides and the time period and update of the results of the statistical analysis were conducted by Koukis et al. [47] with 1200 landslides from 1949 to 1995, by Koukis et al. [48] with 1300 landslides from 1950 to 2004 and by Sabatakakis et al. [49] with 1635 landslides up to 2010.

The landslide inventory map of Greece [50] and the above works present the landslide locations, but no additional qualitative or quantitative characteristics of the landslides, or information on their causes and time of occurrence. As a result of this approach, the marked landslides could not be associated with specific triggering events, e.g., earthquakes, rainfall or even human activity. For the ETL in Greece in general and for the Cephalonia in particular, the maps of the above studies are not significantly useful as it is not possible to determine whether the eight recorded landslides on the island were caused by earthquakes or rainfall or another triggering factor.

Furthermore, Papadopoulos and Plessa [17] collected data on 47 landslides induced by earthquakes from 1650 to 1995 with magnitude ranging from 5.3 to 7.9. Based on the ETL data, they presented magnitude-distance relations for ETL in Greece, which are compatible with curves proposed worldwide. They also examined the spatial distribution of ETL, the size distribution, and the maximum landslide distance from the epicenter. However, the collected data was not sufficient or accurate to be used for studying other parameters, such as the area affected by landslides and correlation with earthquake magnitude, as well as the minimum intensity for generating ETL. According to Papadopoulos and Plessa [17], the historical descriptions of ETL in Greece are usually not precise enough to provide a reliable basis for a modern scientific analysis. Furthermore, after examining hundreds of documents, they supported that a database with data from the 16th century onwards can be exploited.

From these 47 landslides presented by Papadopoulos and Plessa [17], only 4 were generated in the seismically active area of the Ionian Islands: a rockslide in Zakynthos from the October 30, 1840, Ms = 6.7 earthquake; a rockfall in Keri (southern Zakynthos) from the April 17, 1893, Ms = 6.4 earthquake; a rockfall in Lefkada from the 27 November 1914, Ms = 6.3 earthquake; a rockfall in the settlement of Vassiliki in southern Lefkada from the April 22, 1948, Ms = 6.5 earthquake.

In addition to the above studies, the Emergency Events Database (EM-DAT) of the Center of Research on the Epidemiology of Disasters (CRED; [51]) has also been taken into account. Based on a search for disasters induced by geophysical hazards in Europe from 1920 to present, it is concluded that the EM-DAT comprises records of ground movements caused by the 12 August 1953 and the early 2014 Cephalonia earthquakes and the 14 August 2003 Lefkada earthquake. However, there is no further information on the sites of these landslides and their qualitative and quantitative properties.

From the aforementioned, the importance of the present study is ascertained by highlighting the historical and recent ETL in Cephalonia and by presenting all available qualitative and quantitative characteristics of these phenomena on the island and the results of their analysis.

3.2. Methodology

Landslide inventories are essential for predicting future landslides generation on the basis of past conditions [52]. In order to compile a landslide inventory, various mapping approaches can be applied comprising visual interpretation of aerial and satellite imagery and field surveys and mapping, supported by investigation of contemporary sources and post-event field survey reports [52,53]. The resulting landslide inventories are classified into (i) geomorphological inventories, resulted from respective features of the study area, (ii) event inventories, associated with a specific causative event including mainly earthquakes and rainfalls, (iii) multi-temporal inventories of an area for different time periods and (iv) historical inventories for respective time periods and events [52,53].

We conducted a re-examination of the available scientific literature for historical earthquakes of Cephalonia Island with emphasis on their environmental effects and especially on the ETL. Due to the limited related information in the existing scientific literature (e.g., [41,42,54]), we expanded our search to contemporary sources, including not only post-event field survey reports (e.g., the report of the Institute of Geology and Underground Research [55]), but also daily press databases comprising newspapers of local and national circulation.

The newspaper databases comprised (Table 1):

- The Press Museum of the Peloponnese—Epirus—Ionian Islands Daily Newspaper Editors Association [56];
- The Digital Historical Archive of the Lambrakis Press Group [57];
- The Digital Library of Newspapers and Magazines of the National Library of Greece [58];
- The Digital Library of the Greek Parliament [59].

The Press Museum was founded by the Association of Editors of Daily Newspapers of Peloponnese-Epirus-Islands, was based in Patras (Northwestern Peloponnese) in 1956 and has been operating since 28 May 1957. Its purpose is the collection and preservation of newspapers, magazines, manuscripts and other publications from previous centuries to present, which constitute knowledge and information sources about the history of the area. Through this website, we were given the opportunity to search for newspaper clippings about the August 1953 earthquakes. In particular, four newspapers were found (Table 1), which provided important information about the environmental effects triggered by the August 1953 earthquakes:

- ETHNIKOS KIRIX (in Greek: Εθνικός Κήρυξ; lit. National Herald, abbreviated by EK)
- I IMERA (in Greek: H Ημέρα; lit. The Day, abbreviated by IM)
- NEOLOGOS PATRON (in Greek: Νεολόγος Πατρών; lit. Patras Neologist, abbreviated by NP)
- PELOPONNISOS (in Greek: Πελοπόννησος; lit. Peloponnese, abbreviated by PL)

The Digitized Historical Archive of the Lambrakis Press Group comprises black and white as well as colored newspapers and magazines, volumes and microfilms in various dimensions (from A2 to A4) from 1922 to present. From this archive, we managed to retrieve related information from two newspapers of national circulation (Table 1), which provided information about the August 1953 earthquake sequence:

- TO VIMA (Greek: TO BHMA, lit: The Tribune, abbreviated by TV)

- TA NEA (Greek: TA NEA, lit: The News, abbreviated by TN)

Table 1. Contemporary sources comprising newspapers and magazines of local and national circulation used in this study for revisiting landslides induced by historical and recent earthquakes generated in Cephalonia.

Archives	Newspapers	Studied Period
The Press Museum of the Peloponnese—Epirus—Ionian Islands Daily Newspaper Editors Association [56]	ETHNIKOS KIRIX	10–24 August 1953
	I IMERA	11–29 August 1953
	NEOLOGOS PATRON	11–29 August 1953
	PELOPONNISOS	11–29 August 1953
The Digital Historical Archive of the Lambrakis Press Group [57]	TO VIMA	10–25 August 1953
	TA NEA	10–25 August 1953
The Digital Library of Newspapers and Magazines of the National Library of Greece [58]	ANAMORFOSIS	4 February 1867
	AVGI	4 February 1867
	MERIMNA	4 February 1867
	ELEFTHERIA	11–28 August 1953
	EMPROS	11–23 August 1953
The Digital Library of the Greek Parliament [59]	ZIZANION	24 January 1912
	TAXYDROMOS	12–30 August 1953

The Digital Library of Newspapers and Magazines of the National Library of Greece includes digitized archives of newspaper sheets. Through the website of the digital library, we searched information for the post-earthquake period from two newspapers (Table 1):

- ANAMORFOSIS (Greek: Αναμόρφωσις, lit: Reform, abbreviated by AN)
- AVGI (Greek: Αυγή, lit: Dawn, abbreviated by AV)
- MERIMNA (Greek: Μέριμνα, lit: Concern, abbreviated by ME)
- ELEFTHERIA (Greek: Ελευθερία, lit: Freedom, abbreviated by EL)
- EMPROS (Greek: Εμπρός, lit: Ahead, abbreviated by EM)

The information retrieved from the AN, AV and ME newspapers refers to the 1867 earthquake and from the EL and EM to the seismic sequence of 1953.

The Greek Parliament Library houses one of the richest collections of Greek and foreign newspapers and magazines from the 18th century to present. Due to the sensitivity of the material and the large number of visitors, the Library has been utilizing digitization and microfilming since the 1980s. Today, about 25,000 microfilms have been digitized to make their content accessible to all. We managed to retrieve related information from three newspapers of national circulation (Table 1):

- ELPIS (Greek: Ελπίς, lit: Hope, abbreviated by EL).
- ZIZANIO (Greek: Ζιζάνιο, lit: Pest, abbreviated by ZN).
- Taxydromos (Greek: Ταχυδρόμος, lit: The Postman, abbreviated by TD).

The information retrieved from the EL and ZN refers to the 1912 earthquake and from TD to the 1953 seismic sequence.

Furthermore, we also retrieved information included in several papers of a local journal titled "I Kefalonitiki Proodos" (English: "The Cephalonian Progress", abbreviated by KP), which referred to the 1867 earthquake and the 1953 sequence [60,61].

The related information for the recent earthquakes has been retrieved from already published scientific papers depicting the impact of the earthquakes on the natural environment [10,62,63] and from official reports of post-event field surveys [64]. Mavroulis and Lekkas [10] revised the August 1953 earthquake sequence, which comprises the most destructive events in the recent history of Greece. In particular, the latter study reconstructed a complete picture of the primary and secondary effects on the environment of the Cephalonia, Ithaki and Zakynthos Islands induced by the mainshock of 12 August, and its large foreshocks that occurred on 9 and 11 August. Eleftheriou and Mouyiaris [64] reported detailed macroseismic observations of the earthquake in Argostoli, Sami, Lixouri and Fiskardo and their surroundings and presented environmental effects triggered by the earthquake comprising rockfalls and ground failures related to liquefaction phenomena. Lekkas and Mavroulis [62,63] presented the environmental effects triggered by the early 2014 Cephalonia earthquakes (Mw = 6.1 on January 26 and Mw = 5.9 on February 3) along with the ESI-07 intensities based on the qualitative and quantitative information of the generated coseismic and secondary phenomena.

The majority of the extracted information was also verified during field surveys and campaigns held in Cephalonia after their extraction from the aforementioned contemporary sources.

3.3. Historical and Recent ETL in Cephalonia Island

Landslides in Cephalonia have been triggered by earthquakes generated in 1636, 1658, 1767, 1867, 1912, 1953, 1983 and 2014 (Figure 4). The focal parameters of these earthquakes are presented in the following Table 2.

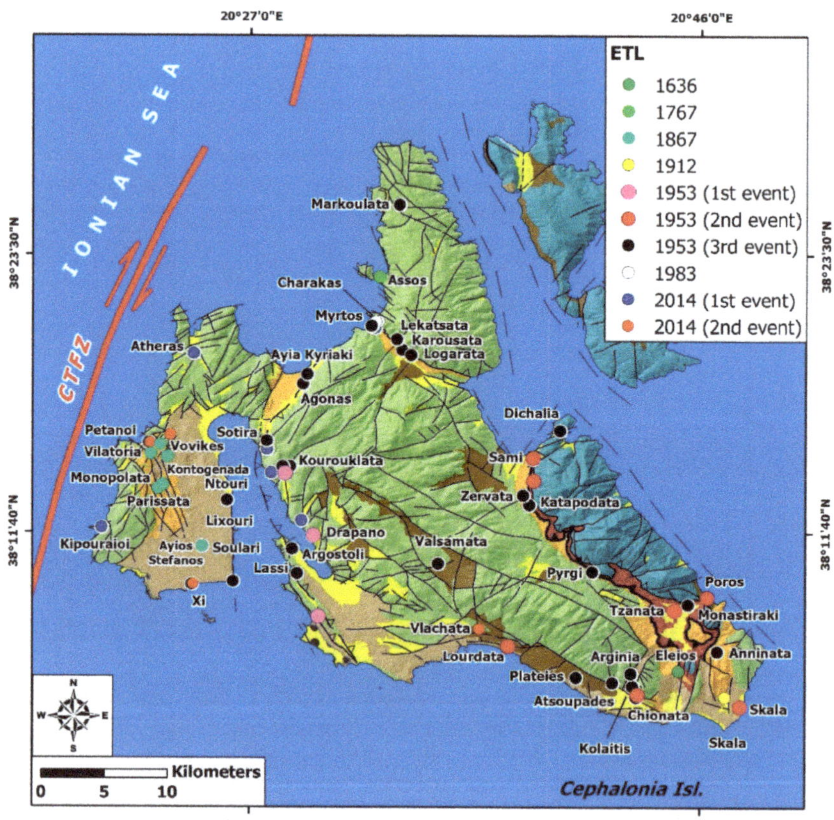

Figure 4. The neotectonic map of Cephalonia along with the historical and recent ETL in the island. Legend of the geological formations in Figure 3.

Table 2. Focal parameters and impact on the local population of the studied earthquakes that triggered landslides in Cephalonia Island.

Lat	Lon	Year	Month	Day	Mw	Io	Fatalities	Injured	Sources [1]
37.928	20.885	1636	9	30	6.62	IX	520	1500	[41,54]
38.2	20.42	1658	8	24	6.7	IX	20	-	[41,54]
38.268	20.464	1767	7	22	6.65	X	330	-	[41,54]
38.233	20.424	1867	2	4	7.15	X	224	-	[41,54]
38.10	20.50	1912	1	24	6.1	X	8	-	[40,54]
38.24	20.80	1953	8	9	5.9	-			[40,54]
38.35	20.74	1953	8	11	6.6	-	455	2412	[40,54]
38.13	20.74	1953	8	12	7.0	X+			[40,54]
38.07	20.25	1983	1	17	6.7	VI	-	-	[40,54]
38.2113	20.4425	2014	1	26	6.1	VII	-	-	[65,66]
38.2716	20.4198	2014	2	3	5.9	VIII	-	-	[65,66]

[1] Sources: Refs. [40,54,65] for epicenter coordinates, occurrence date and magnitude, Ref. [41] for intensities, fatalities and injured people, and [66] for intensities.

3.3.1. The 30 September 1636 Earthquake

The 30 September 1636 earthquake caused severe structural damage to all buildings of Cephalonia, resulting in 520 fatalities and about 1500 injured people [41,42,67]. Heavier damage was reported in the southern part of Cephalonia, where many villages were completely destroyed. Regarding the environmental effects, the earthquake affected the southeastern part of the island. Failures, comprising slides and rockfalls, were triggered along the south facing southern slopes of Aenos Mt. Unstable limestone boulders were detached from slopes in Eleios area (Figure 4) and rolled down resulting in effects on vegetation, buildings and livestock. In particular, a part of the forest on the slopes was destroyed [42], buildings collapsed after crushing and animals were buried under debris.

3.3.2. The 24 August 1658 Earthquake

This earthquake affected the western part of Cephalonia [41,42]. It caused the destruction of 500 houses, resulting in 20 fatalities [67–69]. As regards the induced earthquake environmental effects, Tsitselis [69] referred to the triggering of landslides in the coastal eastern part of Paliki peninsula (Lixouri) resulting in submergence of two coastal sites. Furthermore, ETL were induced, resulting in the collapse of a church founded on the top of a hill. Unfortunately, there is no more information available for determining the exact location of the induced phenomena.

3.3.3. The 22 July 1767 Earthquake

Cephalonia was also affected by the 22 July 1767 earthquake [41,42,67,70,71]. Lixouri, in the eastern part of Paliki peninsula, was completely destroyed, Argostoli, in the northern part of the eponymous peninsula, was less affected, while Sami, in the eastern part of the island, and Fiskardo, in the northern part of Erissos peninsula, were extensively damaged as all churches and monasteries were razed to the ground [72–74]. As a result, 303 fatalities were reported, with 50 of them in Paliki peninsula [69]. Concerning the ETL, rockfalls were generated in Assos area, located in the central-western part of Erissos peninsula [75,76] (Figure 4).

3.3.4. The 4 February 1867 Earthquake

This earthquake is the largest that ever hit the Ionian Sea. The greatest damage was observed in the western part of Cephalonia, especially in Paliki peninsula, where Lixouri and the surrounding villages as well as residential areas in Thinia valley were completely destroyed ([67]; AN, 04.02.1867; AV, 06.02.1867; ME, 07.02.1867). As a result, 3200 houses were destroyed and 2600 were damaged, resulting in 224 fatalities [42].

Concerning ETL, rockfalls and slides were reported in Paliki peninsula. Large volume of clay material failed resulting in a large ground crack with depth of 8 m and width of 3 m in Ayios Stefanos area in southern Paliki [76]. In the same area, and particularly in Soulari village, landslides were also induced during the 2014 earthquake [62]. In addition, debris slides and rockfalls were triggered by the 1867 earthquake in Parissata, Monopolata, Vilatoria, Vovikes and Kontogenada villages, all in the central western part of Paliki peninsula, without causing structural damage [60,68] (Figure 4).

3.3.5. The 24 January 1912 Earthquake

The 1912 earthquake destroyed what had remained intact during the great 1867 earthquake. The human losses were minimal compared to the 1867 disaster, comprising seven fatalities in Asprogerakas village, located between Eleios and Anninata villages in southeastern Cephalonia, attributed to house collapse and one in Skala village from falling debris. In Skala area, rockfalls were reported and the mobilized material ended up in the adjacent roads and fields (ZN, 28.01.1912). At a distance of 400 m from Skala village, large limestone boulders were detached from the slopes, rolled down and ended up to adjacent road resulting in traffic disruption (ZN, 28.01.1912) (Figure 4).

3.3.6. The August 1953 Seismic Sequence

The August 1953 seismic sequence comprised the mainshock on 12 August and its large foreshocks on 9 and 11 August. All earthquakes triggered landslides in Cephalonia with considerable impact on the local population and on the natural and the built environment [10]. All affected sites described below are presented in Figure 4.

- The 9 August 1953 earthquake

This earthquake induced rockfalls in the coastal quarries of Argostoli town and along the western slopes of Ayia Dynati Mt (EK, 10.08.1953; NP, 11.08.1953; PL, 11.08.1953; EM, 11.08.1953).

- The 11 August 1953 earthquake

This seismic event generated rockfalls and slides (TV, 12.08.1953; IM, 12.08.1953; NP, 12.08.1953; PL, 12.08.1953). Rockfalls were generated in Sami area (NP, 12.08.1953), in Tzanata village (PL, 12.08.1953) in Lourdata village (TV, 12.08.1953; IM, 12.08.1953) and along road slopes leading from Sami to Argostoli (TV, 12.08.1953), from Argostoli to Eleios area (IM, 12.08.1953) and along the road network leading to Chionata, Skala and Poros villages (TV, 12.08.1953).

The unstable material was accumulated on adjacent roads, resulting in traffic and communication disruption. In addition, the detached limestone boulders crushed buildings in residential areas, resulting in heavy structural building damage ranging from partial to total collapse and subsequently increased related fatalities and injuries. A characteristic example of such impact is Tzanata village, which suffered rockfall-induced complete building destruction, uprooting of trees and 14 fatalities (PL, 12.08.1953).

- The 12 August 1953 earthquake

The mainshock of the 1953 earthquake sequence induced landslides in several sites of Cephalonia. The generated landslides are classified into detachment of boulders from abrupt slopes and consequent rockfalls and slides. The majority of them were rockfalls, which affected the adjacent infrastructures including the road network and buildings in villages and towns in the earthquake-affected area. The generated landslides are presented below along with the impact on the local population and on the built environment.

Starting from northern Cephalonia, rockfalls were generated along a hill adjacent to Markoulata village, located in the northern part of Erissos peninsula [55]. In the southern part of the same fault block, rockfalls occurred along a gorge between Karousata and Lekatsata villages, causing destruction of olive trees, as well as in Logarata and Karousata villages founded along the southern slopes of Kalon Mt, resulting in damage to buildings.

In the northwestern part of the transition zone between the Erissos peninsula in the north and the fault block of Aenos Mt in the south, Myrtos beach, one of the most impressive sites in Cephalonia, was also affected by rockfalls. A limestone boulder was detached from the slopes and ended up in the adjacent bay, while a church founded on the slope was completely destroyed by the generated rockfalls (EM, 14.08.1953).

Rockfalls were also triggered along the eastern margin of Thinia valley, which has been formed along the transition between the northwestern Ayia Dynati Mt located east and the northeastern Paliki peninsula located west. The reported landslides affected the upper part of Agonas village [55], the coastal Ayia Kyriaki area, where debris reached the sea, and along the slopes over Sotira beach (TX, 16.08.1953).

Rockfalls were also triggered along the abrupt slopes surrounding Kourouklata village [55] located south of the aforementioned sites. It was the worst affected village as it was completely destroyed by fallen boulders (PL, 16.08.1953). Similar phenomena were reported from north of Drapano village. The unstable blocks ended up in Argostoli Gulf without impact on residential areas [61].

Several areas were also affected by ETL in the eastern part of the fault block of Aenos Mt. Rockfalls were generated in Sami, Zervata and Katapodata villages along the western front of Avgo Mt (TN, 12.08.1953; IM, 13.08.1953; TX, 14.08.1953, 19.08.1953; [55]), in Dichalia village in the northern slopes of Avgo Mt [55], in Pyrgi village in the eastern part of Roudi Mt (TX, 19.08.1953), in Valsamata village west of Roudi Mt (TX, 14.08.1953), in Tzanata and Monastiraki villages in the southern part of the Atros Mt (TX, 14.08.1953; IM, 14.08.1953; [55]), and in Atsoupades, Kolaitis, Arginia, Chionata and Plateies villages along the southern front of Aenos Mt (IM, 13.08.53; IGUR, 1954). Similar phenomena were also induced in the southeastern end of Cephalonia, in particular in Skala and Anninata villages (TX, 14.08.1953; NP, 14.08.1953; TN, 19.08.1953).

The Argostoli and Paliki peninsula were also affected by rockfalls. In Argostoli area, rockfalls were generated along the slopes of kilns close to the town (IM, 14.08.1953, 16.08.1953) and in Faraon hill, located in Lassi area (NP, 14.08.1953, 15.08.1953; PL, 14.08.1953; IM, 15.08.1953). In Paliki, rockfalls were generated in Ntouri (Paliokastro) hill, located north of Lixouri, as well as at a cape located at the southeastern end of the peninsula. Moreover, landslides were observed along slopes in the road network leading from Sami to Argostoli and from Argostoli to Livadi, located north of Ntouri site.

The ETL had considerable impact on the local population and the natural and built environments. They affected vegetation, caused damage to infrastructures including the road network and to buildings in residential areas resulting in increasing fatalities and injuries.

Rockfalls caused damage to olive groves along the abrupt slopes of the gorge in Karousata and Lekatsata [55], in Agonas area [55] and in Ntouri area of Paliki as well as to olive groves and vineyards in Zervata area (TX, 14.08.1953). The vegetation acted as protection barrier for the residential areas, for example, in the case of Agonas, where olive trees protected its lower part from rolling boulders and subsequent destruction [55].

As regards the building damage, the ETL caused heavy structural damage in several villages and led to more fatalities and injured people. Characteristic examples of such impact are the villages of Tzanata (PL, 12.08.1953), Chionata (IM, 13.08.53), Anninata (TX, 14.08.1953) and Plateies [55], where rockfalls claimed the life of residents and increased the number of injuries, Kourouklata (PL, 16.08.1953; [55]), Valsamata (TX, 14.08.1953) and Kolaitis [55], which were totally destroyed by rockfalls, and Zervata (TX, 14.08.1953), Katapodata [55] and Atsoupades [55], where buildings suffered extensive damage by rockfalls.

Rockfalls affected several parts of the road network in Cephalonia. The main effect was the accumulation of the mobilized material on the adjacent roads resulting in temporary traffic disruption and communication breakdown between the affected residential areas. In particular, traffic disruption due to proximity to high steep slopes and to rockfalls triggered by the 12 August mainshock and the foreshocks on 9 and 11 August was reported for the road leading from Argostoli to northern Cephalonia (NP, 11.08.1953; PL, 11.08.1953; EM,

11.08.1953), from Argostoli to Sami (TV, 12.08.1953), from Argostoli to Eleios-Pronnoi area (IM, 12.08.1953) and from Sami to the surrounding villages (NP, 12.08.1953). In addition, roads connecting villages in several parts of the island were also affected. Characteristic examples are the roads in Pyrgi village (TX, 19.08.1953), in Sami and Zervata village (TN, 12.08.1953; IM, 13.08.1953; TX, 14.08.1953, 19.08.1953), in Atsoupades village (IM, 13.08.53) and in Skala area (TX, 14.08.1953) among others.

3.3.7. The 17 January 1983 Earthquake

The 1983 Ms = 7.0 earthquake [37] caused slight damage to Cephalonia, due to the fact that its epicenter was located offshore between Cephalonia and Zakynthos Islands. It mainly generated non-structural damage to buildings with reinforced concrete frame and infill walls in Lixouri, Argostoli, Sami and Fiskardo areas [64]. The largest aftershock of M = 6.2 occurred on 23 March [37], but the M = 6.0 aftershock that was generated the same day as the mainshock resulted in larger damage than the mainshock [64].

Along the road to Fiskardo, at Charakas site, rockfalls were triggered along the sub-vertical artificial road cut [64] (Figure 4). Limestone boulders and debris from the slope were concentrated on the adjacent road and resulted in traffic disruption. Downstream of Charakas road, on the steep coastal slope, rockfalls were also generated, resulting in sea water turbidity [64] (Figure 4).

3.3.8. The Early 2014 Cephalonia Sequence

The early 2014 Cephalonia sequence comprised two earthquakes generated on 26 January (Mw = 6.1) and 3 February (Mw = 5.9), with considerable impact on western Cephalonia [62,63]. Among other effects, the 2014 earthquakes triggered landslides mainly in Paliki peninsula (Atheras, Petanoi, Kipouraeoi, Soullari and Xi areas), along the margins of Thinia valley (Agonas and Ayia Kyriaki areas), in Myrtos coastal area, along slopes in the western part of Ayia Dynati Mt and in the southern part of Argostoli peninsula (Vlachata) [62,63]. It is significant to note that such failures were not generated in Erissos peninsula. All affected sites described below are presented in Figure 4, while representative views of the triggered effects and the affected sites are presented in Figure 5.

In the Myrtos area, two limestone boulders were detached from the eastern steep slope, they rolled down and stopped in the beach at about 100 m from the shoreline (Figure 5a–c). During rolling down the boulders resulted in destruction of the road leading from the top of the slope to its base. Furthermore, extensive rockfalls and rockslides were also triggered along the coastal slopes close to the beach, particularly in Charakas (Figure 5a), a site which was also affected by the 1953 and 1983 Cephalonia earthquake [55,64].

Rockfalls and slides were generated along the margins of Thinia valley, mainly in the steep slopes over Agonas village and in Ayia Kyriaki coastal area in its eastern margin and secondarily along slopes adjacent to the road leading to Zola village in its western margin (Figure 5d).

Rockfalls also affected several parts of the road network leading from Argostoli town to Livadi area located northwards. Unstable material was mobilized by the earthquake along steep road slopes and accumulated on the road asphalt surface resulting in some cases to temporary traffic disruption.

Extensive landslides were triggered on slopes of Paliki peninsula by the first earthquake on 26 January. Rockfalls were generated in its northern part, where a building in Atheras village was crushed by a rolling boulder and partially collapsed (Figure 5e,f), fortunately without casualties and injuries. During the second earthquake on 3 February, boulders in the same site were again detached from the slopes, but they did not reach the village.

Figure 5. *Cont.*

Figure 5. The 2014 ELT in Cephalonia: (**a**) The landslide-prone Myrtos coastal area. (**b**) Water turbidity in Myrtos Bay after the 26 January 2014 earthquake attributed to rockfalls along the adjacent coastal slopes. (**c**) Rockfalls by the early 2014 Cephalonia earthquakes over the Myrtos beach and at Charakas site. (**d**) The landslide-prone eastern margin of Thinia valley, where the Kontogourata–Agonas fault occurs. Rockfalls and slides triggered in Cephalonia by the early 2014 earthquakes: (**e**,**f**) Unstable limestone boulders in Atheras village in north Paliki crushed trees and a building. (**g**,**h**) Slides in Petanoi coastal area resulting in traffic disruption. (**i**–**k**) Rockfalls close to the Kipouraioi monastery affecting its yard.

Landslides were also triggered along the western coastal part of Paliki, which is characterized by high coastal fault scarps and steep slopes. Characteristic examples were recorded in Petanoi and Kipouraioi coastal areas.

In Petanoi area, the second earthquake triggered a large slide (Figure 5g,h), which reached the beach and partially covered it. In addition, the slide material blocked the road leading to the beach, resulting in traffic disruption for 3 days.

In the Kipouraeoi area, landslides were generated by both 2014 earthquakes along the impressive steep coastal fault scarps (Figure 5i–k). In the area of Kipouraeoi monastery, rockfalls and rockslides occurred along an abrupt slope, at the top of which the monastery is constructed. The crown of the slide was very close to the monastery and affected its yard (Figure 5j,k). Additionally, slides also occurred along road slopes in the area surrounding the monastery.

In the Xi coastal area, in the southern part of Paliki peninsula, small rockfalls were generated by both 2014 earthquakes. In Soulari area, an impressive slide was triggered by the 26 January earthquake on the top of a hill. A large fragment composed of silt and clay moved downhill and formed a large main scarp, while the main body of the slide was split into smaller sections. With the contribution of heavy rainfalls that followed the first earthquake, the main scarp of the slide further collapsed during the second earthquake and the pre-existing cracks were enlarged, resulting in larger gaps in the area.

The list of landslides induced by the 2014 earthquakes also includes the rockfalls triggered in the limestone quarry situated west of Vlachata village. These rockfalls temporarily blocked the road leading from Argostoli to Poros port.

3.4. Maximum Epicentral Distances as a Function of Earthquake Magnitudes

The epicentral distances of the landslides, which were induced by the aforementioned historical and recent earthquakes in Cephalonia, were calculated, taking into account the locations of the landslides, as obtained from the ETL inventory of the island, and the earthquake magnitudes, as obtained: (a) for the 1636, 1767 and 1867 earthquakes from the SHARE European Earthquake Catalogue (SHEEC) 1000–1899 [54], (b) for the 1912, 1953 and 1983 earthquakes from the updated and extended earthquake catalogue for Greece and adjacent areas since 1900 compiled by Makropoulos et al. [40] and (c) for the early 2014 earthquakes from the relocated catalogue for the 2014 Cephalonia aftershock sequence compiled by Papadimitriou et al. [77], Kapetanidis [78] and Sakkas et al. [65]. The maximum epicentral distance of landslides for each earthquake was then determined. The pairs of maximum epicentral distance of landslides and the magnitude of the causative earthquakes were plotted on a graph (Figure 6). The distribution of these points was also compared with the upper bound limit/curve for the maximum observed epicentral distances of disrupted slides and falls proposed by Keefer [7], which was based on a dataset of 40 worldwide earthquakes. This curve represents the maximum distances at which ETL can be expected to be generated for specific Mw values. The graph in Figure 6a shows how the distribution of Cephalonia ETL data generally fit well the Keefer's relationships. In the graph in Figure 6b, together with the points for Cephalonia, the pairs of maximum epicentral distances for landslides and magnitudes for earthquakes in Greece, as presented by Papadopoulos and Plessa [17], are also shown. From these graphs, it is also concluded that outlier events attributed to various triggering factors and local seismic amplification were not recorded in Cephalonia.

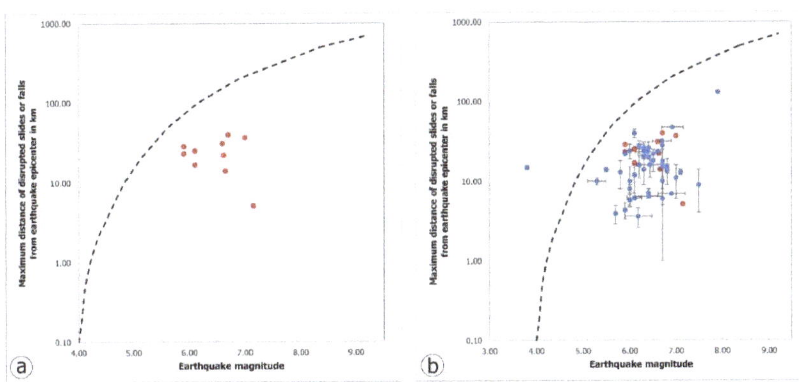

Figure 6. (**a**) Maximum distance from earthquake epicenter of disrupted slides and falls induced by historical and recent earthquakes of different magnitudes generated onshore and offshore Cephalonia versus earthquake magnitude. The dashed line is the upper bound for disrupted slides or falls by Keefer [7]. (**b**) Maximum epicentral distances of landslides triggered by historical and recent earthquakes of different magnitudes versus earthquake magnitude. The red subset comprises data from Cephalonia and the blue subset from around Greece published by Papadopoulos and Plessa [17]. The dashed line is the upper bound for disrupted slides or falls by Keefer [7].

4. ETL Susceptibility Assessment in Cephalonia Based on GIS-Based Analytic Hierarchy Process

4.1. ETL Susceptibility Model

In the context of landslide susceptibility assessment research, 596 thematic parameters have been used so far worldwide, which are classified into five main categories, as shown in the review by Reichenbach et al. [53]: (a) geological parameters, (b) hydrological parameters, (c) parameters related to land use, (d) morphological parameters and (e) other parameters not included in the previous categories.

In the case of the assessment of landslide susceptibility in Cephalonia, the following parameters were used: (a) morphological parameters, including slope, slope aspect and slope curvature; (b) geological parameters, such as lithology of geological formations and distance from tectonic structures (faults and thrusts); (c) seismological parameters, including the peak ground acceleration (PGA); (d) environmental parameters including vegetation, derived from the NDVI index and the distance from the road network; (e) hydrometeorological parameters, such as the spatial distribution of rainfall and the distance from the drainage network (rivers and streams) (Figures 7 and 8; Table 3).

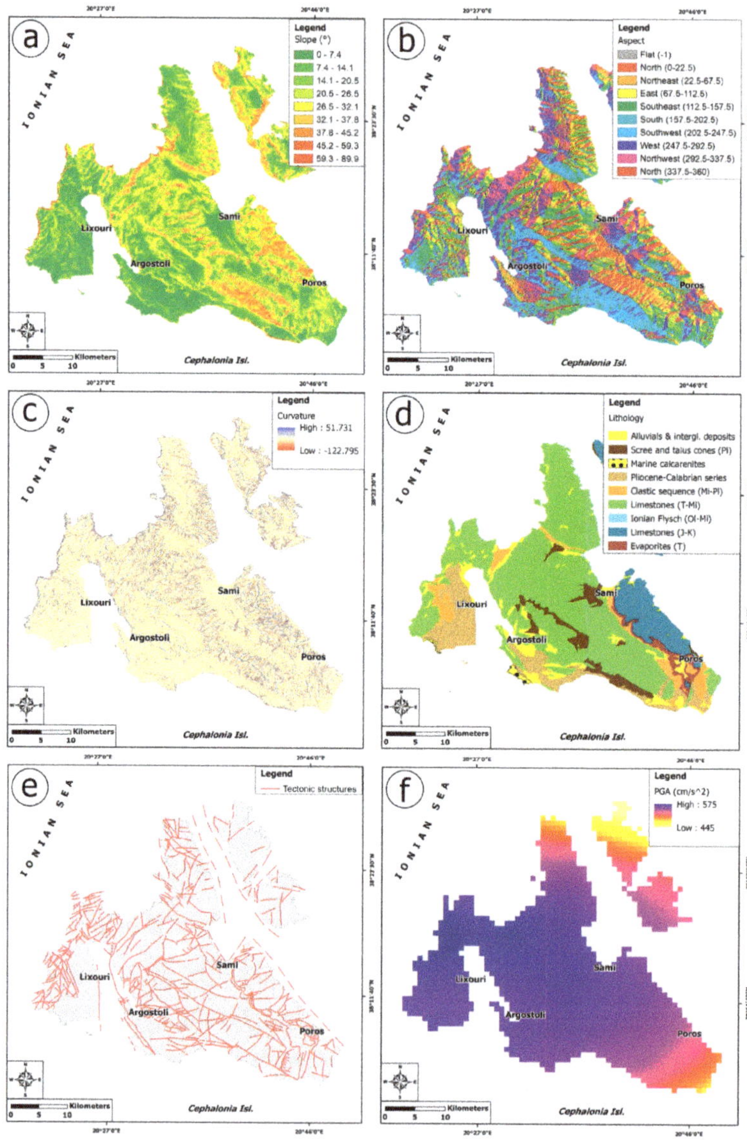

Figure 7. Thematic maps illustrating the morphological, geological and seismological causal factors taken into account for the landslide susceptibility calculations. (**a**) slope, (**b**) aspect, (**c**) curvature, (**d**) lithology, (**e**) tectonic structures and (**f**) PGA.

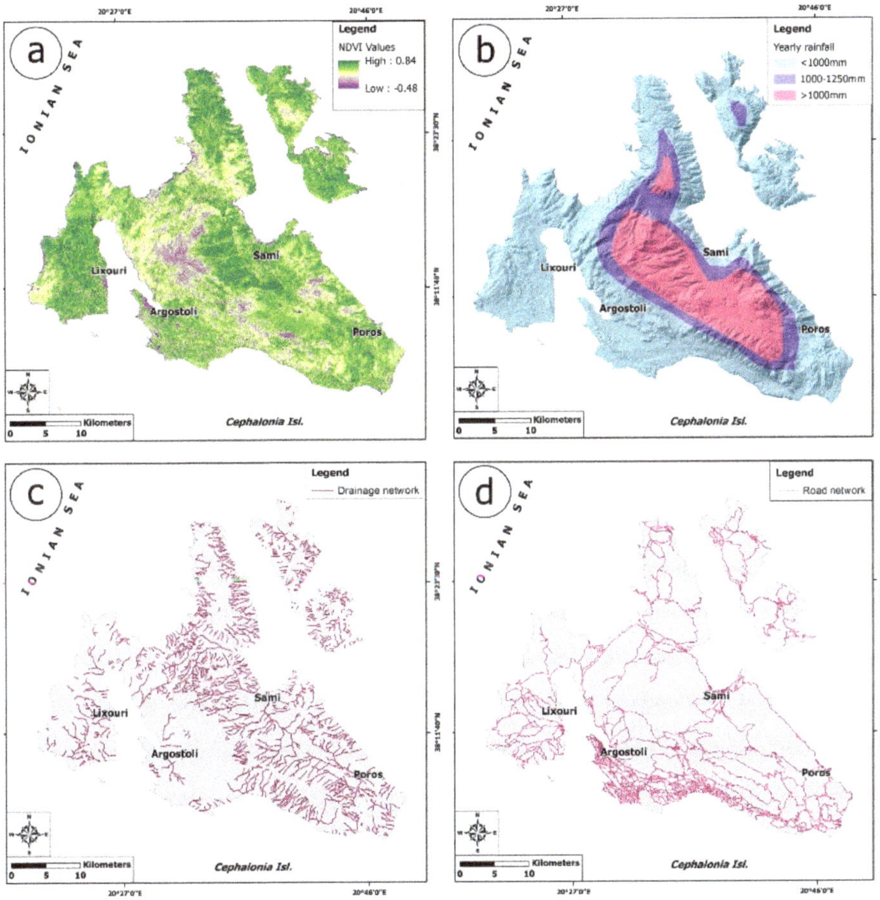

Figure 8. Thematic maps illustrating the environmental and hydrometeorological causal factors taken into account for the landslide susceptibility calculations. (**a**) NDVI, (**b**) yearly rainfall, (**c**) drainage network and (**d**) road network.

Morphological slope controls the balance of restraining and destabilizing load forces acting on a slope [79,80], while forces are developed to maintain a steep rather than a gentle slope. From the statistical analysis of recorded landslides throughout Greece [81], it appears that the 16–30° class is the most susceptible one. Among the morphological parameters, slope is widely used in landslide susceptibility models, as it is the most critical parameter that contributes to the reliable estimation of susceptibility [24,82–84] and to the definition of non-susceptible areas [85].

Slope aspect is a very important parameter as it affects the exposure of the surface to sun and dry winds and indirectly to the flora as well as to the degree of saturation and evapotranspiration of the soil. According to Guzetti et al. [86,87], at northern hemisphere latitudes, where Ionian Islands are located, N- and NW-facing slopes are more prone to failures due to lower temperatures and more shade, factors that favor soil moisture retention.

Table 3. Thematic layers used in the study and their sources.

Layer	Data Type	Data Source
Slope	Grid	DEM of Cephalonia resulted from the TanDEM-X digital data
Aspect	Grid	
Curvature	Grid	
Lithology	Polygon	Neotectonic Map of Gephalonia by Lekkas [32]
Tectonic Structures	Line	Neotectonic Map of Gephalonia by Lekkas [32]
NDVI	Grid	Sentinel (acquisition date 21.04.11)
PGA	Grid	Sakkas et al. [65]
Rainfall	Grid	The Climate Atlas of Greece by the Hellenic National Meteorological Service [88]
Rivers	Line	Topographic maps by the Hellenic Military Geographical Service [89–92]
Roads	Line	Topographic maps by the Hellenic Military Geographical Service [89–92] OpenStreetMap [93]

A concave surface is more likely to suffer landslides, as after prolonged rainfall it can retain and store rainwater for longer time periods, unlike a curved surface. Additionally, a curved slope can delineate the rocky bedrock that is likely to be present in the area. Areas with positive curvature values identify curved surfaces and correspondingly negative values indicate concave surfaces. A zero curvature value indicates a flat surface. The more negative a value is, the higher the probability of landslide, while on the contrary, a surface with a positive curvature value (curved surface) is less likely to experience landslide phenomena.

The maps of slope, aspect and curvature (Figure 7a–c, respectively) were derived from the digital elevation model (DEM) of Cephalonia generated from the TanDEM-X mission, which was based on very high resolution (VHR) synthetic aperture radar (SAR) X-band data [94]. The TanDEM-X global DEM was initially introduced in 2016, based on the processing of SAR data, which were acquired within the years 2011 and 2012, succeeded a spatial DEM resolution of 0.4-arcsecond (ca. 12 m) with the aim of establishing a 2 m relative height accuracy [95–97]. Independently published comparisons prove an of TanDEM-X DEM smaller than ±0.20 m, a root-mean-square error (RMSE) smaller than 1.4 m and an excellent absolute 90% linear height error below 2 m. The assessed accuracy of these data sets is rather excellent and are used quite often for geomorphological and hydrological modeling [98,99].

Concerning the geological parameters used for the assessment of landslide susceptibility, the bedding of formations [100,101], the presence of faults and especially the presence of active faults in seismic active areas [102–106] and local hydrogeological features [24,107] are common in relevant models. The distance to tectonic structures and especially to faults is the most widespread parameter, derived from simple mapping of geo environmental and other data [53,108,109]. Consideration must be given to the fact that the faulted rock along the zone and the surrounding unfaulted rock behave differently during the occurrence of an earthquake, and each fault responds differently to each seismic event, even within the same fault zone.

Regarding the geological parameters used for assessing landslide susceptibility, the presence of scree, talus cones and clastic formations of various age is an aggravating factor for the occurrence of slope failures by earthquakes. Alluvial deposits have little or no influence on the occurrence of these phenomena, while carbonate rocks are susceptible to slides and rockfalls in cases where a steep slope has been formed either by geological processes (e.g., fault zones, incision, erosion, etc.) or by human intervention (e.g., steep road slopes, etc.). Both geological parameters, lithology and tectonic structures, in the case of the landslide susceptibility in Cephalonia were derived from the Neotectonic Map of Cephalonia [32] (Figure 7d,e).

Earthquakes are considered to be a main factor in landslide susceptibility, which is why they are taken into account in relevant studies worldwide, as in Greece [110,111] and Central Asia [112]. In seismic hazard calculations related to landslide susceptibility studies, co-seismic strong ground motion, which is the cause of damage after the occurrence of large earthquakes, is most often expressed in terms of peak ground acceleration (PGA). PGA is often obtained from probabilistic seismic hazard assessment (PSHA). It is mainly preferred as an earthquake metric, as it refers to the overall seismicity of the study area [110]. Another reason is that PGA has been used in many studies in order to determine earthquake-induced displacements [49,113–115].

PSHA in the framework of the present study, in terms of PGA, was adopted by Sakkas et al. [65], who applied the Cornell and McGuire method [116,117] that requires a seismotectonic model. Regarding the latter, the seismogenic crustal area source model of the European Seismic Hazard Model (ESHM), proposed by the Seismic Hazard Harmonization in Europe (SHARE) project [118,119], was used. The required seismicity parameters for PSHA, i.e., the b-value, earthquake rate and maximum magnitude for each zone, were taken from Sakkas et al. [65]. The Modified Gutenberg–Richter (MG-R) model [120,121] was adopted as the earthquake occurrence model. The Ground Motion Prediction Equation (GMPE) of Danciu and Tselentis [122] was used for the prediction of the PGA distribution. It is worth noting that this GMPE, obtained using Greek data, can be considered as reliable, as it has been applied in many PSHA studies in Greece (e.g., [123–125]). The R-CRISIS software [126] was utilized to obtain PGA for a return period of 475 years for Cephalonia Island, considering a rock basement and faulting types corresponding to reverse or strike-slip focal mechanisms (Figure 7f).

In terms of land use, the majority of researchers use combinations of vegetation, land cover and land use from existing maps prepared after visual interpretation of aerial photographs and more recently from automatic or semi-automatic processing of satellite imagery at various scales [127,128] and in different regions, maps showing changes in vegetation or land use [129–132]. The rationale behind incorporating vegetation into landslide susceptibility models is that vegetation depends on slope stability. The normalized difference vegetation index (NDVI) gives a quantitative estimation of the vegetation growth and biomass in the study area and constitutes an index used in landslide susceptibility studies worldwide. NDVI values range from +1.0 to -1.0. Very low NDVI values (for example, 0.1 or less) correspond to no vegetation cover and exposure of the unvegetated area to surface erosion processes and subsequent slope failures [133]. Moderate NDVI values ranging approximately from 0.2 to 0.5 correspond to sparse vegetation, while high NDVI values ranging approximately from 0.6 to 0.9 correspond to dense vegetation [133]. Thus, the lowest NDVI value has the highest impact on the occurrence of landslides. In this study, Sentinel 2 satellite images were used (acquisition date 21.04.11) and the index was calculated by using the red and near infrared spectral bands [134] (Figure 8a), according to the following formula NDVI = (NIR $-$ R)/(NIR + R), where NIR and R are the observed reflectance in the near infrared and red portions of the electromagnetic spectrum, respectively.

Regarding the hydrometeorological parameters, the average annual precipitation, leads to an increase in landslide susceptibility by affecting the water content of rock/soil. In this case, gridded precipitation data were collected from the Hellenic National Meteorological Service [88], based on which the spatial distribution of the average annual rainfall in Cephalonia was modelled in raster form (Figure 8b).

Concerning the environmental parameters and the influence of the drainage network, the distance from drainage network is used according to the following approach: rivers and streams can influence the probability of landslide occurrence, as their erosive processes especially at the base of the slopes and the intense incision across active faults and increase the instability conditions of the slopes and the susceptibility of landslides. As the distance from the branches of the drainage network increases, the susceptibility to landslide events decreases. Rivers and streams in the study area were derived from the topographic maps of the Hellenic Military Geographical Service in 1:50,000 scale [89–92] (Figure 8c).

Due to inadequate drainage and destabilization of hillslopes by undercutting and overloading, landslides are more frequent near the road network [135]. With increasing distance from the road network, the landslide susceptibility in an area decreases. Important information on landslide susceptibility along the road network is obtained from the study of the history of such events. The road network in the study areas was derived from the topographic maps of the Hellenic Military Geographical Service in 1:50,000 scale [89–92], which was then updated by using OpenStreetMap [93] (Figure 8d).

After the factors were selected, they were classified into classes and rank values and standardized ratings were assigned for these classes (Table 4).

In order to evaluate the relative weight of each factor, the Analytic Hierarchy Process (AHP) developed by Saaty [136] was selected as a decision tool. In the frame of the AHP, all factors are compared pairwise by taking into consideration the intensity of their significance and contribution to the generation of ETL, as it is derived from expert opinions, experience and knowledge gained from previous events. In the current study, the SpiceLogic Inc. Analytic Hierarchy Process Software (trial version 3.4.5) [137] was used. In total, 45 pairwise comparisons undertaking an approximate eigenvector calculation were conducted resulting in the pairwise comparison matrix, the consistency ratio and the multi-criteria utility function (Table 5; Figure 9). For the consistency ratio, Saaty [136] suggested that it should be less than or equal to 0.1, to accept the weights. The consistency ratio in this case, as derived from the software, is equal to 0.071, which reveals a suitable and reasonable level of consistency in the pairwise comparison that is a satisfactory value to recognize and estimate the factors weights.

Table 4. Landslide causal factors, classes, rank values, standardized ratings and weights used in the landslide susceptibility assessment in Cephalonia.

Causal Factors	Classes	Rank Values	Standardized Ratings	Weights
Slope	Escarpments, >35°	5	100	0.224
	Steep slopes, 25–35°	4	80	
	Moderately steep slopes, 15–25°	3	60	
	Gentle slopes, 5–15°	2	40	
	Very gentle slopes, <50°	1	20	
PGA	>530 cm/s^2	3	100	0.203
	470–530 cm/s^2	2	67	
	<470 cm/s^2	1	33	
Tectonic Structures	<150 m from thrusts and fault zones	3	100	0.178
	150–300 m from thrusts and fault zones	2	67	
	<50 m from minor faults	1	33	
	All other areas	0	0	

Table 4. *Cont.*

Causal Factors	Classes	Rank Values	Standardized Ratings	Weights
Lithology	Scree and talus cones	6	100	0.136
	Flysch and flysch-type formations	5	83	
	Series with several lithologies	4	97	
	Evaporites	3	50	
	Marine calcarenites	2	33	
	Limestones	1	17	
	Alluvial deposits	0	0	
Roads	<50 m	1	100	0.062
	>50 m	0	0	
Rainfall	>1250 mm	3	100	0.058
	1000–1250 mm	2	67	
	<1000 mm	1	33	
Rivers	<50 m	1	100	0.043
	>50 m	0	0	
Curvature	<−1.5	4	100	0.035
	−1.5–−0.5	3	75	
	−0.5–0	2	50	
	>0	1	25	
NDVI	<0	3	100	0.033
	0–0.1	2	67	
	>0.1	1	33	
Aspect	N and NW facing slopes	3	100	0.028
	W and SW facing slopes	2	67	
	all other directions	1	33	
	Flats	0	0	

Table 5. Pairwise comparison matrix regarding the landslide causative factors as required for applying the AHP method.

	Slope	Aspect	Curvature	Lithology	Tectonic Structures	NDVI	PGA	Rainfall	Rivers	Roads	Weights
Slope	1	4	4	4	2	4	2	5	5	3	0.224
Aspect	0.25	1	1	0.2	0.2	0.5	0.2	0.2	0.5	0.333	0.028
Curvature	0.25	1	1	0.2	0.2	2	0.2	0.5	0.5	0.5	0.035
Lithology	0.25	5	5	1	0.333	5	0.333	5	5	4	0.136
Tectonic Structures	0.5	5	5	3	1	5	0.5	5	5	5	0.178
NDVI	0.25	2	0.5	0.2	0.2	1	0.2	0.5	0.5	0.5	0.033
PGA	0.5	5	5	3	2	5	1	5	5	5	0.203
Rainfall	0.2	5	2	0.2	0.2	2	0.2	1	2	0.5	0.058
Rivers	0.2	2	2	0.2	0.2	2	0.2	0.5	1	0.5	0.043
Roads	0.333	3	2	0.25	0.2	2	0.2	2	2	1	0.062

Consistency Ratio calculated as 0.071. Multi-Criteria Utility Function = 0.22 × [Slope] + 0.03 × [Aspect] + 0.04 × [Curvature] + 0.14 × [Lithology] + 0.18 × [TectonicStructures] + 0.03 × [NDVI] + 0.2 × [PGA] + 0.06 × [Rainfall] + 0.04 × [Rivers] + 0.06 × [Roads].

Next, the thematic layers of the distance to roads, to rivers and to faults were calculated with the use of the Euclidean distance tool included in the Spatial Analyst toolbox of ArcGIS 10.7 [138]. The thematic maps were converted to raster datasets, which were then reclassified based on the aforementioned classes, ranking and standardized ratings. The resulting weighted raster thematic maps were multiplied by the corresponding weights based on the multi-criteria utility function (Table 5) with Raster Calculator tool of the Spatial Analyst toolbox in ArcGIS 10.7. This tool allows researchers to create and execute Map Algebra expressions for outputting a raster. The zonation map is obtained from the sum of the weighted map. Each cell of the map resulted from the aforementioned calculation is characterized by a certain landslide susceptibility index (LSI) value. The final earthquake-triggered landslide susceptibility map of Cephalonia Island comprises seven susceptibility zones (non-susceptible, very low, low, moderate, high, very high, and critically high susceptibility) (Figure 10) according to the Jenks natural breaks classification method [139]. The area of each zone as well as the percentage on the total area of the studied island is presented in Table 6.

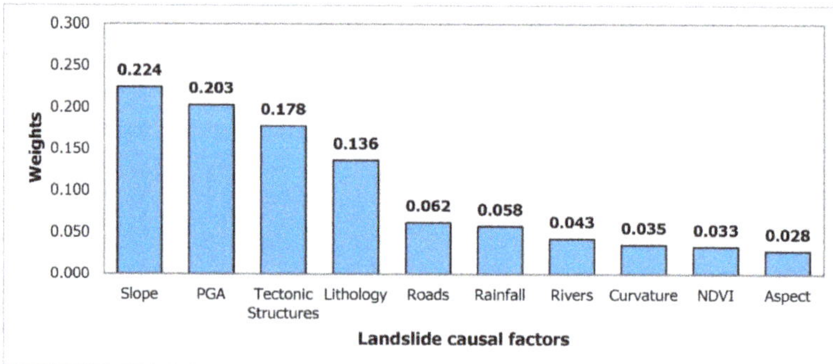

Figure 9. Ranking of weights for the 10 landslide causative factors.

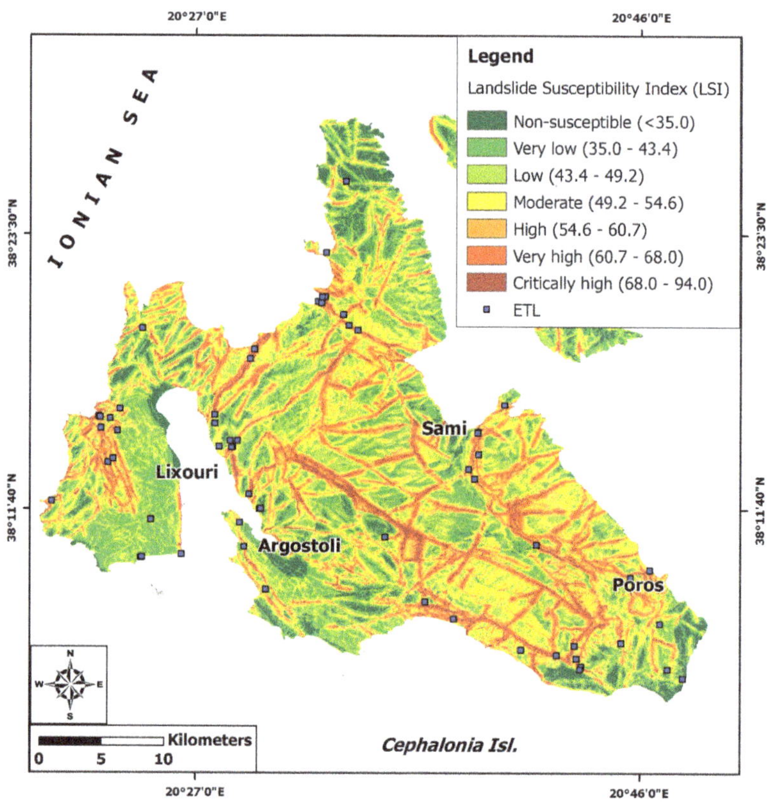

Figure 10. The landslide susceptibility index (LSI) map of Cephalonia Island, against the recorded ETL.

Table 6. Areas of susceptibility map classes.

LSI Class	Area (km^2)	Percentage (%)
No susceptible	86.40	8.46
Very low	210.26	20.60
Low	236.42	23.16
Moderate	199.82	19.57
High	152.78	14.97
Very high	99.68	9.76
Critically high	35.52	3.48
Total	1020.87	100

4.2. Validation of the Model

The creation of the landslide susceptibility map was followed by the verification of the accuracy of the landslide susceptibility model. The validation of the landslide susceptibility results is usually carried out by applying two methods in the frame of the relevant studies. The first approach involves the comparison of the landslide susceptibility map with an inventory map, which includes landslide events in the study area (e.g., [110,111,140–142]). The second approach involves the use of the receiver operating characteristic (ROC; [143–145]). The result of this approach is a success rate resulting from the comparison between the earthquake-triggered landslide grid cells and the earthquake-triggered landslide susceptibility grid (e.g., [146]).

Both approaches were used in this study for Cephalonia. For the first one, the ETL inventory in Cephalonia from 1636 to 2014 was used. The majority of the 67 landslides occurred in areas of high (24/67, 35.82%), very high (16/67, 23.88%) and critically high (15/67, 22.39%) susceptibility (Table 7). In total, 55 out of 67 events (82.09%) occurred in high to critically high susceptibility zones, while 12 occurred in low to medium susceptibility zones (3 in low and 9 in medium susceptibility zones, with a percentage of 4.48% and 13.43%, respectively) (Table 7).

Table 7. Distribution of the recorded historical and recent ETL in the detected landslide susceptibility zones of Cephalonia Island.

Landslide Susceptibility	ETL	Percentage (%)
No	0	0.00
Very low	0	0.00
Low	3	4.48
Moderate	9	13.43
High	24	35.82
Very high	16	23.88
Critically high	15	22.39
Total	67	100.00

For the second approach based on the ROC curve, the curves are obtained by plotting in a graph with both axes having values ranging from 0 to 1. The x-axis comprises values of false positive rate based on the resulting earthquake-triggered landslide susceptibility map and the y-axis comprises values of true positive rate based on the ETL sites. The accuracy of the model is related to the area under the curve (AUC). The AUC ranges from 0.5 to 1.0 [147,148] with a value equal to 1.0 corresponding to a perfect model prediction accuracy. A value found along the diagonal line of the graph implies a 50% probability of predicting the generation of landslides. The larger the value of the AUC, the better the landslide susceptibility model predicts susceptible areas and landslides. In the case of Cephalonia, the ROC graph and the AUC have been derived by using the ArcSDM tools in ArcGIS 10.7.

The AUC was calculated as 0.803 (Figure 11), which reveals fair to good model prediction accuracy.

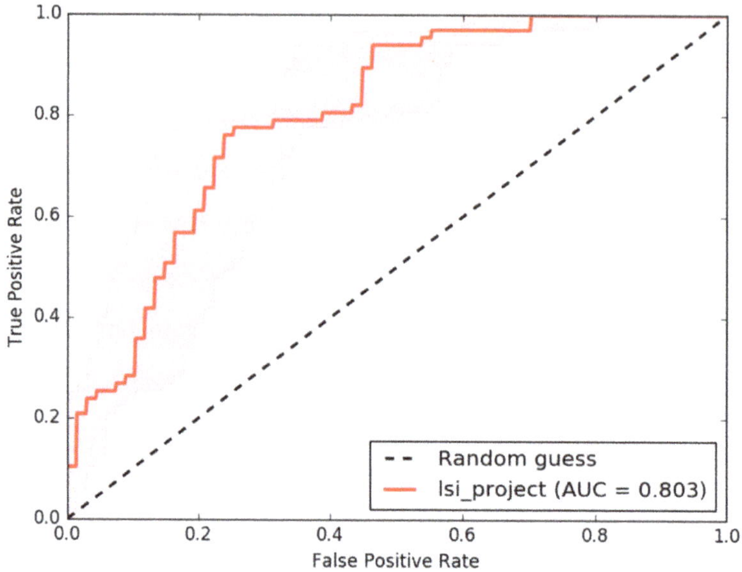

Figure 11. ROC graph and AUC for the landslide susceptibility model applied for Cephalonia.

5. Discussion

This work explores earthquake-triggered landslides (ETL) on the island of Cephalonia, one of the most active areas in the Mediterranean region. The study addresses the gap in the record of past events and examines the ETL susceptibility of the island exploiting the long record that has been created. Given the rich record of the identified ETL phenomena, the island is considered to be particularly prone to slope failures due to earthquakes.

From the implemented inventory, it is concluded that 67 cases were induced by earthquakes in Cephalonia from 1636 to 2014, with most of them being triggered by the 12 August 1953 earthquake (30 cases, 44.78%). The majority of failures have been reported from earthquakes after 1900 (Figure 12). Although historical earthquakes that have affected Cephalonia had magnitudes larger than 5.9, information on environmental effects, despite their important impacts, is limited. This does not mean that these events did not cause such effects, but it is probably due to the fact that the interest of residents, authorities and researchers at the time was mainly focused on damage to buildings which were mainly associated with unpleasant effects on the local population. In addition, the earthquake effects resulted usually in heavy damage of the road network, disrupting transportation between local villages, making it impossible for researchers to record the complete impact of earthquakes on both the natural environment and the building stock. This situation seems to have changed after the first half of the 20th century, as evidenced by the plethora of recorded landslides caused by the 1953, 1983 and 2014 earthquakes.

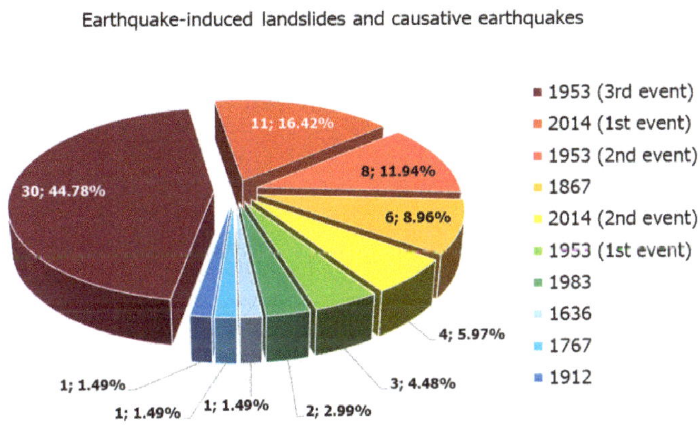

Figure 12. Pie charts showing the number and percentages of landslides triggered in Cephalonia Island by earthquakes from 1636 to 2014. The majority has been generated by the 1953 earthquake sequence.

Based on the rich record of ETL developed, it was possible to validate the result of landslide susceptibility analysis in terms of spatial distribution. In the course of this analysis, we used not only morphological and geological variables as the causative factors, but also seismological, hydrometeorological and environmental parameters, in order to effectively identify and precisely map the most susceptible zones in the island. Both from the landslide susceptibility map individually and from the comparison of the historical and landslide susceptibility maps, it was found that the spatial distribution of the studied effects in Cephalonia is not random. They were mainly generated in certain zones with high to critically high susceptibility to landslide triggering in the case of strong earthquake ground motion. These zones include the following from north to south:

- The northern margins of Pylaros valley, located between Kalon Mt to the north and Ayia Dynati to the south, constitutes one for the most susceptible zones. Along the southern slopes of Kalon Mt, the Ayia Efimia oblique slip reverse fault occurs, which has formed a highly fractured and deformed rock mass, susceptible to slope failure from earthquakes. This situation is highlighted not only by the slope failures generated during the 1953 earthquakes at Lekatsata, Logarata and Karousata villages (Figure 4), but also by similar impressive effects that have taken place further west, in the coastal area of Myrtos bay during the occurrence of the devastating earthquakes of 1953, 1983 and 2014 [10,62,64], which resulted in damage and destruction of adjacent sections of the road network and temporary traffic disruption. It is important to mention that the coastal area of Myrtos is particularly prone to slope failures, caused not only by earthquakes, but also during extreme hydro-meteorological events. A typical example for the later events is the case of the "Ianos" Mediterranean cyclone (medicane) in mid-September 2020, which affected several areas in Greece and especially Lefkada, Cephalonia, Ithaki and Zakynthos Islands, causing flooding, mudflows and debris flows with severe impacts on networks and infrastructure of Cephalonia (Assos area in the Erissos peninsula, and Myrtos and Ayia Efimia areas at the western and eastern ends of the Pylaros valley, respectively; [149,150]).
- The eastern margins of Thinia valley, where the Kontogourata–Agonas oblique slip reverse fault occurs, is also among the most earthquake-triggered landslide-susceptible zones in Cephalonia. This fault places the northwestern part of Aenos Mt on the northeastern part of Paliki peninsula and is responsible for the rugged morphology, dense discontinuities and the decreased cohesion in the area. Along this susceptible zone, landslides were induced by the 1953 and 2014 earthquakes.
- The high steep mountain slopes in the western, eastern and southern parts of Aenos Mt fault block are characterized by high to critically high susceptibility, strongly related

to the eponymous fault zone, which bounds the mountain to its south and west. As a result of the presence and activity of the zone, extensive outcrops of scree and talus cones as well as fragmented limestone of the Paxi unit participated in the geological setting and led to the generation of landslides during the 1636, 1912 and August 1953 earthquakes. In particular, the slope failures of 1953 had considerable impact on settlements in the area, which were founded on the above-mentioned formations, resulting in their partial destruction and human losses.

- Another zone susceptible to ETL is the western coastal part of the Paliki peninsula. This coastal zone has been formed and evolves under the influence of the offshore Cephalonia strike-slip fault and many smaller onshore faults. Their combined action results in the creation of extensive fault scarps and steep slopes, the majority of which are almost vertical and composed by highly fractured, brecciated and almost powdered carbonate formations.
- An important contribution to the susceptibility of Cephalonia to landslides from earthquakes is also made by certain large inactive structures and in particular the thrust of the Ionian unit over the Paxi unit. However, the formations occurring at the front of the thrust appear particularly deformed, a fact that increases the susceptibility of both the overlying Ionian unit and the underlying Paxi unit to failure. Characteristic of this zone is the occurrence of landslides in the wider area of Sami during the 1953 earthquake sequence, with impact on adjacent settlements and parts of the road network.
- The southern part of the Paliki peninsula is locally characterized by high susceptibility, mainly along slopes, which are made up of not very compact mineral deposits. Typical cases of landslides have occurred along slopes in the southern and eastern coastal areas near the town of Lixouri.

Concerning the ETL distribution in the fault blocks of Cephalonia, the most affected in terms of ETL is the fault block of Aenos Mt and the eastern Cephalonia followed by Paliki, Erissos and Argostoli peninsulas.

6. Conclusions

The landslide susceptibility map was classified according to the natural break method into seven classes of critically high, very high, high, moderate, low, very low, and no susceptibility (Figure 10) corresponding to 3.48%, 9.76%, 14.97%, 19.57%, 23.16%, 20.60% and 8.46% of the total study area respectively (Table 6).

Overall, from qualitative assessment of the distribution of the landslides, illustrated in the description of the aforementioned most susceptible zones, it was concluded that four factors (i.e., slope, PGA, tectonic structures and lithology) were associated in a higher degree to the locations where ETL occurred on Cephalonia Island. This supports the difference between this group of parameters and the rest in terms of weighting factors estimated in AHP.

It should be also noted that despite the limited extent of high to critically high susceptible zones in the island (28.21% of its total area based on Table 6) they host 82% of the ETL phenomena included in the inventory. This fact along with the AUC values of 80.3% reveals a fair-to-good accuracy of the landslide susceptibility assessment in the island. Furthermore, the validation results indicate that the geoenvironmental conditions and the contribution of the studied variables to the generation of the landslides were effectively detected and determined.

Such landslide susceptibility assessment approaches are interesting from the scientific viewpoint, but mostly they are useful and necessary for decision making regarding effective and responsible land-use planning and environmental development. The combined use of ETL inventories and susceptibility maps maximizes the information content provided to the involved decision-makers, planners and other stake-holders.

Such applied studies give the opportunity and the possibility to all involved parties from the research community and the risk professionals to further and in detail identify the most important landslide triggering factors, to understand the contribution of several parameters to the generation of landslides to seismically active areas and to interpret the generation mechanisms and processes of landslides. Furthermore, by taking into account the results of such studies, the authorities involved in disaster risk reduction acquire related knowledge in order to adopt preventive measures and site-specific mitigation strategies. These strategies have the potential to mitigate the adverse effects of ETL to the natural environment and the building stock and eliminate and reduce human and economic losses, respectively.

The application of similar approaches is expected to be extensive in the future as urbanization and development have already expanded and will further expand to areas prone to landslides, while the simultaneous presence of landslide causal factors will be continuous and the occurrence of landslide triggering events constantly increasing. Furthermore, it is also expected that the processes applied in the landslide susceptibility models will also be used to other fields of assessment of related hazard and disaster risk reduction.

Author Contributions: Conceptualization, S.M.; methodology, S.M. and M.D.; software, S.M. and M.D.; validation, S.M. and M.D.; formal analysis, S.M. and M.D.; investigation, S.M. and M.D.; resources, N.V. and E.L.; data curation, S.M., M.D., V.K. and I.S.; writing—original draft preparation, S.M., M.D., V.K. and I.S.; writing—review and editing, S.M., M.D., H.K., E.V., V.K., I.S., G.K., E.S., N.V. and E.L.; visualization, S.M. and M.D.; supervision, S.M.; project administration, N.V. and E.L.; funding acquisition, N.V. and E.L. All authors have read and agreed to the published version of the manuscript.

Funding: This research was partially funded by the "Telemachus—Innovative Operational Seismic Risk Management System of the Ionian Islands" project, included in the Priority Axis "Environmental Protection and Sustainable Development" of the Regional Operational Programme "Ionian Islands 2014–2020", grant number MIS 5007986, which is funded by the European Regional Development Fund (ERDF) and National Resources under the National Strategic Reference Framework NSRF 2014–2020.

Institutional Review Board Statement: Not applicable.

Informed Consent Statement: Not applicable.

Data Availability Statement: Data available on request from the corresponding author.

Conflicts of Interest: The authors declare no conflict of interest.

References

1. Michetti, A.M.; Esposito, E.; Guerrieri, L.; Porfido, S.; Serva, L.; Tatevossian, R.; Vittori, E.; Audemard, F.; Azuma, T.; Clague, J.; et al. *Environmental Seismic Intensity Scale 2007—ESI 2007*; Memorie Descrittive Della Carta Geologica d'Italia; Guerrieri, L., Vittori, E., Eds.; Agenzia per la Protezione dell'Ambiente e per i Servizi Tecnici (APAT): Roma, Italy, 2007; pp. 1–41.
2. Marano, K.D.; Wald, D.J.; Allen, T.I. Global earthquake casualties due to secondary effects: A quantitative analysis for improving rapid loss analyses. *Nat. Hazards* **2010**, *52*, 319–328. [CrossRef]
3. Petley, D. Global patterns of loss of life from landslides. *Geology* **2012**, *40*, 927–930. [CrossRef]
4. Kennedy, I.T.R.; Petley, D.N.; Williams, R.; Murray, V. A Systematic Review of the Health Impacts of Mass Earth Movements (Landslides). *PLoS Curr.* **2015**, *30*, 7. [CrossRef] [PubMed]
5. Nowicki Jessee, M.A.; Hamburger, M.W.; Allstadt, K.; Wald, D.J.; Robeson, S.M.; Tanyas, H.; Hearne, M.; Thompson, E.M. A global empirical model for near-real-time assessment of seismically induced landslides. *J. Geophys. Res. Earth Surf.* **2018**, *123*, 1835–1859. [CrossRef]
6. Bird, J.F.; Bommer, J.J. Earthquake losses due to ground failure. *Eng. Geol.* **2014**, *75*, 147–179. [CrossRef]
7. Keefer, D.K. Landslides caused by earthquakes. *Geol. Soc. Am.* **1984**, *95*, 406–421. [CrossRef]
8. Marui, H.; Nadim, F. Landslides and Multi-Hazards. In *Landslides—Disaster Risk Reduction*; Sassa, K., Canuti, P., Eds.; Springer: Berlin/Heidelberg, Germany; pp. 435–450.
9. Shieh, C.-L.; Wang, C.-M.; Lai, W.-C.; Tsang, Y.-C.; Lee, S.P. The composite hazard resulted from Typhoon Morakot in Taiwan. *J. JSECE* **2009**, *62*, 61–65. [CrossRef]
10. Mavroulis, S.; Lekkas, E. Revisiting the Most Destructive Earthquake Sequence in the Recent History of Greece: Environmental Effects Induced by the 9, 11 and 12 August 1953 Ionian Sea Earthquakes. *Appl. Sci.* **2021**, *11*, 8429. [CrossRef]

11. Hansen, A.; Franks, C.A.M. Characterization and mapping of earthquake triggered landslides for seismic zonation. State-of-the-art Paper. In Proceedings of the Fourth International Conference on Seismic Zonation, Stanford, CA, USA, 26–29 August 1991; Volume 1, pp. 149–195.
12. Keefer, D.K. Landslides generated by earthquakes: Immediate and long-term effects. In *Treatise on Geomorphology*; Shroder, J., Owen, L.A., Eds.; Academic Press: San Diego, CA, USA, 2013; Volume 5, pp. 250–266.
13. Rodriguez, C.; Bommer, J.; Chandler, R. Earthquake-induced landslides: 1980–1997. *Soil Dyn. Earthq. Eng.* **1999**, *18*, 325–346. [CrossRef]
14. Hancox, G.T.; Perrin, N.D.; Dellow, G.D. *Earthquake-Induced Landsliding in New Zealand and Implications for MM Intensity and Seismic Hazard Assessment*; Institute of Geological & Nuclear Sciences Client Report 43601 B; Lower Hutt: Wellington, New Zealand, 1997.
15. Hancox, G.T.; Perrin, N.D.; Dellow, G. Recent studies of historical earthquake-induced landsliding, ground damage, and MM intensity in New Zealand. *Bull. N. Z. Soc. Earthq. Eng.* **2002**, *35*, 59–95. [CrossRef]
16. Rosser, B.; Dellow, S.; Haubrock, S.; Glassey, P. New Zealand's National Landslide Database. *Landslides* **2017**, *14*, 1949–1959. [CrossRef]
17. Papadopoulos, G.A.; Plessa, A. Magnitude–distance relations for earthquake-induced landslides in Greece. *Eng. Geol.* **2000**, *58*, 377–386. [CrossRef]
18. Prestininzi, A.; Romeo, R. Earthquake-induced ground failures in Italy. *Eng. Geol.* **2000**, *58*, 387–397. [CrossRef]
19. Martino, S.; Prestininzi, A.; Romeo, R.W. Earthquake-induced ground failures in Italy from a reviewed database. *Nat. Hazards Earth Syst. Sci.* **2014**, *14*, 799–814. [CrossRef]
20. Rondoyanni, T.; Sakellariou, M.; Baskoutas, J.; Christodoulou, N. Evaluation of active faulting and earthquake secondary effects in Lefkada Island, Ionian Sea, Greece: An overview. *Nat. Hazards* **2012**, *61*, 843–860. [CrossRef]
21. Mavroulis, S.; Stanota, E.-S.; Lekkas, E. Evaluation of environmental seismic intensities of all known historical and recent earthquakes felt in Zakynthos Island, Greece using the Environmental Seismic Intensity (ESI 2007) scale. *Quat. Int.* **2019**, *532*, 1–22. [CrossRef]
22. EAK. *Greek Seismic Code*; Earthquake Planning & Protection Organization: Athens, Greece, 2003; Volume 72, p. 7.
23. Sakkas, G.; Kouskouna, V.; Makropoulos, K. Seismic hazard analysis in the Ionian Islands using macroseismic intensities. *Hell. J. Geosci.* **2010**, *45*, 239–247.
24. Carrara, A.; Cardinali, M.; Detti, R.; Guzzetti, F.; Pasqui, V.; Reichenbach, P. GIS techniques and statistical models in evaluating landslide hazard. *Earth Surf. Proc. Land.* **1991**, *16*, 427–445. [CrossRef]
25. Renz, C. *Die Vorneogene Stratigraphie der Normal-Sedimentaren Formationen Griechenlands*; Institute for Geology and Subsurface Research: Athens, Greece, 1955; p. 637.
26. Aubouin, J. Contribution a l' étude géologique de la Grèce septentrionale, les confins de l' Epire et de la Thessalie. *Ann. Géol. Pays Hellén.* **1959**, *10*, 1–525.
27. Aubouin, J.; Dercourt, J. Zone preapulienne, zone ionienne et zone du Gavrovo en Peloponncse occidentale. *Bull. Soc. Géol. Fr.* **1962**, *4*, 785–794. [CrossRef]
28. Georgiadou-Dikaioulia, E. The Neogene of Kephallinia. Ph.D. Thesis, University of Athens, Athens, Greece, 1967.
29. BP Co. *The Geological Results of Petroleum Exploration in Western Greece*; Institute of Geology and Subsurface Research: Athens, Greece, 1971.
30. BP Co.; University of Munich; Migiros, G. *Geological Map of Greece, "Cephalonia Island (Northern and Southern Part)", 1:50,000 Scale*; Institute of Geology and Mineral Exploration: Athens, Greece, 1985.
31. Underhill, J.R. Triassic evaporites and Plio-Quaternary diapirism in western Greece. *J. Geol. Soc.* **1988**, *145*, 269–282. [CrossRef]
32. Lekkas, E. *"Cephalonia—Ithaki" Sheet; Scale 1:100,000, Neotectonic Map of Greece*; National and Kapodistrian University of Athens: Athens, Greece, 1996.
33. Lekkas, E.; Danamos, G.; Mavrikas, G. Geological Structure and Evolution of Kefallonia and Ithaki Islands. *Bull. Geol. Soc. Greece* **2001**, *34*, 11–17. [CrossRef]
34. Sorel, D. Etude Néotectonique des Îles Ioniennes de Céphalonie et de Zante et de l' Elide Occidentale (Gréce). Ph.D. Thesis, Orsay, Université Paris Sud, Paris, France, 1976; p. 200.
35. Lagios, E.; Sakkas, V.; Papadimitriou, P.; Damiata, B.N.; Parcharidis, I.; Chousianitis, K.; Vassilopoulou, S. Crustal deformation in the Central Ionian Islands (Greece): Results from DGPS and DInSAR analyses (1995–2006). *Tectonophysics* **2007**, *444*, 119–145. [CrossRef]
36. Hollenstein, C.H.; Müller, M.D.; Geiger, A.; Kahle, H.-G. Crustal motion and deformation in Greece from a decade of GPS measurements 1993–2003. *Tectonophysics* **2008**, *449*, 17–40. [CrossRef]
37. Scordilis, E.M.; Karakaisis, G.F.; Karacostas, B.G.; Panagiotopoulos, D.G.; Comninakis, P.E.; Papazachos, B.C. Evidence for transform faulting in the Ionian sea: The Cephalonia island earthquake sequence of 1983. *Pure Appl. Geophys. PAGEOPH* **1985**, *123*, 388–397. [CrossRef]
38. Louvari, E.; Kiratzi, A.; Papazachos, B. The Cephalonia Transform Fault and its extension to western Lefkada Island (Greece). *Tectonophysics* **1999**, *308*, 223–236. [CrossRef]
39. Sachpazi, M.; Hirn, A.; Clément, C.; Haslinger, F.; Laigle, M.; Kissling, E.; Charvis, P.; Hello, Y.; Lépine, J.-C.; Sapin, M.; et al. Western Hellenic subduction and Cephalonia Transform: Local earthquakes and plate transport and strain. *Tectonophysics* **2000**, *319*, 301–319. [CrossRef]

40. Makropoulos, K.; Kaviris, G.; Kouskouna, V. An updated and extended earthquake catalogue for Greece and adjacent areas since 1900. *Nat. Hazards Earth Syst. Sci.* **2012**, *12*, 1425–1430. [CrossRef]
41. Papazachos, B.; Papazachou, K. *The Earthquakes of Greece*; Ziti Publications: Thessaloniki, Greece, 2003; p. 286.
42. Ambraseys, N. *Earthquakes in the Mediterranean and Middle East, a Multidisciplinary Study of Seismicity up to 1900*; Cambridge University Press: Cambridge, UK, 2009; p. 970. [CrossRef]
43. Papadimitriou, E.E. Mode of Strong Earthquake Recurrence in the Central Ionian Islands (Greece): Possible Triggering due to Coulomb Stress Changes Generated by the Occurrence of Previous Strong Shocks. *Bull. Seism. Soc. Am.* **2002**, *92*, 3293–3308. [CrossRef]
44. Bonatis, P.; Akinci, A.; Karakostas, V.; Papadimitriou, E.; Kaviris, G. Near-fault broadband ground motion simulation applications at the Central Ionian Islands, Greece. *Pure Appl. Geophys.* **2021**, *20*, 10. [CrossRef]
45. Koukis, G.; Ziourkas, C. Slope instability phenomena in Greece: A statistical analysis. *Bull. Int. Assoc. Eng. Geol.* **1991**, *43*, 47–60. [CrossRef]
46. Koukis, G.; Tsiambaos, G.; Sabatakakis, N. Slope movements in Greek territory: A statistical approach. In Proceedings of the 7th International Congress of IAEG, Lisboa, Portugal, 5–9 September 1994; Oliveira, R., Rodrigues, L.F., Coelho, A.G., Cunha, A.P., Eds.; A. A. Balkema: Rotterdam, The Netherlands, 1994; pp. 4621–4628.
47. Koukis, G.; Tsiambaos, G.; Sabatakakis, N. Landslides movements in Greece: Engineering geological characteristics and environmental consequences. In Proceedings of the International Symposium of Engineering Geology and Environment, Athens, Greece, 23–27 June 1997; A. A. Balkema: Rotterdam, The Netherlands, 1997; Volume 1, pp. 789–792.
48. Koukis, G.; Sabatakakis, N.; Nikolau, N.; Loupasakis, C. Landslide Hazard Zonation in Greece. In *Landslides*; Sassa, K., Fukuoka, H., Wang, F., Wang, G., Eds.; Springer: Berlin/Heidelberg, Germany, 2005. [CrossRef]
49. Sabatakakis, N.; Koukis, G.; Vassiliades, E.; Lainas, S. Landslide susceptibility zonation in Greece. *Nat. Hazards* **2013**, *65*, 523–543. [CrossRef]
50. Koukis, G.; Andronopoulos, V.; Rozos, D.; Kinigalaki, M.; Tzitziras, A.; Pogiatzi, E.; Garivaldi, A. *Geotechnical Map of Greece, Scale 1:500,000*; Institute of Geology and Mineral Exploration: Athens, Greece, 1993.
51. Centre for Research on the Epidemiology of Disasters (CRED). EM-DAT—The International Disaster Database. Available online: https://public.emdat.be/ (accessed on 15 November 2021).
52. Guzzetti, F.; Mondini, A.C.; Cardinali, M.; Fiorucci, F.; Santangelo, M.; Chang, K.T. Landslide inventory maps: New tools for an old problem. *Earth-Sci. Rev.* **2012**, *112*, 42–66. [CrossRef]
53. Reichenbach, P.; Rossi, M.; Malamud, B.D.; Mihir, M.; Guzzetti, F. A review of statistically-based landslide susceptibility models. *Earth-Sci. Rev.* **2018**, *180*, 60–91. [CrossRef]
54. Stucchi, M.; Rovida, A.; Gomez Capera, A.A.; Alexandre, P.; Camelbeeck, T.; Demircioglu, M.B.; Gasperini, P.; Kouskouna, V.; Musson, R.M.W.; Radulian, M.; et al. The SHARE European earthquake catalogue (SHEEC) 1000–1899. *J. Seismol.* **2013**, *17*, 523–544. [CrossRef]
55. Institute of Geology and Underground Research (IGUR). *The Devastating Earthquakes of the Ionian Islands in August 1953: Proposals for the Reconstruction of Cities, Towns and Villages of the Earthquake-Affected Islands*; Monograph; Institute of Geology and Underground Research: Athens, Greece, 1954; p. 76.
56. The Press Museum of the Peloponnese—Epirus—Islands Daily Newspaper Editors Association. Available online: http://www.mouseiotipou.gr/arxeion-xml/pages/esiepi/internet/intro (accessed on 10 January 2020).
57. The Digital Historical Archive of the Lambrakis Press Group. Available online: http://premiumarchives.alteregomedia.org/Login.aspx (accessed on 20 January 2020).
58. The Digital Library of Newspapers and Magazines of the National Library of Greece. Available online: http://efimeris.nlg.gr/ns/main.html (accessed on 5 February 2020).
59. The Digital Library of the Greek Parliament. Available online: https://library.parliament.gr/Portals/6/pdf/digitalmicrofilms.pdf (accessed on 10 February 2020).
60. Pentogalou, V.G. The destructive earthquake of 1867 in Paliki and its similarities with the earthquakes of January—February of 2014. *I Kefalonitiki Prood.* **2014**, *10*, 11–14.
61. Petratos, P. The earthquakes of 1953—Argostoli Prison. The forgotten offer of political prisoners. *Kefalonitiki Prood.* **2013**, *8*, 10–14.
62. Lekkas, E.L.; Mavroulis, S.D. Earthquake environmental effects and ESI 2007 seismic intensities of the early 2014 Cephalonia (Ionian Sea, Western Greece) earthquakes (January 26 and February 3, Mw 6.0). *Nat. Hazards* **2015**, *78*, 1517–1544. [CrossRef]
63. Lekkas, E.L.; Mavroulis, S.D. Fault zones ruptured during the early 2014 Cephalonia Island (Ionian Sea, Western Greece) earthquakes (January 26 and February 3, Mw 6.0) based on the associated co-seismic surface ruptures. *J. Seismol.* **2016**, *20*, 63–78. [CrossRef]
64. Eleftheriou, A.; Mouyiaris, N. *Macroseismic Reconnaissance in the Area of Cephalonia-Zakynthos (Earthquakes 17 & 19-1-83)*; Monograph; Institute of Geological and Mineral Research: Athens, Greece, 1983.
65. Sakkas, V.; Kapetanidis, V.; Kaviris, G.; Spingos, I.; Mavroulis, S.; Diakakis, M.; Alexopoulos, J.D.; Kazantzidou-Firtinidou, D.; Kassaras, I.; Dilalos, S.; et al. Seismological and Ground Deformation Study of the Ionian Islands (w. Greece) during 2014–2018, a Period of Intense Seismic Activity. *Appl. Sci.* **2022**, *12*, 2331. [CrossRef]

66. Mavroulis, S.; Alexoudi, V.; Grambas, A.; Lekkas, E.; Carydis, P. The January-February 2014 Cephalonia (Ionian Sea, western Greece) earthquake sequence: Damage pattern on buildings. In Proceedings of the 16th World Conference of Earthquake Engineering, Santiago, Chile, 9–13 January 2017; WCEE Online Proceedings; Paper No 414. National Information Centre of Earthquake Engineering, IIT Kanpur: Kanpur, India, 2017.
67. Schmidt, J. *Study about the Kefalonia Earthquake of 23 January 1867*; Ethniko Typogr.: Athens, Greece, 1867; 30p.
68. Partsch, J. *Kephallenia und Ithaka*; Eine Geographische Monographie: Gotha, Germany, 1890; p. 108.
69. Tsitselis, H.A. *Kefalliniaka Symmikta*; Argostoli: Cephalonia, Greece, 1904.
70. Kouskouna, V.; Makropoulos, K.C.; Tsiknakis, K. Contribution of historical information to a realistic seismicity and hazard assessment of an area. The Ionian Islands earthquakes of 1767 and 1769: Historical investigation. In *Materials of CEC Project "Review of Historical Seismicity in Europe"*; Albini, P., Moroni, A., Eds.; CNR, Istituto di Ricerca sul Rischio Sismico: Milano, Italy, 1992; Volume 1, pp. 195–206.
71. Makropoulos, K.C.; Kouskouna, V. The Ionian Islands earthquakes of 1767 and 1769: Seismological aspects. Contribution of historical information to a realistic seismicity and hazard assessment of an area. In *Materials of CEC Project "Review of Historical Seismicity in Europe"*; Albini, P., Moroni, A., Eds.; CNR, Istituto di Ricerca sul Rischio Sismico: Milano, Italy, 1994; Volume 2, pp. 27–36.
72. Scrofani, X. *Voyage en Gréce Fait en 1794 et 1795*; Treuttel et Würtz: Paris-Strasburg, France, 1801; Volume 3.
73. Vergotis, P. *The 23 January 1867 Earthquake*; Typography of Kefallinia: Argostoli, Greece, 1867; 15p.
74. Katramis, N. *Philological Analects of Zakinthos*; Avgi Publications: Zakynthos, Greece, 1880.
75. Albini, P.; Ambraseys, N.N.; Monachesi, G. Material for the Investigation of the seismicity of the Ionian Islands between 1704 and 1766. In *Historical Investigation of European Earthquakes*; Albini, P., Moroni, A., Eds.; CNR, Istituto di Ricerca sul Rischio Sismico: Milano, Italy, 1994; Volume 2, pp. 11–26.
76. Papathanassiou, G.; Valkaniotis, S.; Ganas, A. Evaluation of the macroseismic intensities triggered by the January/February 2014 Cephalonia, (Greece) earthquakes based on ESI-07 scale and their comparison to 1867 historical event. *Quat. Int.* **2017**, *451*, 234–247. [CrossRef]
77. Papadimitriou, P.; Voulgaris, N.; Kouskouna, V.; Kassaras, I.; Kaviris, G.; Pavlou, K.; Karakonstantis, A.; Bozionelos, G.; Kapetanidis, V. The Kefallinia Island earthquake sequence January–February 2014. In Proceedings of the Second European Conference on Earthquake Engineering and Seismology (2ECEES), Istanbul, Turkey, 24–29 August 2014.
78. Kapetanidis, V. Spatiotemporal Patterns of Microseismicity for the Identification of Active Fault Structures using Seismic Waveform Cross-Correlation and Double-Difference Relocation. Ph.D. Thesis, Department of Geophysics-Geothermics, Faculty of Geology and Geoenvironment, National and Kapodistrian University of Athens, Athens, Greece, 2017.
79. Taylor, D.W. *Fundamentals of Soil Mechanics*; John Wiley & Son: New York, NY, USA, 1948; p. 700.
80. Wu, W.; Sidle, R.C. A distributed slope stability model for steep forested basins. *Water Resour. Res.* **1995**, *31*, 2097–2110. [CrossRef]
81. Bliona, M. Development of a Database of Landslides in Greece. Master's Thesis, Engineering Geology Lab, Faculty of Applied Geology and Geophysics, Department of Geology, University of Patras, Patras, Greece, 2008.
82. Fabbri, A.G.; Chung, C.-J.F.; Cendrero, A.; Remondo, J. Is prediction of future landslides possible with a GIS? *Nat. Hazards* **2003**, *30*, 487–503. [CrossRef]
83. Budimir, M.E.A.; Atkinson, P.M.; Lewis, H.G. A systematic review of landslide probability mapping using logistic regression. *Landslides* **2015**, *12*, 419–436. [CrossRef]
84. Qiu, C.; Su, L.; Zou, Q.; Geng, X. A hybrid machine-learning model to map glacier-related debris flow susceptibility along Gyirong Zangbo watershed under the changing climate. *Sci. Total Environ.* **2022**, *818*, 151752. [CrossRef] [PubMed]
85. Marchesini, I.; Ardizzone, F.; Alvioli, M.; Rossi, M.; Guzzetti, F. Non-susceptible landslide areas in Italy and in the Mediterranean region. *Nat. Hazards Earth Syst. Sci.* **2014**, *14*, 2215–2231. [CrossRef]
86. Guzzetti, F.; Peruccacci, S.; Rossi, M.; Stark, C.P. Rainfall thresholds for the initiation of landslides in central and southern Europe. *Meteorol. Atmos. Phys.* **2007**, *98*, 239–267. [CrossRef]
87. Guzzetti, F.; Peruccacci, S.; Rossi, M.; Stark, C.P. The rainfall intensity–duration control of shallow landslides and debris flows: An update. *Landslides* **2008**, *5*, 3–17. [CrossRef]
88. Hellenic National Meteorological Service. *The Climate Atlas of Greece (1971–2000): Precipitation*; Hellenic National Meteorological Service (HNMS): Athens, Greece, 2021; Available online: http://climatlas.hnms.gr/ (accessed on 5 January 2022).
89. Hellenic Military Geographical Service. *Topographic Map "Atheras" Sheet, 1:50,000 Scale*; HMGS: Athens, Greece, 1973.
90. Hellenic Military Geographical Service. *Topographic Map "Argostolion" Sheet, 1:50,000 Scale*; HMGS: Athens, Greece, 1973.
91. Hellenic Military Geographical Service. *Topographic Map "Ag. Eirini" Sheet, 1:50,000 Scale*; HMGS: Athens, Greece, 1973.
92. Hellenic Military Geographical Service. *Topographic Map "Ithaki" Sheet, 1:50,000 Scale*; HMGS: Athens, Greece, 1976.
93. OpenStreetMap. Available online: https://www.openstreetmap.org/ (accessed on 5 January 2022).
94. Esch, T.; Taubenböck, H.; Roth, A.; Heldens, W.; Felbier, A.; Schmidt, M.; Mueller, A.; Thiel, M.; Dech, S. TanDEM-X mission-new perspectives for the inventory and monitoring of global settlement patterns. *J. Appl. Remote Sens.* **2012**, *6*, 061702. [CrossRef]
95. Krieger, G.; Moreira, A.; Fiedler, H.; Hajnsek, I.; Werner, M.; Younis, M.; Zink, M. TanDEM-X: A Satellite Formation for High-Resolution SAR Interferometry. *IEEE Trans. Geosci. Remote Sens.* **2007**, *45*, 3317–3341. [CrossRef]

96. Krieger, G.; Zink, M.; Bachmann, M.; Bräutigam, B.; Schulze, D.; Martone, M.; Rizzoli, P.; Steinbrecher, U.; Antony, W.J.; De Zan, F.; et al. TanDEM-X: A radar interferometer with two formation-flying satellites. *Acta Astronaut.* **2013**, *89*, 83–98. [CrossRef]
97. Wessel, B.; Huber, M.; Wohlfart, C.; Marschalk, U.; Kosmann, D.; Roth, A. Accuracy assessment of the global TanDEM-X Digital Elevation Model with GPS data. *ISPRS J. Photogramm. Remote Sens.* **2018**, *139*, 171–182. [CrossRef]
98. Huber, M.; Osterkamp, N.; Marschalk, U.; Tubbesing, R.; Wendleder, A.; Wessel, B.; Roth, A. Shaping the Global High-Resolution TanDEM-X Digital Elevation Model. *IEEE J. Sel. Top. Appl. Earth Obs. Remote Sens.* **2021**, *14*, 7198–7212. [CrossRef]
99. Pertiwi, A.P.; Roth, A.; Schaffhauser, T.; Bhola, P.K.; Reuß, F.; Stettner, S.; Kuenzer, C.; Disse, M. Monitoring the Spring Flood in Lena Delta with Hydrodynamic Modeling Based on SAR Satellite Products. *Remote Sens.* **2021**, *13*, 4695. [CrossRef]
100. Clerici, A.; Perego, S.; Tellini, C.; Vescovi, P.A. procedure for landslide susceptibility zonation by the conditional analysis method. *Geomorphology* **2002**, *48*, 349–364. [CrossRef]
101. Ruff, M.; Czurda, K. Landslide susceptibility analysis with a heuristic approach in the Eastern Alps (Vorarlberg, Austria). *Geomorphology* **2008**, *94*, 314–324. [CrossRef]
102. Gökceoglu, C.; Aksoy, H. Landslide susceptibility mapping of the slopes in the residual soils of the Mengen region (Turkey) by deterministic stability analyses and image processing techniques. *Eng. Geol.* **1996**, *44*, 147–161. [CrossRef]
103. Saha, A.K.; Gupta, R.P.; Sarkar, I.; Arora, M.K.; Csaplovics, E. An approach for GIS based statistical landslide susceptibility zonation—with a case study in the Himalayas. *Landslides* **2005**, *2*, 61–69. [CrossRef]
104. Ladas, I.; Fountoulis, I.; Mariolakos, I. Using GIS and Multicriteria decision analysis in landslide susceptibility mapping—A case study in Messinia prefecture area (SW Peloponnesus, Greece). *Bull. Geol. Soc. Greece* **2007**, *40*, 1973–1985. [CrossRef]
105. Ladas, I.; Fountoulis, I.; Mariolakos, I. Large scale landslide susceptibility mapping using GIS-based weighted linear combination and multicriteria decision analysis—A case study in Northern Messinia (SW Peloponnesus, Greece). In Proceedings of the 8th Congress of the Hellenic Geographical Society, Athens, Greece, 4–7 October 2007; Hellenic Geographical Society: Athens, Greece, 2007; pp. 93–102.
106. He, Y.; Beighley, R.E. GIS-based regional landslide susceptibility mapping: A case study in southern California. *Earth Surf. Proc. Land.* **2008**, *33*, 380–393. [CrossRef]
107. Neuhäuser, B.; Terhorst, B. Landslide susceptibility assessment using "weights-ofevidence" applied to a study area at the Jurassic escarpment (SW-Germany). *Geomorphology* **2007**, *86*, 12–24. [CrossRef]
108. Binaghi, E.; Luzi, L.; Madella, P.; Pergalani, F.; Rampini, A. Slope instability zonation: A comparison between certainty factor and fuzzy Dempster–Shafer approaches. *Nat. Hazards* **1998**, *17*, 77–97. [CrossRef]
109. Süzen, M.L.; Doyuran, V. Data driven bivariate landslide susceptibility assessment using geographical information systems: A method and application to Asarsuyu catchment, Turkey. *Eng. Geol.* **2004**, *71*, 303–321. [CrossRef]
110. Sakkas, G.; Misailidis, I.; Sakellariou, N.; Kouskouna, V.; Kaviris, G. Modeling landslide susceptibility in Greece: A weighted linear combination approach using analytic hierarchical process, validated with spatial and statistical analysis. *Nat. Hazards* **2016**, *84*, 1873–1904. [CrossRef]
111. Karpouza, M.; Chousianitis, K.; Bathrellos, G.D.; Skilodimou, H.D.; Kaviris, G.; Antonarakou, A. Hazard zonation mapping of earthquake-induced secondary effects using spatial multi-criteria analysis. *Nat. Hazards* **2021**, *109*, 637–669. [CrossRef]
112. Li, F.; Torgoev, I.; Zaredinov, D.; Li, M.; Talipov, B.; Belousova, A.; Kunze, C.; Schneider, P. Influence of Earthquakes on Landslide Susceptibility in a Seismic Prone Catchment in Central Asia. *Appl. Sci.* **2021**, *11*, 3768. [CrossRef]
113. Newmark, N.M. Effects of earthquake on dams and embankments. *Geotechnique* **1965**, *15*, 139–160. [CrossRef]
114. Ambraseys, N.N.; Menu, J.M. Earthquake-induced ground displacements. *Earthq. Eng. Struct. Dyn.* **1988**, *16*, 985–1006. [CrossRef]
115. Ambraseys, N.; Srbulov, M. Attenuation of earthquake-induced ground displacements. *Earthq. Eng. Struct. Dyn.* **1994**, *23*, 467–487. [CrossRef]
116. Cornell, C.A. Engineering Seismic Risk Analysis. *Bull. Seismol. Soc. Am.* **1968**, *58*, 1583–1606. [CrossRef]
117. McGuire, R.K. *FORTRAN Computer Program for Seismic Risk Analysis. Open-File Report 76-67*; U.S. Department of the Interior, U.S. Geological Survey: Reston, VA, USA, 1976. [CrossRef]
118. Giardini, D.; Wössner, J.; Danciu, L. Mapping Europe's Seismic Hazard. *EOS* **2014**, *95*, 261–262. [CrossRef]
119. Woessner, J.; Danciu, L.; Giardini, D.; Crowley, H.; Cotton, F.; Grünthal, G.; Valensise, G.; Arvidsson, R.; Basili, R.; Demircioglu, M.B.; et al. The 2013 European Seismic Hazard Model: Key components and results. *Bull. Earthq. Eng.* **2015**, *13*, 3553–3596. [CrossRef]
120. Cornell, C.A.; Vanmarke, E.H. The major influences on seismic risk. In Proceedings of the 3rd World Conference on Earthquake Engineering, Santiago, Chile, 13–18 January 1969.
121. Sornette, D.; Sornette, A. General theory of the modified Gutenberg-Richter law for large seismic moments. *Bull. Seismol. Soc. Am.* **1999**, *89*, 1121–1130. [CrossRef]
122. Danciu, L.; Tselentis, G.-A. Engineering Ground-Motion Parameters Attenuation Relationships for Greece. *Bull. Seismol. Soc. Am.* **2007**, *97*, 162–183. [CrossRef]
123. Tselentis, G.-A.; Danciu, L. Probabilistic seismic hazard assessment in Greece-Part 1: Engineering ground motion parameters. *Nat. Hazards Earth Syst. Sci.* **2010**, *10*, 25–39. [CrossRef]
124. Pavlou, K.; Kaviris, G.; Kouskouna, V.; Sakkas, G.; Zymvragakis, A.; Sakkas, V.; Drakatos, G. Minor seismic hazard changes in the broader area of Pournari artificial lake after the first filling (W. Greece). *Results Geophys. Sci.* **2021**, *7*, 100025. [CrossRef]

125. Kaviris, G.; Zymvragakis, A.; Bonatis, P.; Sakkas, G.; Kouskouna, V.; Voulgaris, N. Probabilistic seismic hazard assessment for the broader Messinia (SW Greece) region. *Pure Appl. Geophys.* **2022**, *179*, 551–567. [CrossRef]
126. Ordaz, M.; Salgado-Gálvez, M.A.; Giraldo, S. R-CRISIS: 35 years of continuous developments and improvements for probabilistic seismic hazard analysis. *Bull. Earthq. Eng.* **2021**, *19*, 2797–2816. [CrossRef]
127. Lee, C.-T.; Huang, C.-C.; Lee, J.-F.; Pan, K.-L.; Lin, M.-L.; Dong, J.-J. Statistical approach to storm event-induced landslides susceptibility. *Nat. Hazards Earth Syst. Sci.* **2008**, *8*, 941–960. [CrossRef]
128. Mondini, A.C.; Guzzetti, F.; Reichenbach, P.; Rossi, M.; Cardinali, M.; Ardizzone, F. Semi-automatic recognition and mapping of rainfall induced shallow landslides using satellite optical images. *Remote Sens. Environ.* **2011**, *115*, 1743–1757. [CrossRef]
129. Pontius Jr, R.G.; Schneider, L. Land-cover change model validation by a ROC method for the Ipswich watershed, Massachusetts, USA. *Agric. Ecosyst. Environ.* **2001**, *85*, 239–248. [CrossRef]
130. Glade, T. Landslide occurrence as a response to land use change: A review of evidence from New Zealand. *Catena* **2003**, *51*, 297–314. [CrossRef]
131. Pontius, R.G., Jr.; Hao, C. *GEOMOD Modeling*; Clark University: Worcester, MA, USA, 2006; Available online: https://www2.clarku.edu/~{}rpontius/pontius_chen_2006_idrisi.pdf (accessed on 10 January 2022).
132. Reichenbach, P.; Busca, C.; Mondini, A.C.; Rossi, M. The influence of land use change on landslide susceptibility zonation: The Briga catchment test site (Messina, Italy). *Environ. Manag.* **2014**, *54*, 1372–1384. [CrossRef] [PubMed]
133. United States Geological Survey. NDVI, the Foundation for Remote Sensing Phenology. Available online: https://www.usgs.gov/special-topics/remote-sensing-phenology (accessed on 15 January 2021).
134. Tucker, C.J. Red and photographic infrared linear combinations for monitoring vegetation. *Remote Sens. Environ.* **1979**, *8*, 127–150. [CrossRef]
135. Irigaray, C.; Lamas, F.; El Hamdouni, R.; Fernández, T.; Chacón, J. The importance of the precipitation and the susceptibility of the slopes for the triggering of landslides along the roads. *Nat. Hazards* **2000**, *21*, 65–81. [CrossRef]
136. Saaty, T.L. *The Analytic Hierarchy Process*; McGraw-Hill: New York, NY, USA, 1980.
137. SpiceLogic Inc. Analytic Hierarchy Process Software (Trial Version 3.4.5). Available online: https://www.spicelogic.com/Products/ahp-software-30/ (accessed on 15 January 2022).
138. ESRI. ArcGIS 10.7. Available online: https://desktop.arcgis.com/en/arcmap/10.7/get-started/setup/arcgis-desktop-quick-start-guide.htm (accessed on 15 January 2022).
139. Jenks, G.F. Generalization in Statistical Mapping. *Ann. Assoc. Am. Geogr.* **1963**, *53*, 15–26. [CrossRef]
140. Semlali, I.; Ouadif, L.; Bahi, L. Landslide susceptibility mapping using the analytical hierarchy process and GIS. *Curr. Sci.* **2019**, *116*, 773–779. [CrossRef]
141. Boroumandi, M.; Khamehchiyan, M.; Nikoudel, M.R. Using of Analytic Hierarchy Process for Landslide Hazard Zonation in Zanjan Province, Iran. In *Engineering Geology for Society and Territory—Volume 2*; Lollino, G., Giordan, D., Crosta, G.B., Corominas, J., Azzam, R., Wasowski, J., Sciarra, N., Eds.; Springer International Publishing: Cham, Switzerland, 2015. [CrossRef]
142. Cignetti, M.; Godone, D.; Giordan, D. Shallow landslide susceptibility, Rupinaro catchment, Liguria (northwestern Italy). *J. Maps* **2019**, *15*, 333–345. [CrossRef]
143. Gárcía-Rodriguez, M.J.; Malpica, J.A.; Benito, B.; Díaz, M. Susceptibility assessment of earthquake triggered landslides in El Salvador using logistic regression. *Geomorphology* **2008**, *95*, 172–191. [CrossRef]
144. Ozdemir, A.; Altural, T. A comparative study of frequency ratio, weights of evidence and logistic regression methods for landslide susceptibility mapping: Sultan Mountains, SW Turkey. *J. Asian Earth Sci.* **2013**, *64*, 180–197. [CrossRef]
145. Wu, X.; Niu, R.; Ren, F.; Peng, L. Landslide susceptibility mapping using rough sets and back-propagation neural networks in the Three Gorges, China. *Environ. Earth Sci.* **2013**, *70*, 1307–1318. [CrossRef]
146. Chung, C.J.F.; Fabbri, A.G. Validation of spatial prediction models for landslide hazard mapping. *Nat. Hazards* **2003**, *30*, 451–472. [CrossRef]
147. Fawcett, T. An introduction to ROC analysis. *Pattern Recognit. Lett.* **2006**, *27*, 861–874. [CrossRef]
148. Nandi, A.; Shakoor, A. A GIS-based landslide susceptibility evaluation using bivariate and multivariate statistical analyses. *Eng. Geol.* **2010**, *110*, 11–20. [CrossRef]
149. Lekkas, E.; Nastos, P.; Cartalis, C.; Diakakis, M.; Gogou, M.; Mavroulis, S.; Spyrou, N.-I.; Kotsi, E.; Vassilakis, E.; Katsetsiadou, K.-N.; et al. Impact of Medicane "IANOS" (September 2020). *Newsl. Environ. Disaster Cris. Manag. Strateg.* **2020**, *20*, 1–140. [CrossRef]
150. Vassilakis, E.; Konsolaki, A.; Petrakis, S.; Kotsi, E.; Fillis, C.; Lozios, S.; Lekkas, E. Quantification of Mass Movements with Structure-from-Motion Techniques. The Case of Myrtos Beach in Cephalonia, after Ianos Medicane (September 2020). *Bull. Geol. Soc. Greece Spec. Publ.* **2021**, *8*, 16–20.

applied sciences

MDPI

Article

Reconstructing the 26 June 1917 Samoa Tsunami Disaster

Laura Sischka [1], Cyprien Bosserelle [2], Shaun Williams [2,*], Josephina Chan Ting [3], Ryan Paulik [2], Malcolm Whitworth [1], Lameko Talia [4] and Paul Viskovic [5]

1 School of the Environment, Geography and Geosciences, University of Portsmouth, Portsmouth PO1 3QL, UK; laura.sischka@myport.ac.uk (L.S.); malcolm.whitworth@port.ac.uk (M.W.)
2 NIWA Taihoro Nukurangi, Ōtautahi Christchurch 8440, New Zealand; cyprien.bosserelle@niwa.co.nz (C.B.); ryan.paulik@niwa.co.nz (R.P.)
3 Disaster Management Office & National Emergency Operations Centre, Ministry of Natural Resources and Environment, Private Bag, Apia WS1338, Samoa; josephina.chanting@mnre.gov.ws
4 Geosciences Section, Ministry of Natural Resources and Environment, Private Bag, Apia WS1338, Samoa; lameko.talia@mnre.gov.ws
5 GNS Science, Avalon, Lower Hutt 5011, New Zealand; p.viskovic@gns.cri.nz
* Correspondence: shaun.williams@niwa.co.nz

Abstract: The 1917 Samoa tsunamigenic earthquake is the largest historical event to impact this region. Over a century later, little is known about the tsunami magnitude and its implications for modern society. This study reconstructs the 1917 tsunami to understand its hazard characteristics in the Samoan region and assesses the risk implications of tsunamis sourced from different locations along the subduction zone bend of the Northern Tonga Trench (NTT). We model the event from its origin to produce outputs of tsunami inundation extent and depth at spatially flexible grid resolution, which are validated using available runup observations and Apia harbour tide gauge records. We then combine the inundation model with digital distributions of buildings to produce exposure metrics for evaluating the likely impacts on present-day coastal assets and populations if a similar tsunami were to occur. Results exhibit recorded and modelled wave arrival time discrepancies in Apia harbour of between 30–40 min, with runup underestimated in southeast Upolu Island compared with the rest of the country. These differences could reflect complexities in the tsunami source mechanism that are not represented in our modelling and require further investigation. Nevertheless, our findings suggest that if a characteristic 1917-type event were to occur again, approximately 71% of exposed people would reside in Savai'i. Overall, this study provides the first detailed inundation model of the 1917 tsunami that supports an appreciation of the regional risk to local tsunamis sourced at the subduction zone bend of the NTT in Samoa.

Keywords: tsunami inundation; historical records; hazard risk exposure; Pacific; BG-Flood; RiskScape

Citation: Sischka, L.; Bosserelle, C.; Williams, S.; Ting, J.C.; Paulik, R.; Whitworth, M.; Talia, L.; Viskovic, P. Reconstructing the 26 June 1917 Samoa Tsunami Disaster. *Appl. Sci.* **2022**, *12*, 3389. https://doi.org/10.3390/app12073389

Academic Editors: Spyridon Mavroulis and Efthymios Lekkas

Received: 28 February 2022
Accepted: 24 March 2022
Published: 26 March 2022

1. Introduction

More than 700 million people live in island states, most of them developing countries, and low-lying areas at the coast are under constant risk from tsunamis, storm surges, or severe fluctuations of sea levels. In the Pacific, small island developing states (SIDS) represent a collection of remote island communities with developing economies that are often at elevated risk from climate change, sea level rise, coastal erosion, and both natural and anthropogenic hazards [1,2]. It is often the case that due to their geography and the relatively small size of these islands, a significant proportion of the population, infrastructure, and commercial and industrial activity are concentrated in low lying areas, typically in a strip close to the coasts, which render them at considerable risk from coastal inundation from tsunamis [3].

The central south Pacific region is frequently affected by tsunamis generated from earthquakes centred on the Tonga Kermadec Trench, including 39 events between 1837 and 2009 that included the 1917 and 2009 tsunamis that affected the Samoan islands [4]. These two nearly identical events suggest that tsunamis in this region are relatively common, and, as a consequence, the SIDS in this area are exposed not only to global tsunamigenic events, but also to frequent locally derived tsunamis triggered by earthquakes, volcanic eruptions, and submarine landslides [5]. Indeed, the recent submarine volcanic eruption of Hunga-Tonga Hunga-Ha'apai (HTHH) in Tonga on 14 January 2022 [6,7], and the resulting tsunami along with potential inferred predecessors (e.g., [8]), illustrate that there is a high degree of residual risk to these islands from local tsunami generating events.

Local tsunamis with less than 30 min impact time are extremely hazardous to the island communities, due to the limited warnings and response times between the triggering event (earthquake or eruption, for example), and the tsunami wave making landfall. Consequently, when trying to understand current tsunami risks and the nature and extent of exposure to these islands, it is common to use data or records from historical events to constrain the likely intensity (inundation extent and flow depths, for example), and then use this within a scenario-based context to understand the present-day exposure if such an event was to occur today.

The focus of this study is to reconstruct the 1917 tsunami that struck the islands of Samoa, which was the second most deadly tsunamigenic event on record to affect this region after the fatal 2009 tsunami that devastated the southeast coast of Upolu [9,10] (Figure 1). Although the 2009 event provides a benchmark to help plan for and mitigate the impacts of local tsunamis sourced at the Northern Tonga Trench (NTT) in Samoa, areas that experienced little or no impact give rise to a public perception that these areas are at minimal threat from tsunamis. This is particularly the case for coastal areas in Savai'i island to the west of Upolu. However, very little is known about the impacts of the 1917 tsunami predecessor, despite the earthquake source magnitude being the largest ever recorded in this region [11].

Figure 1. Location of the 1917 earthquake epicentre (star) and 2009 earthquakes (circles) relative to Samoa.

The 1917 earthquake epicentre was located along the transform segment of the NTT margin about 100 km south of west Savai'i, and approximately 150 km west of the 2009 earthquake. Preliminary modelling by Okal et al. (2011) [12] suggests a focusing of tsunami flux northwards toward the west and southern areas of Savai'i. This implies that these areas would have been severely impacted compared with what they had experienced in the 2009 event. Paradoxically, the evidence presented in [12] does not seem commensurate with the lack of documented impacts for the event, which led them to hypothesize that the scale of impacts was masked by the impacts of the 1918 influenza pandemic, which occurred approximately 1 yr later. The 1918 pandemic is Samoa's deadliest disaster in history, which saw the loss of between 20 and 25% of the country's population at the time [13,14], most of whom were adults or knowledge holders. This is a gap in our current understanding of the scale and magnitude of the 1917 tsunami disaster within the context of regional hazard risk resilience planning. To help elucidate this enigma, we model the tsunami from source to inundation and use this scenario to evaluate the exposure characteristics of a similar event-type on present-day distributions of buildings and people.

We provide an overview of the geographical and historical event context in Section 2, and describe the data and analytical methods in Section 3. Findings of the analysis are presented in Section 4, which includes a comparison with observations of the 2009 event. The uncertainties and implications of our results in understanding the regional hazard characteristics that tsunamigenic-earthquakes occurring on different segments of the NTT have on the distribution of exposed areas are discussed in Section 5, with conclusions and suggestions for future research provided in Section 6.

2. Geographical and Historical Context

Samoa consists of two main islands, Upolu and Savai'i, with several smaller inhabited and uninhabited islands between them (e.g., Manono and Apolima), as well as east and south of Upolu (e.g., Fanuatapu, Namu'a, Nu'utele, Nu'ulua, and Nu'usafe'e) (Figure 1). Comprising part of a larger archipelago encompassing the geologically younger islands of American Samoa to the east, the island chain originated from hotspot volcanic activity and is fringed by coral reefs [15–17]. The geology of Samoa largely consists of mafic material (e.g., basalt and gabbro), due to its oceanic intraplate volcanic hotspot origins [18]. Savai'i is the bigger island with an area of 1820 km^2, whereas Upolu has an area of 1114 km^2 and accommodates over 67% of the total population of approximately 200,000 people [19]. Apia, the capital of Samoa, is located in the central north of Upolu.

In 1902, a temporary geophysical observatory was established in Apia on the then-German administered island of Upolu, whereby it was initially set up to obtain baseline earth observations to compare with the British and German south polar expeditions of 1902–1903. Meteorological instruments and seismographs were installed in 1902 and magnetic instruments in 1905. These enabled studies in geomagnetism, seismology, meteorology, tidal variations, and atmospheric electricity and were so productive that in 1908 the observatory was established on a permanent basis.

In August 1914, troops of the New Zealand Expeditionary Force seized control of German-controlled Samoa and the observatory. Its operations were much curtailed during the World War 1 (WW1) years, but the German Director (G. Angenheister) continued observatory operations until it was formerly taken over by the New Zealand government in 1921 [20–22]. New Zealand administration of the observatory continued until the handover in 1963, shortly after Samoa achieved political independence and control of the observatory.

Between 1917 to 1919, four tidal waves were recorded by the observatory on continuously recording tidal gauges that correlated with four earthquakes recorded by the seismograph installed at the observatory, with observations reported in [23].

Situated approximately 100 km north of the Tonga Trench, Samoa is exposed to a range of local geophysical hazards (e.g., earthquakes, volcanoes, landslides, tsunami). For example, the subaerial volcanic eruption from 1905 to 1911 on northeast Savai'i caused displacement/relocation of affected villagers to neighbouring Upolu [24] and generated several small tsunamis during this period, with the most damaging occurring in 1907 [4]. Indeed, the recent 2009 complex earthquake sequence [25] and consequent tsunami that resulted in severe casualties and livelihood destruction in southeast Upolu reinforces this vulnerability [26,27].

The lesser known predecessor to the 2009 event, the 26 June 1917 UTC (local time in 1917 = UTC-11) earthquake and tsunami that originated in a proximal source region northwest of the 2009 epicentre (Figure 1), is arguably considered the largest earthquake to have occurred in this region in terms of magnitude (i.e., *Mw* 8.3 compared with *Mw* 8.1 for the 2009 earthquake sequence) [11,28]. However, the scale of impacts from the resulting tsunami appear to have paled in comparison with the devastation observed in the 2009 event (e.g., [12,29,30]). Although anecdotal records indicate that the 1917 tsunami inundation had flooded several villages and caused damage to buildings and infrastructure (e.g., Satupaitea in southeast Savai'i and Lotofaga in southeast Upolu) [4,31,32], there are virtually no accounts of any casualties. Available modern interpretations assume at least two people lost their lives based on generic descriptions of damage recorded after the event (e.g., [32]). Here, we use our inundation model for the 1917 tsunami along with present-day patterns of inundation exposure as a proxy to discuss and offer alternative views to help elucidate this enigma.

3. Methods and Data

The methods used in this study are described in two sections: (1) tsunami modelling, which describes the process from the initial earthquake to the benchmarking process and inundation on land; and (2) tsunami exposure and damage analysis used in quantifying the hazard exposure on present-day buildings and population.

3.1. *Tsunami Modelling*

3.1.1. Model Setup and Configuration

This tsunami modelling analysis adapts a similar approach used by Bosserelle et al. (2020) [27] to model the inundation of the 2009 event. For generation of the tsunami from the initial earthquake through to propagation and inundation, the BG-Flood software was used. BG Flood (Block-adaptive on Graphics processing unit Flood model) is suited for the simulation of flooding and/or inundation caused by rivers, rain, tides, or tsunamis. It is based on the formulation of Basilisk as well as on the memory structure on the GPU of block uniform quadtree of Vacondio et al. (2017) [33]. The block uniform quadtree structure enables various resolutions with the same memory size but with different physical sizes [33].

For tsunami initialization, we used the earthquake parameters for the 1917 event presented by [12] to configure the source model. To start generating the initial earthquake, the following values were needed: fault parameters (strike, dip, rake, slip), the dimensions of the rupture (length and width), the hypocenter of the earthquake (coordinates and depth), and the timing of the rupture. The rupture length was 150 km with a rupture width of 50 km. The earthquake epicentre was located at 15.13° S and 173.28° W on 26 June 1917 at 05:49 UTC (i.e., 25 June 1917 at 18:49 local time) at a depth of 10 km. After applying vertical deformation to the fault, the initial water displacement was calculated as a theoretical visco-elastic fault displacement using the formulation of [34].

Figure 2 displays the water displacement after the 1917 earthquake. For the tsunami modelling, a fault length of 150 km was used to match the runup observations. As the model uses an adaptive grid (more than one resolution), three output NetCDF files were created with resolutions of 10 m, 20 m, and 40 m.

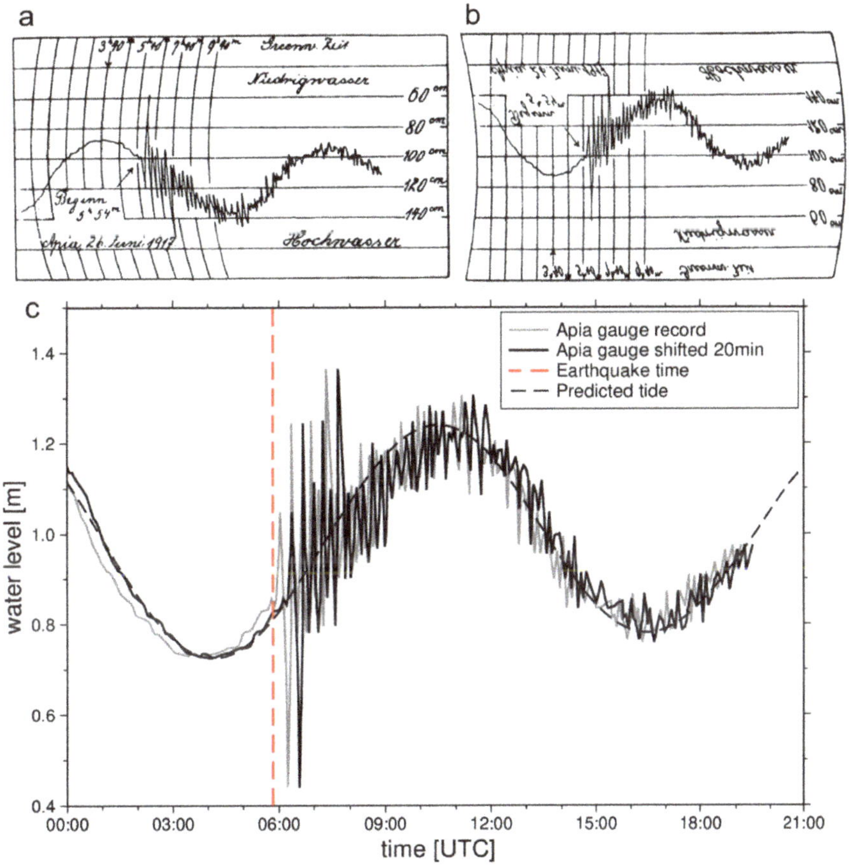

Figure 2. Tide gauge records for the 1917 tsunami in Apia harbour. (**a**) Original maregram taken from Angenheister (1920) [23]; (**b**) inverted and reprojected maregram for digitization; (**c**) digitized tide record (grey solid line) and predicted tide (black dashed line). The maregram shows the fluctuations of the sea in Apia harbour arriving only a few minutes after the earthquake. The earthquake occurred at 05:49 UTC (dashed red line), with the first noticeable sea level peak at 06:03 UTC. The maregram record prior to the earthquake better matched the predicted tide when shifting the record by 20 min (black solid line), which also produces more consistent arrival time for the tsunami.

3.1.2. Tide Gauge and Runup Observations

Available tide gauge records from Apia harbour as well as runup observations from different parts of Upolu and Savai'i were used to validate the inundation modelling and subsequent assessment of present-day exposure and impacts in the runup zone. Tide gauge readings for the 1917 event measured in Apia harbour were digitized using the analog maregraph provided in [23]. The maregram in Figure 2 shows the fluctuations of the tsunami waves within the harbour. The digitized maregram generally matches the predicted tide at Apia (predicted by analyzing tide constituent from recent tide record). However, the tide time reference given in [23] cannot be reconciled with an expected 30–40 min travel time for the tsunami to reach Apia. Therefore, either the tide time reference or earthquake time/location is inaccurate. Moving the tide time reference given in [23] to 20 min later improves the correlation between predicted and measured tide and resolves the arrival time inconsistency. Although there is no clear evidence that the tide time reference from [23] is incorrect, it appears to be the simplest explanation of the inconsistency, and it still provides the first indication of the likely tsunami arrival time in Apia.

Runup observations derived from historical records of eyewitness accounts documented in [4,30,31] provided benchmarks to infer the extent of wave runup onto land. These observations were digitized to help validate the tsunami runup modelling.

3.2. Tsunami Exposure and Damage Analysis

3.2.1. Building and Population Exposure Data

Buildings on Savaii and Upolu that were located within the maximum tsunami inundation extent were remotely digitized from aerial and Google satellite imagery captured between 2016 and 2020. Buildings were manually digitized in GIS software, using roof outlines to create a vector polygon layer. Physical and non-physical attributes including use category and construction frame were assigned to each building object (Table 1). Samoan building construction frame typologies defined by [9] were attributed to features based on their size (i.e., outline area), roof shape, and use category. In the absence of resources such as Google street view to visually validate use category and construction frame, these attributes were confirmed by local engineers and disaster risk management experts. The outline area (m^2) for confirmed buildings was calculated in GIS software (Esri, Redlands, CA, USA).

Table 1. Summary of attributes represented in the building exposure data.

Primary Attribute	Secondary Attribute	Metric or Value
Construction Frame	Masonry, Steel, Reinforced Concrete, Timber	Text
Usually Resident Population	-	Floating
Outline Area	-	m^2
Use Category	Commercial; Community; Education; Fale; Hotel, Resort; Industrial; Outbuilding; Religious; Residential Dwelling; Tourist Fale	Text

Samoa's usually resident population was obtained from the 2016 national census [13]. Descriptive statistics for 'usually-resident population' at their residence on census day are aggregated and publicly available at national, district, and village levels. Here, we apply usually resident population at village levels (V_{Pop}) to determine a residential building-object population rate (BP_{Rate}) as follows:

$$VB_{area} = \sum_{j=1}^{n_i} BA_i \tag{1}$$

$$VBP_{Rate} = VP_j \ / VB_{area} \tag{2}$$

$$BP_{Rate} = VBP_{Rate} \ / BA_i \tag{3}$$

where BA_i is the outline area (m^2) for a residential building located in village j, with n_i, the number of residential buildings within village j. The variable VB_{area} is the total residential building outline area within village j. Residential building-object VBP_{Rate} is the per m^2 residential building population based on the usually resident population (VP) of village j.

3.2.2. Building Fragility Model

Fragility functions relate tsunami hazard intensity (e.g., flow depth) to the conditional probability of a building reaching or exceeding a given damage state [35]. Here, physical building damage is measured from empirical fragility curves representing Samoan buildings damaged in the 2009 SPT [9]. The fragility curves apply a cumulative lognormal function for 'timber', 'masonry', and 'reinforced concrete' construction frame buildings to determine the conditional probability (0–1) of "light", "minor", "moderate", "severe", and "collapse" damage states (DS) being reached or exceeded for a maximum tsunami inundation depth (Table 2). In the absence of representative fragility curves for some building typologies, 'masonry' curves are applied for 'steel' construction frame buildings, and DS1 and DS2 fragility curves are applied for timber and reinforced concrete building typologies.

Table 2. Building fragility model parameters applied in this study.

Damage State (DS)		Construction Frame						Damage Description
		Timber		Masonry		Reinforced Concrete		
		μ	σ	μ	σ	μ	σ	
DS0	None			-				None
DS1	Light	-		0.29	0.46	-		Non-structural damage only
DS2	Minor	-		0.46	0.4	-		Significant non-structural and minor structural damage
DS3	Moderate	1.15	0.38	1.28	0.35	1.38	0.56	Significant structural and non-structural damage
DS4	Severe	1.26	0.4	1.86	0.41	3.45	0.54	Irreparable structural damage, will require demolition
DS5	Collapse	1.62	0.28	2.49	0.4	7.3	0.94	Complete structural collapse

3.2.3. Tsunami Inundation Exposure and Damage Model

A deterministic model is applied to quantify the present-day building and population exposure as well as damage from tsunami inundation. To this end, we use RiskScape, an open-source software that provides a multi-hazard risk modelling framework for deterministic analysis of tsunami impacts [36]. Here, a deterministic model 'pipeline' is developed to analyze the exposure and damage based on the tsunami model, exposure inventory, and fragility model components described in Sections 3.2.1 and 3.2.2. These components formed the 'input data' for the model pipeline used, which sequences a series of steps and step-functions to sample and analyze deterministic tsunami impacts (Figure 3).

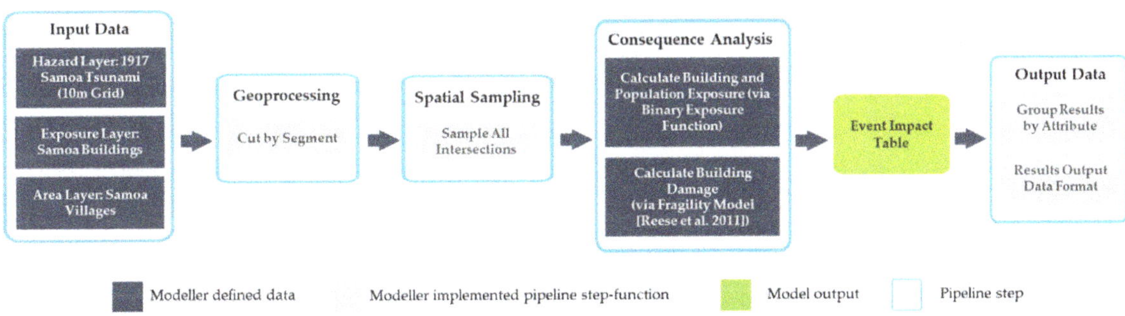

Figure 3. A schematic representation of the RiskScape model pipeline steps and functions applied in this study.

The 1917 event input hazard data layer represented at the adaptive grid resolutions 10 m, 20 m, and 40 m were segmented using a geoprocessing step-function (i.e., 'cut by segment') to extract all tsunami inundation grid cells within the exposure data layer (building outlines). Extracted grid cells were then spatially sampled (i.e., 'all intersections') to determine the maximum inundation flow depth ($MaxD$) at each building location. The consequence analysis applies $MaxD$ to determine: (1) building and population exposure to tsunami inundation; and (2) building damage state. Individual building exposure (Bld_{exp}) to inundation is quantified using a simple binary function:

$$Bld_{exp} = \begin{cases} 1, \ MaxD < 0\ m \\ 0, \ MaxD \geq 0\ m \end{cases} \quad (4)$$

When inundation is present or not at a building location, the corresponding binary value is assigned to the building in the 'event impact table' (EIT). Where inundation is not present (i.e., '0'), no damage (DS0) is assumed. Where inundation is present (i.e., '1'), conditional probability (i.e., 0–1) of damage states DS1 to DS5 based on fragility curves from [9] is calculated in response to the independent variable *MaxD*. Fragility curves scripted in Python using nested statements apply a lognormal function for each curve based on the dependant variables shown in Table 2 for the corresponding building construction frame in Table 1. The conditional probability determined from each fragility curve is then reported in the EIT for each building exposed to tsunami inundation.

The resulting EIT contains tsunami exposure and damage information for model output reporting. In this study, the EIT includes attributes, hazard intensity (i.e., *MaxD*), exposure (i.e., Bld_{exp}), and damage state information for each building object. The 'output data' pipeline step-function 'group results by attribute' is applied here to report descriptive statistics of model results. Building 'count' and population 'sum' exposure to tsunami inundation is enumerated and reported at national and village scales and by hazard intensity (flow depth) bins of 0.5 m. Building damage states are also reported by building count for 0.1 conditional probability bins between 0 and 1. The step-function 'results output file format' outputs this information as spatial file formats (e.g., GIS shapefile, comma-separated value) for national and sub-national spatial analyses of present-day building and population exposure, as well as damage from the 1917 tsunami event.

4. Results

4.1. Tsunami Inundation and Validation

Figure 4 shows the modelled maximum tsunami wave heights for the 1917 tsunami event. Of particular note is that the southwest side of Savai'i is mostly affected where wave heights of 2 m appear to have impacted the coast. Interestingly, the arrival of the simulated wave in Apia harbour suggests that it took approximately 34 min travel time to this location. This is consistent with the observed tide gauge record (Figure 5) when using a 20 min shift in the reference time.

Figures 6 and 7 show which parts of Upolu and Savai'i were most affected by the tsunami, which are generally consistent with available runup observations [4,31] as well as sedimentary evidence presented in [30]. Inundation on Savai'i mostly affects the south western side of the island with much higher flow depths compared with eastern parts of the island including Upolu. Northern Upolu appears unaffected except in areas near Apia. Observed runup points derived from historical records identified in Savai'i (Figure 6) and Upolu (Figure 7) highlight the limited runup observations available for this event.

4.2. Damage to Present-Day Buildings and Population Exposure

If a characteristic 1917-tsunami event scenario were to occur in the near future, we estimate that approximately 2295 buildings would be affected by the inundation (based on present-day building stock). Most exposed buildings on both Savai'i and Upolu are subjected to flow depths >0.0 m to $\leq$0.5 m (Figure 8). As flow depth increases, the number of buildings in each category decreases. However, it is worth noting that 206 buildings on Savai'i are exposed to flow depths >3.0 m, which generally means these buildings are most likely to experience moderate to severe damage. Construction frames made of timber have a higher probability of suffering from severe damage or undergoing complete collapse (Figure 8b).

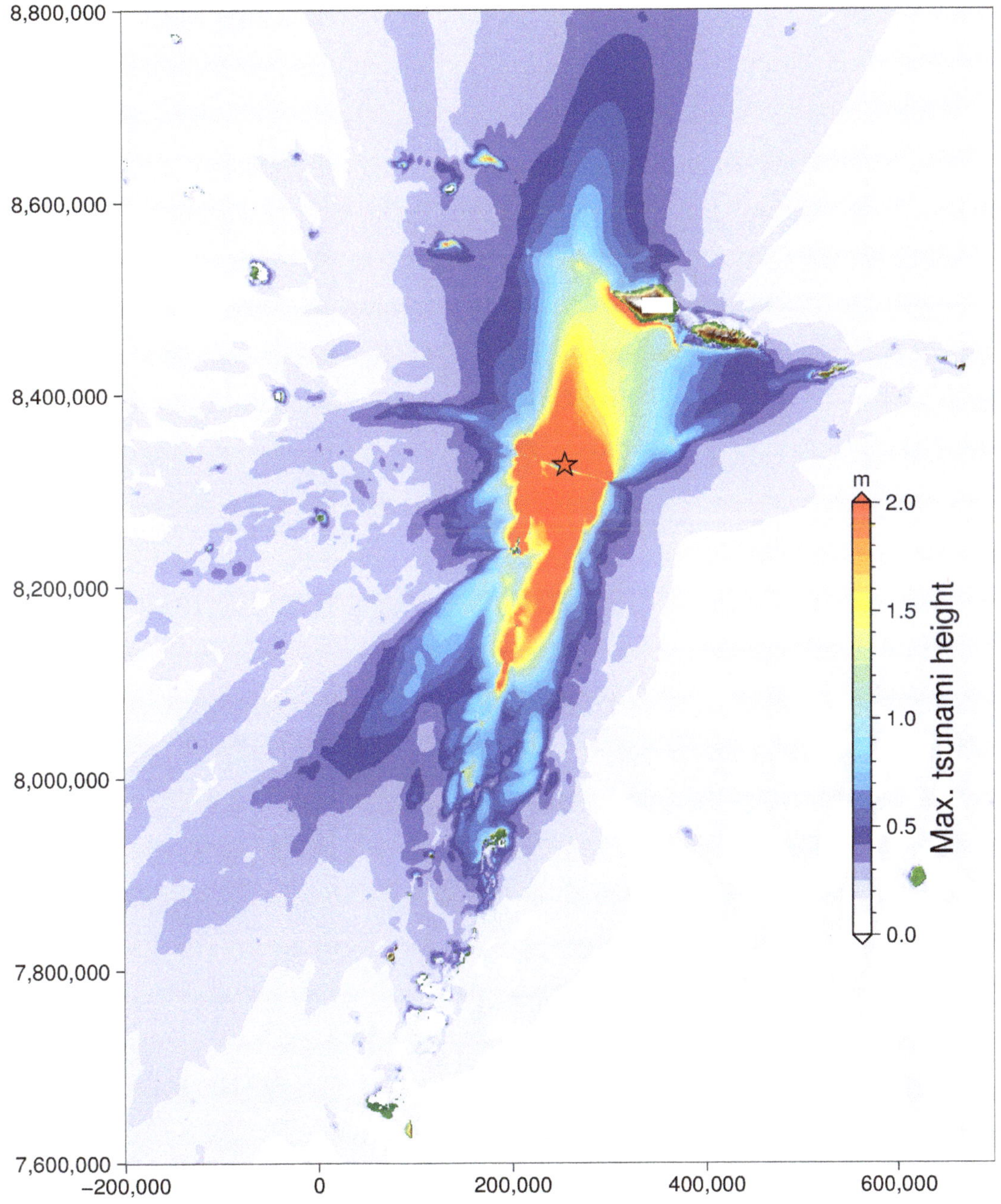

Figure 4. Maximum wave height offshore for the modelled 1917 tsunami.

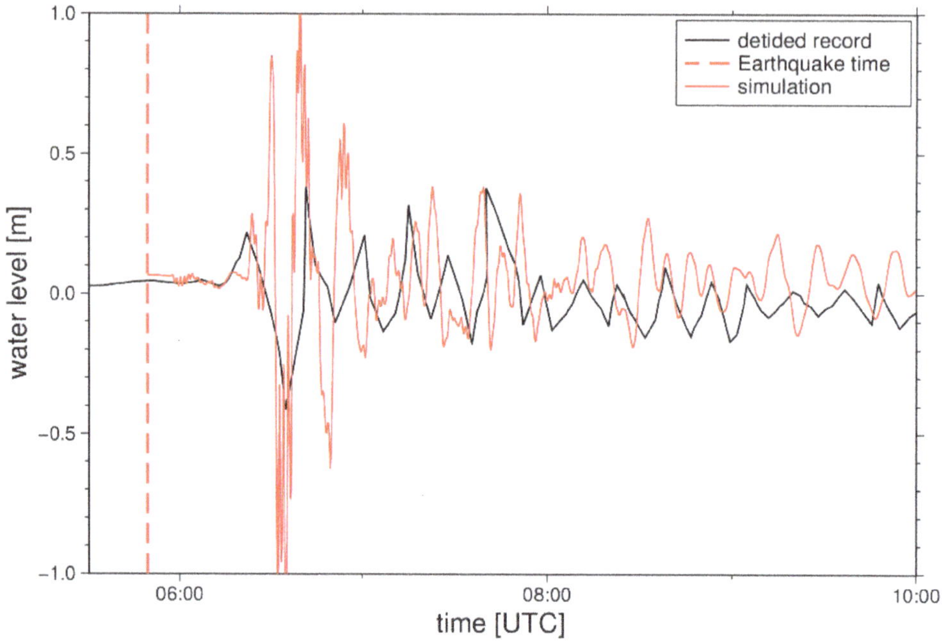

Figure 5. Detided water level recorded in Apia (black line) and simulated (red line). Earthquake time is also given.

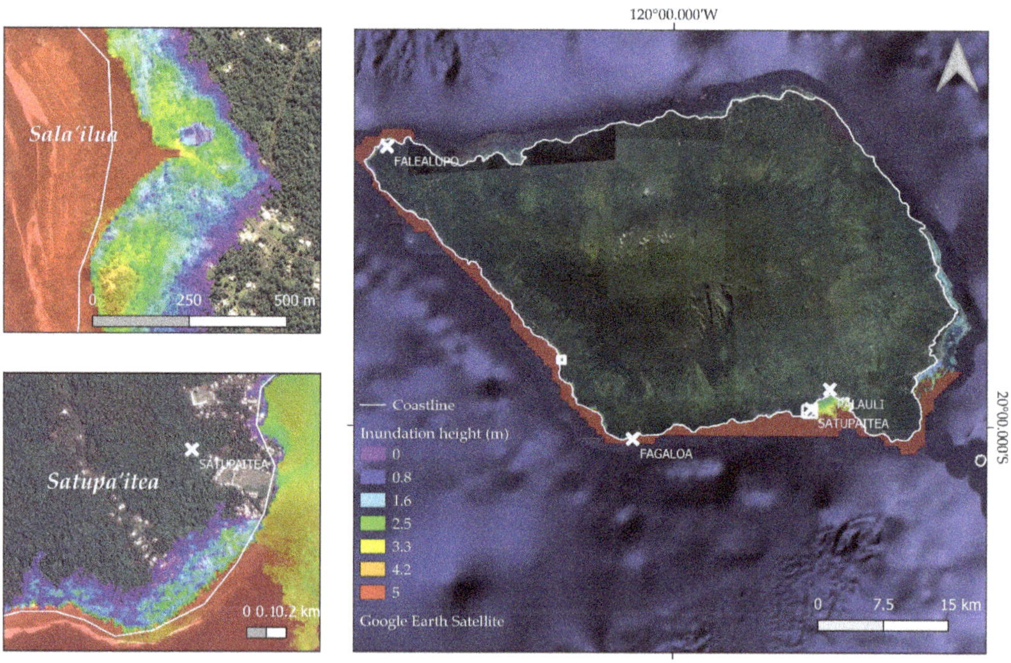

Figure 6. Inundation on Savai'i in close-ups (10 m horizontal resolution). The map shows the 10 m resolution run-up on Savai'i (**a**) and close-up of inundation in Satuiatua (**b**), and in Palauli and Satupa'itea (**c**).

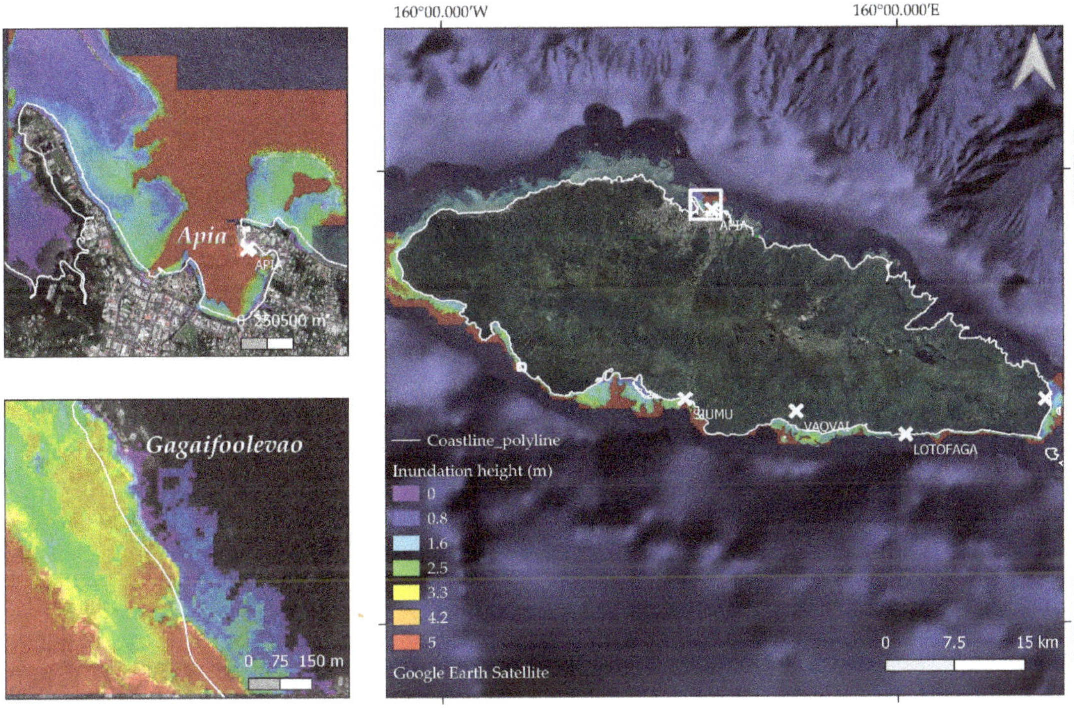

Figure 7. Inundation on Upolu in close-ups (10 m res.). The map shows the 10 m resolution run-up on Upolu (**a**) and in Apia (**b**), with the most severe inundation in Upolu close to Gagaifo'olevao (**c**).

The total number of people living in residential buildings within the modelled 1917 tsunami inundation zone are exhibited in Figure 9. In total, approximately 7919 people across 1074 residential buildings (71% of which are in Savai'i) would be affected by the tsunami, which amounts to approximately 4% of the total population in 2016 [37].

It is worth noting that on Savai'i, 13% of the affected inhabitants live in buildings that are estimated to sustain damage states of DS5 (i.e., complete building collapse). This is particularly the case for the district of Palauli West. On Upolu, more than half of the affected population live in buildings estimated to sustain damage states DS0 (i.e., no damage). The more inundated part of Upolu in the west of the island exposes approximately 77 people who live in buildings likely to sustain damage state DS3 or greater.

4.3. Comparison with the 2009 Tsunami

The 2009 event is the most devastating tsunami to have affected Samoa in recent history. Occurring on 29 September, at 06:48 a.m. local time, two earthquakes only minutes apart shook the ground and caused large waves to travel quickly through the ocean, which caused major destruction and loss of life in less than 30 min after the earthquake rupture [26,38,39]. As the 1917 event can be considered a historical predecessor to the 2009 event in terms of source region [12], a comparison between the two events is made.

Apart from the obvious differences in instrumental monitoring quality in 1917 compared with 2009 [26,40], the main differences between the two events in terms of the distribution of affected coast are highlighted in Figure 10. The main energy beam for the 1917 tsunami was focused towards west and south Savai'i, whereas for the 2009 event, energy was focused towards American Samoa and southeast Upolu, which reflects the proximal epicentral locations of the generating earthquakes, respectively, which were about 150 km apart.

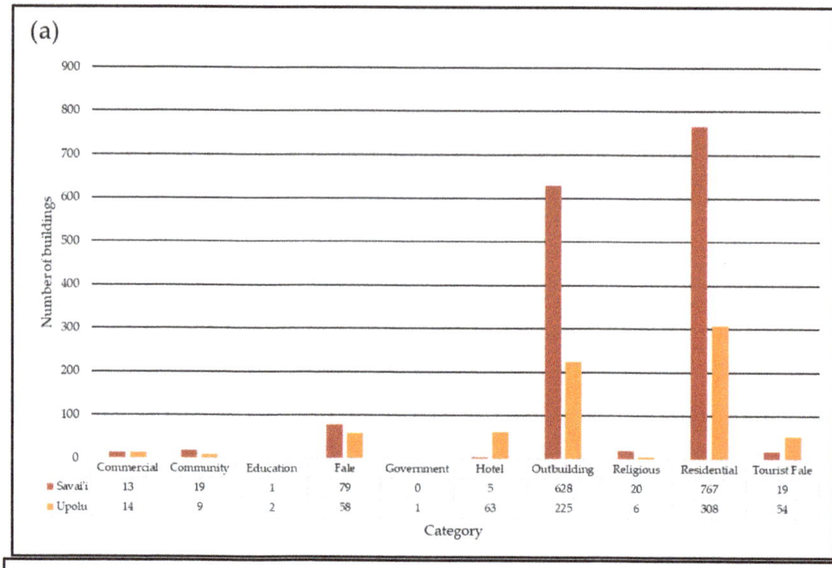

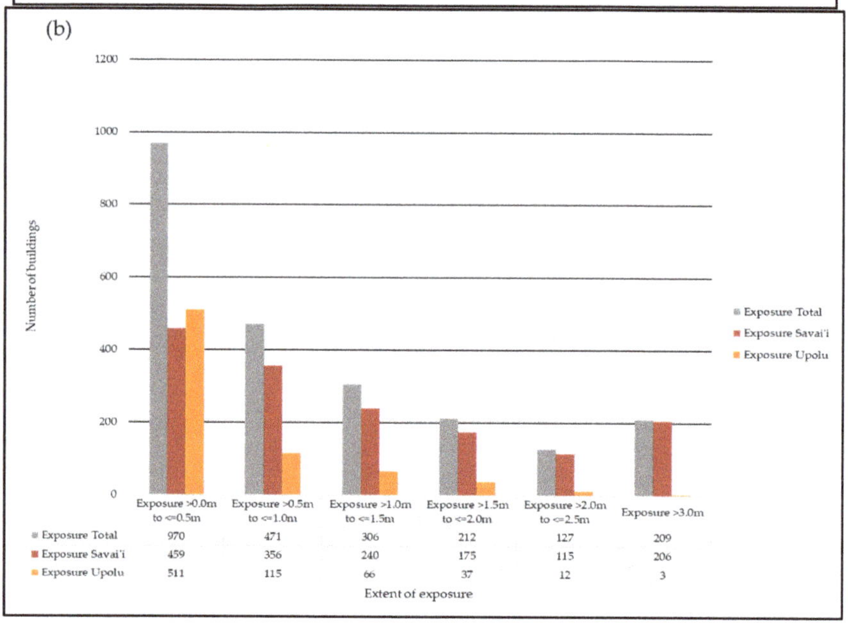

Figure 8. *Cont.*

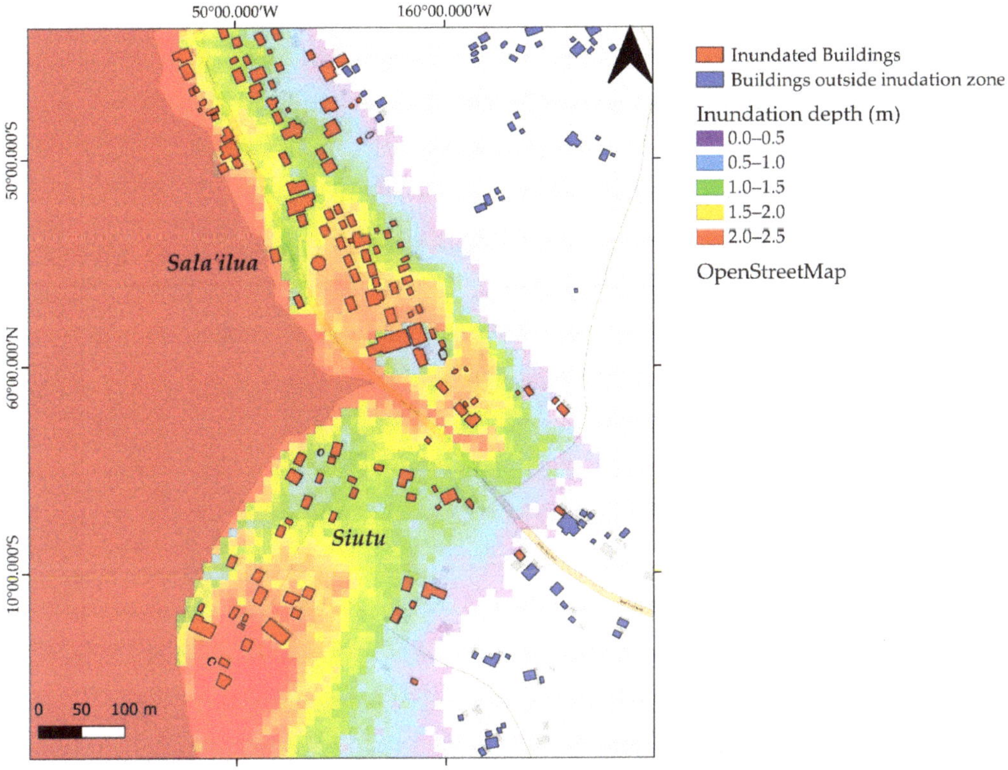

Figure 8. (**a**) Present-day buildings categories on Savai'i and Upolu. (**b**) Exposure of buildings in Savai'i and Upolu to the 1917 modelled scenario event. (**c**) Close up of Sala'ilua village on southwest Savai'i showing the modelled tsunami inundation and exposed buildings.

These hazard characteristics help to explain the observed variations in exposure impact distributions between the modelled 1917 and benchmark 2009 events. A larger proportion of people would be exposed in Savai'i island (71% of the total exposure) compared with Upolu island if a characteristic 1917-type event were to occur. In contrast, coastal areas in southeast and east Upolu were most severely affected in the 2009 event, which reflects the epicentral location of the generating earthquake at the subducting bend of the NTT terminus and main direction of tsunami flux east/northeast towards southeast Upolu. The 1917 epicentre 150 km to the west at the transform segment of the NTT results in the main direction of tsunami flux northward towards Savai'i. This implies that the distribution of the relative exposure of elements at risk to NTT-sourced tsunamis depends on the location of earthquake origin along the subduction zone bend of the NTT.

Notwithstanding the spatial differences in exposure relative to the location of tsunami origin at the NTT, of note is the time of day that each event occurred. Both events struck either in the morning (2009 tsunami) or in the evening (1917 tsunami) during times when the residential population was not at maximum capacity. Had they occurred in the middle of the night, for example, then the scale of human losses might have been significantly greater in each event; this is a key consideration in resilience planning for future NTT sourced tsunamis.

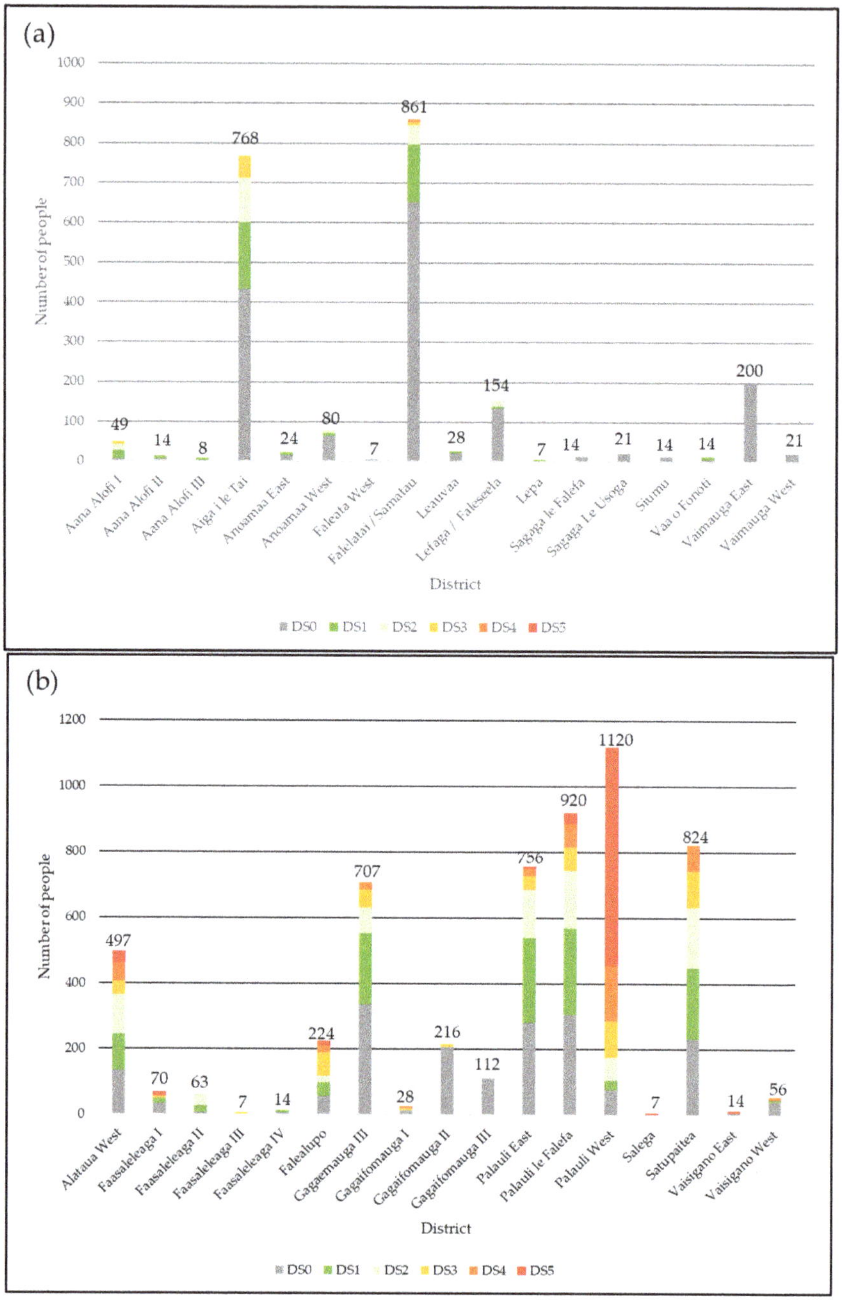

Figure 9. Number of people affected in residential buildings in the six damage states on: (**a**) Upolu, and (**b**) Savai'i.

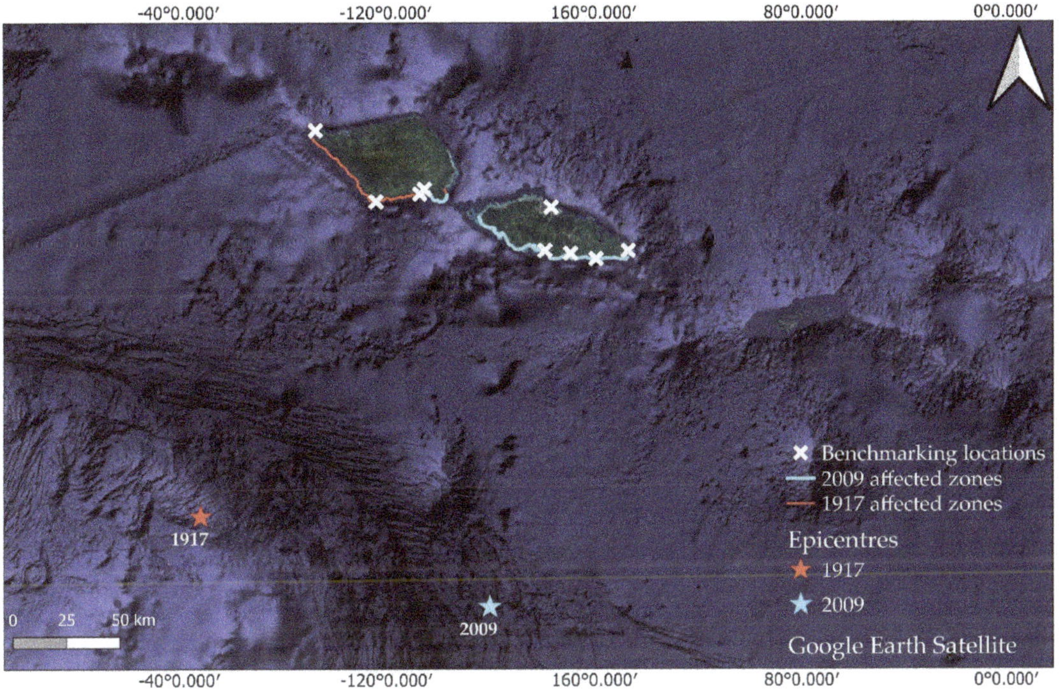

Figure 10. Epicentres of the 1917 and 2009 earthquakes and corresponding extents at the coast that were predominantly affected by each respective tsunami according to the models. The red crosses indicate the 1917 runup benchmarks shown in Figures 5 and 6 showing locations that should be inundated.

5. Discussion

Our results suggest that the south coast of Savai'i including parts of south Upolu were affected by the 1917 tsunami. A comparison of the historic evidence with the modelled inundation shows some inconsistencies. For example, observations of runup in the village of Lotofaga in southeast Upolu indicate that half of the village was submerged [4,31], which suggests significantly greater inundation extent and flow depth in Lotofaga than what our model reproduces. That is, the modelled inundation suggests that Lotofaga was not severely impacted. This inconsistency likely reflects the simplicity in our tsunami source model in the light of recent evidence of complex behaviour at 'subduction-zone bends', as is the case at the northern Tonga Trench [26].

In addition, records of the maregram of Apia harbour indicated initial water level fluctuations only 5–10 min after the earthquake compared with approximately 34 min modelled wave arrival time in Apia. The observed fluctuations of 5–10 min in Apia after earthquake initiation is highly unlikely if considering that the tsunami was generated/influenced by the earthquake source alone, which was located further west of the 2009 epicentre and farther away from Apia. However, the maregram accurately recorded the normal tides at the time of the event, which were within 14 min of tide predictions. This suggests that either the recorded timing of the tsunami or earthquake were inaccurate, or that there is more complexity in the tsunami source mechanism than what is currently captured in our modelling. For example, the possibility of complex near-simultaneous faulting source comparable with that seen for the 2009 event [25,38,39] has not been assessed in this study. It is also probable that the earthquake could have caused co-seismic submarine landsliding close to or along the north coast of Upolu, which might explain the early fluctuations observed in Apia harbour that cannot be accounted for using our current earthquake source mechanism (e.g., [41]). Further investigations are needed to unravel this dilemma.

Of particular note is the comparison in total number of casualties between the two events. In the 2009 tsunami, 146 people lost their lives [26] and several thousand were displaced, compared with only two casualties inferred from damage descriptions in the 1917 event [32]. In contrast, the extent of inundation inferred from our modelling of the 1917 tsunami, particularly on Savai'i, suggests that damage to property, threat to life, and displacement would have been more severe than what is currently inferred from the historical records. It is plausible that loss of life may have been minimal based on the modern analogy presented by the devastating Hunga-Tonga Hunga-Ha'apai volcano-generated tsunami where three casualties along with major destruction to property, homes, and businesses, and displacement of approximately 1500 people were observed [42]. However, it is equally plausible that loss of life and human displacement far exceeded that which has been documented in historical records.

Nevertheless, the hazard risk patterns presented in this study when compared with the 2009 event indicate that the 1917 tsunami would have been more severe on the island of Savai'i compared with Upolu. However, the absence of verifiable records pertaining to loss of life in the 1917 event suggest that either: (1) inundation from the 1917 event, on balance, was less destructive than the inundation caused by the 2009 event; (2) the occurrence of the 1917 tsunami in the evening meant that people who felt the earthquake shaking might have been more aware of the potential tsunami threat and self-evacuated, which helped to minimize/avoid loss of life; or (3) potential casualties from this event were simply not accurately reported and documented in the historical literature. The latter likely reflects a probable association with the 1918 influenza pandemic that overwhelmed Samoa and resulted in the loss of close to 25% of the population, most of them adults and knowledge holders [13,14]. Coupled with the impact of post-WW1 colonialism in Samoa and a shift in political administration could have resulted in the inaccurate reporting of total loses in the event. An example where this is potentially evident is the significant mis-representation of post-colonial population decline in Samoa of up to 80–90% compared with previous estimates of only 20–50% [43].

6. Conclusions

This study aimed to reconstruct the 1917 tsunami in Samoa and assess the impacts of inundation on present-day buildings and population exposure. Our findings show variable consistency between modelled-to-observed event reconstructions, which are exemplified by the inconsistency in the wave arrival time in Apia and underestimation of inundation extent/intensity in southeast Upolu. The observed discrepancies are probably due to: (1) earthquake source model and geometry configuration; (2) instrumental seismic and/or tide gauge record uncertainties for Samoa in 1917 that might explain the 20 min anomaly in the tidal reference time; (3) limited records of runup observations for validation; (4) uncertainties in potential source and/or co-seismic mechanisms that might have exacerbated the observed characteristics of the tsunami (e.g., in Apia harbour). These uncertainties represent areas for further modelling investigation.

Notwithstanding these discrepancies, our modelling provides a first-order estimation to better quantify the magnitude of impacts for the 1917 tsunami inundation in Samoa that can support scenario-based hazard risk assessment. Our modelling suggests that the scale of impacts, in particular on Savai'i and with regards to potential casualties and human displacement, likely exceeded that which was recorded in historical records. However, it is equally plausible that, although the extent of property damage and displaced peoples was likely severe, the casualty rate may have been low comparable to the death toll observed in the January 2022 Hunga-Tonga Hunga-Ha'apai tsunami in Tonga.

Nevertheless, comparison between the 1917 and 2009 events suggests that the extent of present-day exposure distribution around the two main islands of Samoa from local tsunamis originating at the Northern Tonga Trench is highly influenced on the earthquake epicentre and location/orientation of co-seismic displacement. That is, Savai'i Island is more exposed to tsunamis originating along the western segment of the NTT (e.g., 1917 event), compared with Upolu in the east, which exhibits greater exposure to outer-rise events originating along the east NTT segment.

Author Contributions: Conceptualization, L.S., C.B., S.W., R.P., J.C.T., M.W., L.T. and P.V.; methodology, C.B., S.W., R.P. and L.S.; software, C.B., R.P., L.S., S.W. and M.W.; validation, L.S., C.B., S.W., R.P., J.C.T., M.W., L.T. and P.V.; formal analysis, L.S., C.B., S.W., R.P., J.C.T., M.W., L.T. and P.V.; investigation, L.S., C.B., S.W., R.P. and P.V.; resources, S.W. and M.W.; data curation, L.S., C.B., S.W., M.W., J.C.T. and L.T.; writing—original draft preparation, L.S., S.W., C.B., R.P., M.W. and P.V.; writing—review and editing, L.S., C.B., S.W., R.P., J.C.T., M.W., L.T. and P.V.; visualization, L.S., C.B. and R.P.; supervision, S.W., M.W. and J.C.T.; project administration, L.S., S.W. and M.W.; funding acquisition, S.W., L.S., M.W., P.V., J.C.T. and L.T. All authors have read and agreed to the published version of the manuscript.

Funding: The APC was funded by NIWA Taihoro Nukurangi Project No: CARH2206.

Institutional Review Board Statement: Not applicable.

Informed Consent Statement: Not applicable.

Data Availability Statement: Baseline topography, bathymetry, and exposure datasets used in this study as well as raw results are accessible via formal request to the Samoa Ministry of Natural Resources and Environment. The BG-Flood hydrodynamics modelling software is available on the GitHub: https://github.com/CyprienBosserelle/BG_Flood (accessed on 1 June 2021), and the RiskScape multi-hazard impacts modelling software is available at: https://riskscape.org.nz/ (accessed on 28 June 2021).

Acknowledgments: This research was supported via collaboration between NIWA Taihoro Nukurangi Natural Hazards National Science Centre (S.W.; C.B. and R.P.), the University of Portsmouth (L.S. and M.W.), the Samoa Ministry of Natural Resources and Environment (J.C.T. and L.T.) and GNS Te Pu Ao National Earthquake Information Database (P.V.). Three anonymous reviewers are thanked for the helpful comments which significantly improved the paper.

Conflicts of Interest: The authors declare no conflict of interest.

References

1. Shultz, J.M.; Cohen, M.A.; Hermosilla, S.; Espinel, Z.; McLean, A. Disaster Risk Reduction and Sustainable Development for Small Island Developing States. *Disaster Health* **2016**, *3*, 32–44. [CrossRef] [PubMed]
2. Martyr-Koller, R.; Thomas, A.; Schleussner, C.-F.; Nauels, A.; Lissner, T. Loss and damage implications of sea-level rise on Small Island Developing States. *Curr. Opin. Environ. Sustain.* **2021**, *50*, 245–259. [CrossRef]
3. Williams, S.; Paulik, R.; Weaving, R.; Bosserelle, C.; Chan Ting, J.; Wall, K.; Simi, T.; Scheele, F. Multiscale Quantification of Tsunami Hazard Exposure in a Pacific Small Island Developing State: The Case of Samoa. *GeoHazards* **2021**, *2*, 63–79. [CrossRef]
4. Pararas-Carayannis, G.; Dong, B. *Catalog of Tsunamis in the Samoan Islands*; International Tsunami Information Center: Honolulu, HI, USA, 1980.
5. Goff, J.; Dominey-Howes, D. The 2009 South Pacific Tsunami. *Earth-Sci. Rev.* **2011**, *107*, 5–7. [CrossRef]
6. Gusman, A.R.; Roger, J. Hunga Tonga—Hunga Ha'apai Volcano-Induced Sea Level Oscillations and Tsunami Simulations. *GNS Sci. Webpage* **2022**. [CrossRef]
7. Klein, A. Tongan volcano erupts. *New Sci.* **2022**, *253*, 7. [CrossRef]
8. Lavigne, F.; Morin, J.; Wassmer, P.; Weller, O.; Kula, T.; Maea, A.V.; Kelfoun, K.; Mokadem, F.; Paris, R.; Malawani, M.N.; et al. Bridging Legends and Science: Field Evidence of a Large Tsunami that Affected the Kingdom of Tonga in the 15th Century. *Front. Earth Sci.* **2021**, *9*, 748755. [CrossRef]
9. Reese, S.; Bradley, B.A.; Bind, J.; Smart, G.; Power, W.; Sturman, J. Empirical building fragilities from observed damage in the 2009 South Pacific tsunami. *Earth-Sci. Rev.* **2011**, *107*, 156–173. [CrossRef]
10. Paulik, R.; Williams, S.; Simi, T.; Bosserelle, C.; Ting, J.C.; Simanu, L. Evaluating building exposure and economic loss changes after the 2009 South Pacific Tsunami. *Int. J. Disaster Risk Reduct.* **2021**, *56*, 102131. [CrossRef]
11. Kanamori, H. Importance of Historical Seismograms for Geophysical Research. In *Historical Seismograms and Earthquakes of the World*; Lee, W.H.K., Meyers, H., Shimazaki, K., Eds.; Academic Press: San Diego, CA, USA, 1988; pp. 16–33, ISBN 9780124408708. Available online: https://resolver.caltech.edu/CaltechAUTHORS:20141216-125156671 (accessed on 10 November 2021).
12. Okal, E.A.; Borrero, J.C.; Chagué-Goff, C. Tsunamigenic predecessors to the 2009 Samoa earthquake. *Earth-Sci. Rev.* **2011**, *107*, 128–140. [CrossRef]

13. Tomkins, S.M. The Influenza Epidemic of 1918-19 in Western Samoa. *J. Pac. Hist.* **1992**, *27*, 181–197. Available online: https://www.jstor.org/stable/25169127 (accessed on 27 November 2021). [CrossRef]
14. Alofaituli, B.T. The 1918 Influenza Epidemic in Sāmoa and the Sāmoa Chruch (LMS). *J. Samoan Stud.* **2018**, *8*, 34–44. Available online: https://journal.samoanstudies.ws/2019/05/23/the-1918-influenza-epidemic-in-samoa-and-the-samoa-church-lms/ (accessed on 8 August 2021).
15. Hart, S.R.; Coetzee, M.; Workman, R.K.; Blusztajn, J.; Johnson, K.T.M.; Sinton, J.M.; Steinberger, B.; Hawkins, J.W. Genesis of the Western Samoa seamount province: Age, geochemical fingerprint and tectonics. *Earth Planet. Sci. Lett.* **2004**, *227*, 37–56. [CrossRef]
16. Koppers, A.A.P.; Russell, J.; Jackson, M.G.; Konter, J.; Staudigel, H.; Hart, S.R. Samoa reinstated as a primary hotspot trail. *Geology* **2008**, *36*, 435–438. [CrossRef]
17. Koppers, A.A.P.; Russell, J.A.; Roberts, J.; Jackson, M.G.; Konter, J.G.; Wright, D.J.; Staudigel, H.; Hart, S.R. Age systematics of two young en-echelon Samoan volcanic trails. *Geochem. Geophys. Geosyst.* **2011**, *12*, Q07025. [CrossRef]
18. Kear, D.; Wood, B.L. The Geology and Hydrology of Western Samoa. In *New Zealand Geological Survey Bulletin No. 63*; New Zealand Depart of Scientific and Industrial Research: Wellington, New Zealand, 1959.
19. Samoa Bureau of Statistics. Population & Demographic Indicator Summary. Available online: https://www.sbs.gov.ws/population (accessed on 28 October 2021).
20. Angenheister, G.G. Geschichte des Samoa-Observatoriums von 1902 bis 1921. In *Zur Geschichte der Geophysik*; Springer: Berlin, Germany, 1974; pp. 43–66.
21. Hatherton, T. *Geophysics Division, DSIR 1951–1976: An Account of Geophysical Studies in the Department of Scientific and Industrial Research*; Department of Scientific and Industrial Research: Wellington, New Zealand, 1980; p. 45.
22. Tomlinson, L.A. Observatories in New Zealand and The South Pacific. In *Encyclopedia of Geomagnetism and Paleomagnetism*; Gubbins, D., Herrero-Bervera, E., Eds.; Springer: Dordrecht, The Netherlands, 2007. [CrossRef]
23. Angenheister, G. Vier Erdbeben und Flutwellen im Pazifischen Ozean, beobachtet am Samoa-Observatorium, 1917–1919. *Nachr. Königlichen Ges. Wiss. Göttingen Math. Phys. Kl.* **1920**, *1920*, 201–204. Available online: https://eudml.org/doc/59086 (accessed on 30 May 2021).
24. Németh, K.; Cronin, S.J. Volcanic structures and oral traditions of volcanism of Western Samoa (SW Pacific) and their implications for hazard education. *J. Volcanol. Geotherm. Res.* **2009**, *186*, 223–237. [CrossRef]
25. Lay, T.; Ye, L.; Wu, Z.; Kanamori, H. Macrofracturing of Oceanic Lithosphere in Complex Large Earthquake Sequences. *J. Geophys. Res. Solid Earth* **2020**, *125*, e2020JB020137. [CrossRef]
26. Okal, E.A.; Fritz, H.M.; Synolakis, C.E.; Borrero, J.C.; Weiss, R.; Lynett, P.J.; Titov, V.V.; Foteinis, S.; Jaffe, B.E.; Liu, P.L.-F.; et al. Field Survey of the Samoa Tsunami of 29 September 2009. *Seism. Res. Lett.* **2010**, *81*, 577–591. [CrossRef]
27. Bosserelle, C.; Williams, S.; Cheung, K.F.; Lay, T.; Yamazaki, Y.; Simi, T.; Roeber, V.; Lane, E.; Paulik, R.; Simanu, L. Effects of Source Faulting and Fringing Reefs on the 2009 South Pacific Tsunami Inundation in Southeast Upolu, Samoa. *J. Geophys. Res. Oceans* **2020**, *125*, e2020JC016537. [CrossRef]
28. Engdahl, E.; Villasenor, A. 41 Global seismicity: 1900–1999. *Int. Geophys.* **2002**, *81*, S0074–S6142.
29. NCEI/WDS Global Significant Earthquake Database, 2150 BC to Present. Available online: https://www.ncei.noaa.gov/access/metadata/landing-page/bin/iso?id=gov.noaa.ngdc.mgg.hazards:G012153;view=iso (accessed on 20 January 2021).
30. Williams, S.; Titimaea, A.; Bosserelle, C.; Simanu, L.; Prasetya, G. Reassessment of Long-Term Tsunami Hazards in Samoa Based on Sedimentary Signatures. *Geosciences* **2020**, *10*, 481. [CrossRef]
31. Samoa 1917 Earthquake, Search Results. Available online: https://paperspast.natlib.govt.nz/newspapers?items_per_page=10&snippet=true&query=earthquake+samoa (accessed on 1 December 2021).
32. Significant Earthquake Information, 1917 Earthquake—Samoa Islands. Available online: https://www.ngdc.noaa.gov/hazel/view/hazards/earthquake/event-more-info/3087 (accessed on 13 December 2021).
33. Vacondio, R.; Palù, A.; Ferrari, A.; Mignosa, P.; Aureli, F.; Dazzi, S. A non-uniform efficient grid type for GPU-parallel Shallow Water Equations models. *Environ. Modell. Softw.* **2017**, *88*, 119–137. [CrossRef]
34. Okada, Y. Surface deformation due to shear and tensile faults in a half space. *Bull. Seismol. Soc. Am.* **1985**, *75*, 1135–1154. [CrossRef]
35. Tarbotton, C.; Dall'Osso, F.; Dominey-Howes, D.; Goff, J. The use of empirical vulnerability functions to assess the response of buildings to tsunami impact: Comparative review and summary of best practice. *Earth-Sci. Rev.* **2015**, *142*, 120–134. [CrossRef]
36. Paulik, R.; Horspool, N.; Woods, R.; Griffiths, N.; Beale, T.; Magill, C.; Wild, A.; Popovich, B.; Walbran, G.; Garlick, R. RiskScape: A flexible multi-hazard risk modelling engine. *Res. Sq.* **2022**. Vers. 1, 22, preprint. [CrossRef]
37. World Bank. Population, Total—Samoa. Available online: https://data.worldbank.org/indicator/SP.POP.TOTL?locations=WS (accessed on 10 December 2021).
38. Beavan J, Wang X, Holden C, Wilson K, Power W, Prasetya G, Bevis M, Kautoke R Near-simultaneous great earthquakes at Tongan megathrust and outer rise in September 2009. *Nature* **2010**, *466*, 959–963. [CrossRef]
39. Lay, T.; Ammon, C.J.; Kanamori, H.; Rivera, L.; Koper, K.D. The 2009 Samoa-Tonga great earthquake triggered doublet. *Nature* **2010**, *466*, 964–968. [CrossRef]

40. Eiby, G.A. Seismograms made before 1963 at stations in the South-west Pacific. In *Historical Seismograms and Earthquakes of the World*; Lee, W.H.K., Meyers, H., Shimazaki, K., Eds.; Academic Press: San Diego, CA, USA, 1988; pp. 455–461, ISBN 9780124408708. Available online: https://catalogue.nla.gov.au/Record/191994 (accessed on 10 November 2021).
41. Rahiman, T.; Pettinga, J.; Watts, P. The source mechanism and numerical modelling of the 1953 Suva tsunami, Fiji. *Mar. Geol.* **2007**, *237*, 55–70. [CrossRef]
42. World Bank. The 15 January 2022 Hunga Tonga-Hunga Ha'apai Eruption and Tsunami, Tonga: Global Rapid Post Disaster Damage Estimation (GRADE) Report. Available online: https://thedocs.worldbank.org/en/doc/b69af83e486aa652d4232276ad698c7b-0070062022/original/GRADE-Report-Tonga-Volcanic-Eruption.pdf (accessed on 10 February 2022).
43. Code, B. A Secret Pyramid Consumed by the Jungle. January 2020. Available online: https://www.bbc.com/travel/article/20200109-a-secret-pyramid-consumed-by-the-jungle (accessed on 24 September 2021).

applied sciences

MDPI

Article

Developing a Guideline of Unmanned Aerial Vehicle's Acquisition Geometry for Landslide Mapping and Monitoring

Konstantinos G. Nikolakopoulos *, Aggeliki Kyriou and Ioannis K. Koukouvelas

Department of Geology, University of Patras, 265 04 Patras, Greece; a.kyriou@upnet.gr (A.K.); iannis@upatras.gr (I.K.K.)

* Correspondence: knikolakop@upatras.gr

Abstract: Remote sensing data and techniques are widely used for monitoring and managing natural or man-made disasters, due to their timeliness and their satisfactory accuracy. A key stage in disaster research is the detailed and precise mapping of an affected area. The current work examines the relationship that may exist between the acquisition geometry of Unmanned Aerial Vehicle (UAV) campaigns and the topographic characteristics of an investigated area, toward landslide mapping and monitoring that is as accurate as possible. In fact, this work, concerning the systematic research of the acquisition geometry of UAV flights over multiple active landslides, is conducted for the first time and is focused on creating a guideline for any researcher trying to follow the UAV photogrammetric survey during landslide mapping and monitoring. In particular, UAV flights were executed over landslide areas with different characteristics (land cover, slope, etc.) and the collected data from each area were classified into three groups depending on UAV acquisition geometry, i.e., nadir imagery, oblique imagery, and an integration of nadir and oblique imagery. High-resolution orthophotos and Digital Surface Models (DSMs) emerged from the processing of the UAV imagery of each group through structure-from-motion photogrammetry (SfM). Accuracy assessment was carried out using quantitative and qualitative comparative approaches, such as root mean square error calculation, length comparison, and mean center estimation. The evaluation of the results revealed that there is a strong relationship between UAV acquisition geometry and landslide characteristics, which is evident in the accuracy of the generated photogrammetric products (orthophotos, DSMs). In addition, it was proved that the synergistic processing of nadir and oblique imagery increased overall centimeter accuracy.

Keywords: photogrammetry; viewing angle; oblique; nadir; mapping; geomorphology

Citation: Nikolakopoulos, K.G.; Kyriou, A.; Koukouvelas, I.K. Developing a Guideline of Unmanned Aerial Vehicle's Acquisition Geometry for Landslide Mapping and Monitoring. *Appl. Sci.* **2022**, *12*, 4598. https://doi.org/10.3390/app12094598

Academic Editors: Spyridon Mavroulis and Efthymios Lekkas

Received: 21 March 2022
Accepted: 29 April 2022
Published: 2 May 2022

1. Introduction

Remote sensing has emerged as an important and valuable tool for various earth observation applications. In particular, remote sensing data and techniques are widely used for monitoring and managing natural or man-made disasters, due to their timeliness and their satisfactory accuracy [1]. The development of UAVs has opened up new possibilities in hazard assessment and disaster risk management [2–4]. In fact, these low-cost remote sensors have proven their effectiveness in mapping hazards and disasters (earthquakes, floods, landslides, etc.), monitoring human activity during emergencies, and protecting and preserving cultural heritage sites affected by geo-hazards [5–10].

A key stage in disaster research is the detailed and precise mapping of an affected area. In this framework, scientists have developed several new processing methodologies, in which different parameters were assessed in achieving the most efficient results. Specifically, the selection of an appropriate number of ground control points (GCPs) during the georeferencing procedure of the obtained UAV imagery was one of the primary parameters under investigation. Thus, various photogrammetric campaigns, consisting of different combinations of GCPs (varying from 4 to up to 20), were evaluated using root mean square error (RMSE) values extracted from the respective DSMs [11–13]. The results demonstrated a clear influence of the number of GCPs on the accuracy of UAV photogrammetry. Although the selection of about 20 GCPs (defined according to the study area) managed to reduce the RSME by 50%, it was proved that a higher number of GCPs slightly improved the accuracy. Moreover, other studies dealt with analyzing the distribution of GCPs within the study area and the effect of such distribution on the accuracy of the derived DSMs, which decreases as the distance to the nearest GCP increases [14,15]. In general, the accuracy of photogrammetric products increases asymptotically as the number of GCPs increase, until an optimal GCP density is reached.

The collection of GCPs is a time-consuming task. Thus, the development of UAVs with onboard Global Navigation Satellite System Real Time Kinematic (GNSS RTK) positioning—in which the georeferencing of the images does not require GCPs—constitutes a promising solution [16,17]. The main disadvantage of the specific approach is the presence of systematic elevation errors, which emerge from the incorrect determination of the interior orientation parameters estimated during bundle adjustment [18]. A recent study proposed that a combination of two UAV flights at the same altitude, consisting of nadir and oblique imagery, was able to reduce the systematic elevation error to less than 0.03 m [19].

Other approaches focused on evaluating the effect of different scenarios, consisting of multi-view camera combinations, on the accuracy of UAV photogrammetry. Specifically, the combination of non-metric oblique and vertical views, along with an appropriate collection of GCPs, has significantly enhanced the accuracy of the photogrammetric procedure, extracting DSMs comparable to those derived by Light Detection and Ranging (LiDAR) [20]. Indeed, this integration of nadir and oblique imagery has proven its effectiveness in numerous studies, including the reconstruction of the surfaces and the determination of the main geometries in a quarry environment, the geomorphological mapping of landslides, and geotechnical/ hazard mapping [21–23]. In addition, it has been proven that the reconstruction of a topographically complex terrain requires the synergy of oblique and facade images to complement nadir views [24]. This synergistic use significantly improved the geometric accuracy of topographic reconstruction, by approximately 35%.

Although the combination of multi-view UAV imagery is a very successful approach in a variety of applications, nadir images combined with a dense distribution of GCPs constituted an ideal solution for the reconstruction of 3D agricultural surfaces [25]. On the other hand, oblique UAV imagery was considered particularly suitable for the three-dimensional modeling of buildings, cities, or urban settlements, presenting an enhanced achievable accuracy [26]. Moreover, oblique-viewing images can be utilized effectively for the 3D modelling of historical architectures and cultural heritage research, especially in areas characterized by limited accessibility [27]. In addition, oblique images increased by 50% the accuracy of 3D representations of topography in areas with high and steep slopes [28].

The current study aims to create a guideline for researchers who try to follow the UAV photogrammetric survey method for landslide mapping and monitoring, by exploring derived data quality (orthophotos and DSMs) in reference to data acquisition geometry. There are previous studies on the influence of the ground control number to the horizontal or vertical accuracy of orthophotos and DSMs, respectively [12–15]. However, the current study is the first to conduct systematic research on UAV data acquisition methodology applied on multiple active landslides. UAV flights were performed over four landslide areas with different characteristics (land cover, slope, volume, dimensions, etc.), and the collected data from each area were classified into three groups, depending on UAV acquisition geometry: (a) nadir-only images, (b) oblique-only images, and (c) an integration of nadir and oblique images. Moreover, a flat urban area was also surveyed for validation purposes. The processing of UAV imagery in each category was based on structure-from-motion (SfM) photogrammetry, leading to the creation of high resolution orthophotos and digital surface models (DSMs). Accuracy assessment was carried out using quantitative and qualitative comparative approaches, such as root mean square error calculation, length comparison, and mean center estimation. The derived results demonstrate the benefits of combining oblique and nadir images in order to reduce systematic errors and increase the overall accuracy of orthophotos and Digital Surface Models.

2. Materials and Methods

2.1. Case Studies

The selection of research areas was based on the criterion of heterogeneity in terms of slope, land cover, vegetation coverage, etc. Thus, four different landslides were set as case studies. We also surveyed a flat industrial area for validation purposes (Figure 1) (Table 1). In order to eliminate the influence of vegetation height on DSM production, the study areas presented, in general, with low and sparse vegetation, as described in Table 1.

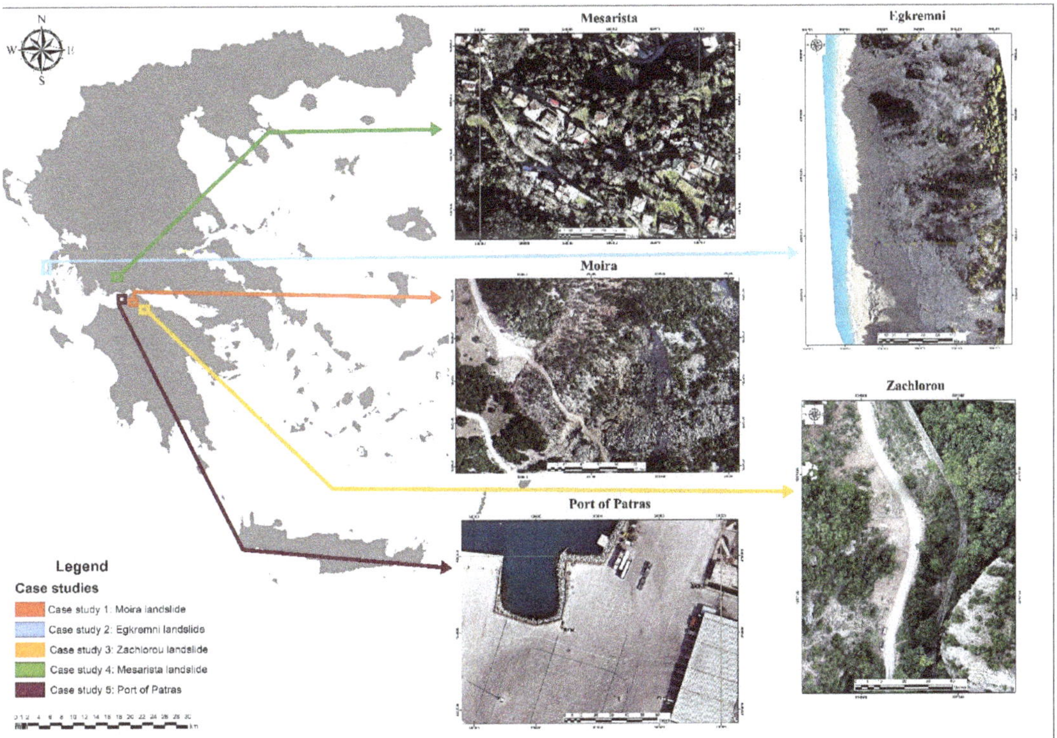

Figure 1. Location of the five case studies within Greece and UAV orthophotos of each area of interest.

Table 1. Selected case studies and characteristics (topography, vegetation).

Case Study ID	Locations	Topography Description	Vegetation
1	Moira	Steep slope, large extent, grassland [1]	Low and sparse vegetation
2	Egkremni	Steep slope, coastal area, grassland	Very sparse vegetation
3	Zachlorou	Steep slope, narrow gorge, grassland	Low vegetation
4	Messarista	Steep slope, urban settlement, narrow roads, densely build-up area	Low vegetation
5	Patras Port	Flat, industrial environment	No vegetation

[1] Based on coordination of information on the environment (CORINE) land cover classification.

The first landslide area is located close to the village of Moira, within Western Greece. The landslide occurred on 20 January 2017 on a mountainous, steep slope, covering an area of approximately 65,569.20 m^2. The landslide material was spread 300 m in length and 300 m in width and it was geologically structured from flysch, limestone, and loose cherts. It was characterized as a complex slide, based on the different lithologies—silicate and/or silicate lithology was presented in the northwestern part, while limestones were detected in the southeast. The occurrence of the landslide was related to geological factors (flysch lithology), as well as to the rapid snow melting. It has been demonstrated that the increase of the water content of clay soils (flysch) reduces the shear strength of soils, acting as a triggering factor for the occurrence of landslides [29].

The destruction of the road connecting the village of Moira with the city of Patras, as well as a significant change in the local landscape, were the main consequences of the landslide.

The second landslide took place in November 2015 on the Egkremni beach on the island of Lefkada; it was classified as a debris slide . The Ionian Islands constitute the northwestern part of the Hellenic arc, which is considered as a site of complex continent-continent to continent-ocean convergent plate margins. The Cephalonia Transform Fault Zone (CTFZ) is recognized as the major tectonic structure of the site and is responsible for the particularly active seismicity in the wider area. In addition, another fault, named Athani, is detected ashore, shaping the west coast of Lefkada island [30,31]. The Athani fault, along with smaller, parallel faults with similar kinematics, form steep slopes, which are an ideal site for the occurrence of landslides during seismic events [32]. On 17 November 2015, a 6.5 magnitude earthquake struck the west coast of the island, causing a series of landslides on the cliffs [33,34].

Another active continental region, in terms of tectonics and seismicity, is Northern Peloponnese, due to the fast-extending rift of the Gulf of Corinth. This third area of interest is located close to the village of Kato Zachlorou within Western Greece. The first mass movements at the specific area occurred in April 2019, while the main event, consisting of rock falls and earth flows, took place on 4 April 2020. The repetitive sliding episodes were strongly associated with the increase of the water content of the clay soils, due to intense rainfall [29].

The fourth area of interest includes a landslide located in a semi-mountainous village of Western Greece named Messarista. Extremely heavy and prolonged rains hit the wider region on 11 December 2021, acting as a main triggering factor for the occurrence of a series of landslides in different places throughout the region, as well as within the village of Messarista. The phenomenon is still ongoing and is being monitored by local authorities, while the landslides were classified as earth slides and earth flows.

Finally, a flat industrial area, covering part of the new port of the city of Patras, was chosen for the validation of the results, due to the past view on photogrammetry, in which flat areas were surveyed using nadir imagery solely.

2.2. Data Collection

UAV imagery was obtained using a DJI Phantom 4, which is equipped with a built-in GNSS system and a CMOS camera (12.4 MP) with 4000 × 3000 resolution. Each case study consisted of three different flight grids (nadir, oblique, and nadir and oblique). During these distinct campaigns, flight characteristics were kept the same (Table 2). Specifically, flights were executed with a 90% along-track overlap and a 75% cross-track overlap. However, the flight acquisition altitude was adapted to the topography of each study area, with the aim of extracting photogrammetric products with spatial resolution between 2.3 and 3.5 cm (Table 2). The collected UAV imagery of each area was classified into three groups of data. The first group included nadir-only images with a 90-degree gimbal pitch angle; the second group consisted of oblique-only images with a 65-degree gimbal pitch angle; and a synergistic use of nadir and oblique imagery was selected for the third group.

Table 2. Characteristics and parameters of UAV campaigns.

Case Study ID	Acquisition Geometry	Number of Photos	Average Flight Altitude (m)	Average GSD (cm)	Along the Track Overlap %	Along the Track Overlap %
Moira	Nadir Oblique Synergistic use	189 174 363	110	3.5	90	75
Egremni	Nadir Oblique Synergistic use	97 84 181	60	2.3	90	75
Zachlorou	Nadir Oblique Synergistic use	60 210 286	70	2.5	90	75
Messarista	Nadir Oblique Synergistic use	65 70 135	110	3.5	90	75
Patras Port	Nadir Oblique Synergistic use	96 85 181	60	2.3	90	75

In addition, GCPs were distributed throughout the survey areas in order to orient and match the aerial imagery to data measured terrestrially. GCPs were collected using a Leica GS08 GNSS receiver. Coded targets were created according to the general recommendations of Agisoft Metashape software [35], and were printed on matte finish plastic boards. These targets were used as GCPs (Figure 2). Furthermore, large rectangular aluminum targets were placed in the survey areas for comparison and validation of the photogrammetric outputs (Figure 3). These aluminum targets had specific dimensions and a hole in each corner, in order to be accurately measured with a GNSS sensor.

Figure 2. Printed GCP pattern, distributed throughout the survey area during UAV campaigns.

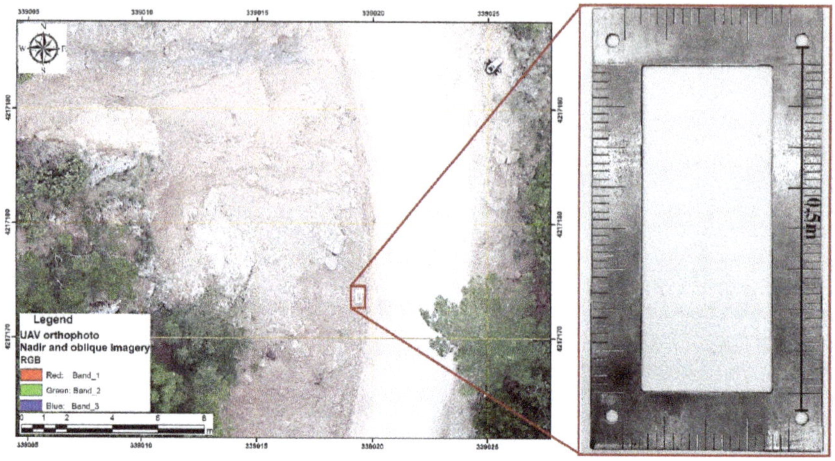

Figure 3. Aluminum targets, as captured in orthophotos during UAV campaigns over the landslide of Zachlorou.

2.3. Methodology

The aim of the current study was to examine whether the geometry of UAV acquisition plays a key role in the accuracy of derived photographic products that are subsequently used as high-precision data for landslide mapping and monitoring. A schematic illustration of the applied methodology is displayed in Figure 4.

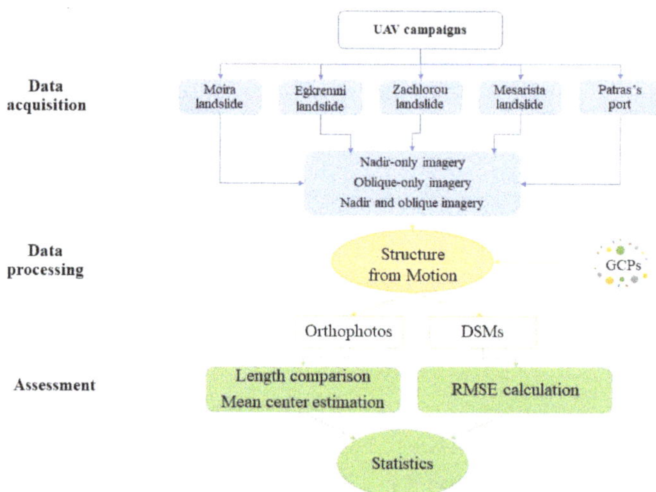

Figure 4. A flowchart showing the processing steps of the research methodology.

In more detail, UAV surveys were conducted over five study areas, presenting different characteristics (topography, land cover, etc.). Three UAV flights, consisting of: (a) nadir-only acquisitions, (b) oblique-only acquisitions, and (c) an integration of nadir and oblique acquisitions, were executed for each case study. The different UAV acquisition trajectories of nadir- and oblique-viewing campaigns over the landslide of Messarista (case study) are depicted in the Figure 5. It is worth mentioning that the collection of the oblique-viewing imagery was implemented as the UAV moved forward and backward, following line paths that were perpendicular to the slope of each study area. Furthermore, photogrammetric GCPs were collected during the UAV surveys.

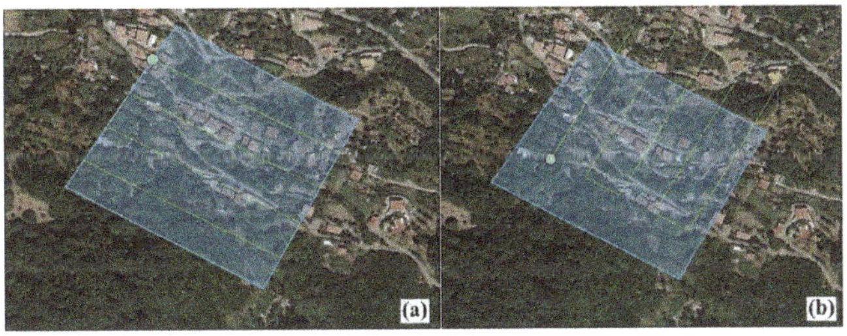

Figure 5. UAV acquisitions over the landslide of Messarista. (**a**) Nadir-viewing acquisition trajectory; (**b**) Oblique-viewing acquisition trajectory.

The processing of the collected UAV imagery was carried out through SfM photogrammetry. The specific technique contributes to the three-dimensional reconstruction of the topography, using the basic principles of photogrammetry along with computer vision algorithms [36–39]. The main advantage of this functional and low-cost method is that it allows the automatic and simultaneous determination of scene geometry, camera positions, and orientation, without requiring pre-existing known points. Thus, multiple, overlapped and shifted 2D images are transformed into 3D representations using an automatic, feature-matching algorithm.

The photogrammetric processing of UAV imagery took place in Agisoft Metashape software. The calibration and optimization of the camera took place in accordance with Agisoft's default values, as set for the DJI Phantom 4 camera. Internal orientation parameters were estimated automatically, due to the ability of the software to recognize the model of the camera and to specify the appropriate settings.

The high-quality option was selected for the alignment of the images, aiming at a more precise estimation of the camera positions [40]. At the same time, the processing of UAV imagery was implemented using the original image size. The quality option was closely linked to the quality of the topographic reconstruction. Moreover the, ultra-high setting was defined as the parameter during "build dense cloud" and "build mesh" procedures.

Orthophotos and DSMs emerged from the processing of the collected UAV imagery. These products were projected into the Hellenic Geodetic Reference System 1987. Specifically, three orthophotos and three DSMs were created for each study area, corresponding to the three different acquisition geometries (i.e., nadir-viewing images, oblique-viewing images, and an integration of nadir and oblique imagery). The evaluation of the accuracy of the generated orthophotos and DSMs was based on qualitative and quantitative comparative approaches, including RMSE calculation, length comparison, and mean center estimation.

3. Results

3.1. Accuracy Asssessment of Orthophotos

Three high resolution orthophotos were extracted from the photogrammetric processing for each case study, consisting of UAV data from the three different acquisition groups, i.e., (a) nadir-viewing imagery, (b) oblique-viewing imagery, and (c) the integration of nadir and oblique imagery. The aforementioned orthophotos, covering the area of case study 4 (Messarista), are depicted in Figure 6. As can be observed, a visual comparison between the orthophotos created by data for the different groups of geometry acquisitions is a particularly difficult task, due to apparent similarity.

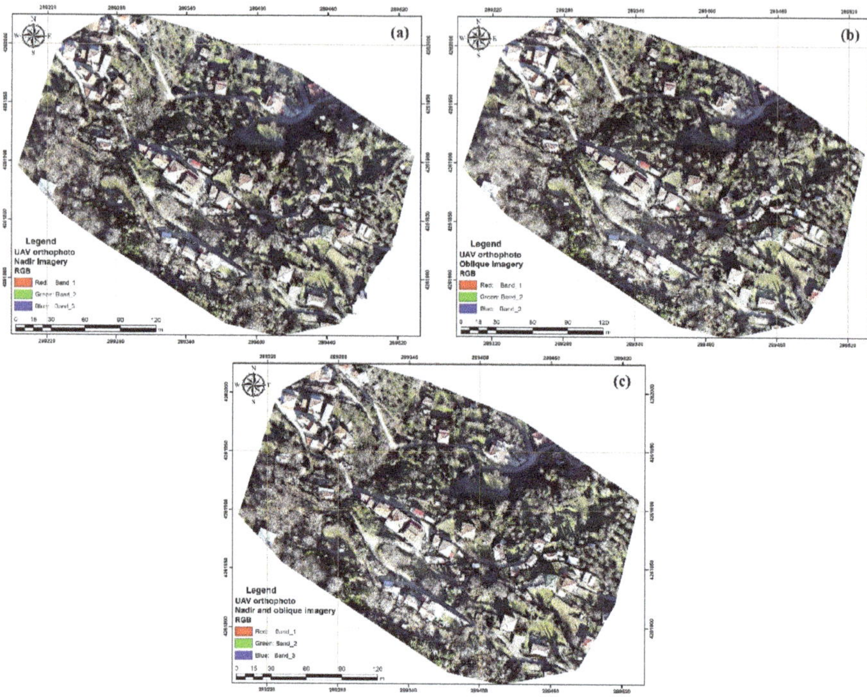

Figure 6. UAV orthophotos of the landslide of Messarista, emerging from the processing of (**a**) nadir-viewing images, (**b**) oblique-viewing images, and (**c**) synergistic use of nadir and oblique images. The scale of the orthophotos was set to 1:1500.

In this context, the evaluation of the accuracy of the derived orthophotos was performed using comparison approaches in an ArcGIS environment. GIS applications have been utilized in respective studies for the characterization of geological features and geological mapping [41]. In this study, GCPs were utilized as precise reference positions. Then, two geodetic lines were shaped from the x-y coordinates of the GCPs of each study area. Figure 7 displays the reference geodetic lines, as formed for each case study. It is important to highlight that we tried to create the geodetic lines in different topographic reliefs, varying from steep slopes to flat regions. Subsequently, we digitized the same lines in all orthophotos, derived from the processing of the UAV imagery of all groups of acquisition geometry. The digitization of the lines was based on the visual identification of the respective GCPs, which were used to shape the reference geodetic lines. The length of each digitized and reference geodetic line were calculated and compared (Table 3). The length variations were significantly small, which proved the high accuracy of the photogrammetric products. In addition, the integration of nadir and oblique viewing acquisitions displayed the smallest length variations in eight out of ten comparisons, demonstrating the positive influence of the specific acquisition geometry on the enhancement of the accuracy of the derived orthophotos. Moreover, oblique-viewing geometry showed a better performance, compared to nadir-viewing acquisitions. In particular, the difference in the length of line 1, in comparison to the corresponding reference line in the case study of Zachlorou, was zero for the orthophoto resulting from the integration of nadir and oblique imagery. In addition, the variation for the nadir-viewing and oblique-viewing orthophotos was estimated at 0.190% and 0.067%, respectively. In the case of Egnemni, the percentage variations in line 2 were calculated at about 0.034% for the orthophotos that emerged from either the synergistic use of nadir and oblique images or the nadir-viewing imagery. In contrast, the corresponding difference for the nadir-viewing orthophoto was obviously larger (i.e., 0.228%).

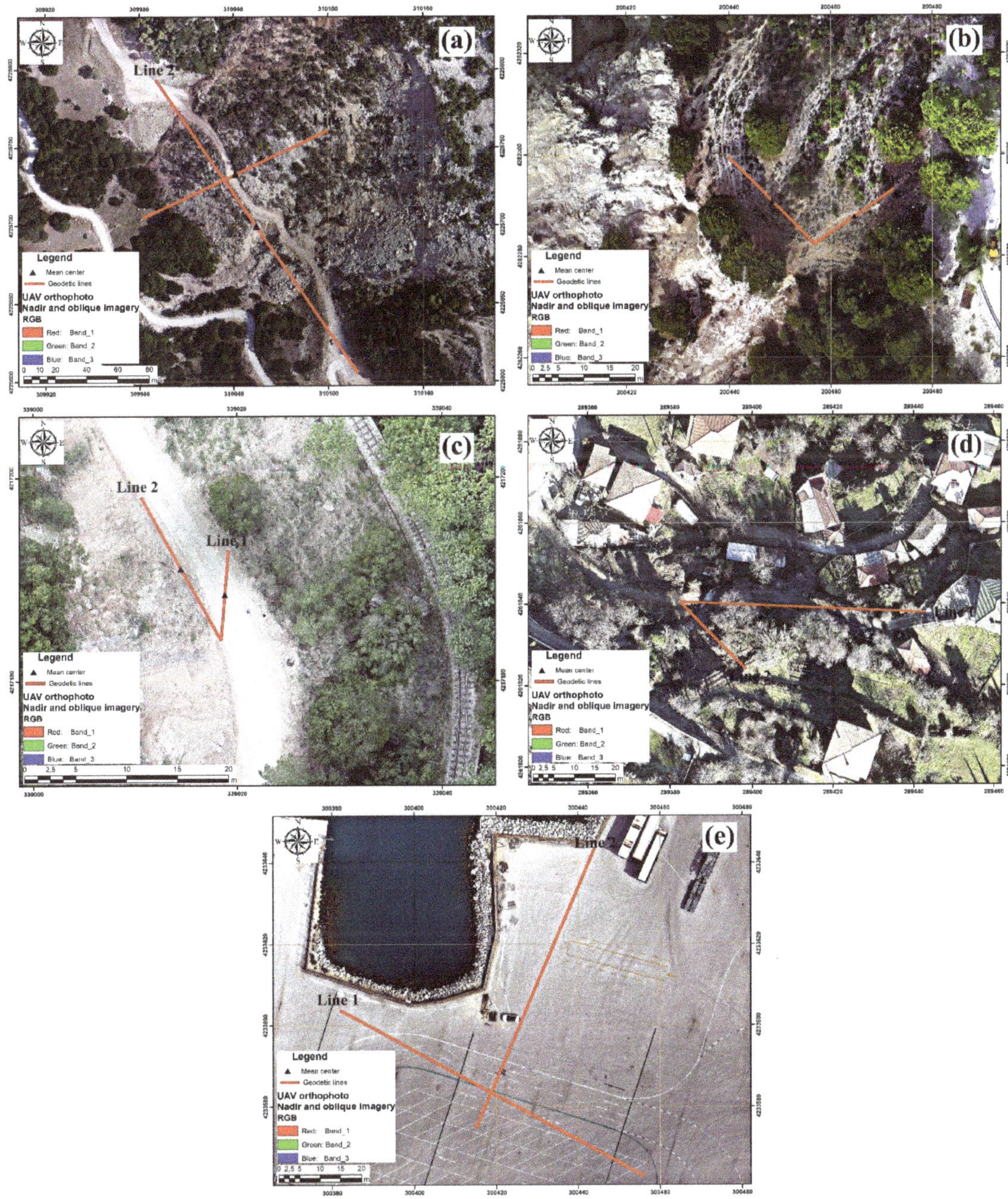

Figure 7. Reference geodetic lines as created in: (**a**) case study 1, (**b**) case study 2, (**c**) case study 3, (**d**) case study 4, and (**e**) case study 5.

Table 3. Comparison of the length of the geodetic lines of each case study in accordance with the respective reference length.

Case Study ID	Acquisition Geometry	Length Line 1 (m)	Difference (m)	Difference %	Length Line 2 (m)	Difference (m)	Difference %
Reference line		226.676			129.881		
Moira	Nadir	226.933	−0.257	−0.113%	129.550	0.331	0.254%
	Oblique	226.745	−0.069	−0.030%	129.937	−0.056	−0.043%
	Synergistic use	226.734	−0.058	−0.029%	129.725	0.156	0.120%
Reference line		32.898			23.221		
Egremni	Nadir	32.861	0.038	0.112%	23.274	−0.054	−0.228%
	Oblique	32.906	−0.007	−0.024%	23.229	−0.008	−0.034%
	Synergistic use	32.868	0.030	0.091%	23.229	−0.008	−0.034%
Reference line		8.938			16.152		
Zachlorou	Nadir	8.921	0.017	0.190%	16.179	−0.027	−0.167%
	Oblique	8.944	−0.006	−0.067%	16.179	−0.027	−0.167%
	Synergistic use	8.938	0.000	0.000%	16.177	−0.025	−0.154%
Reference line		60.524			23.064		
Messarista	Nadir	60.577	−0.053	−0.088%	23.149	−0.085	−0.368%
	Oblique	61.002	−0.478	−0.790%	23.299	−0.235	−1.018%
	Synergistic use	60.539	−0.015	−0.014%	23.097	−0.033	−0.143%
Reference line		84.240			74.220		
Patras Port	Nadir	84.559	−0.319	−0.379%	74.492	−0.272	−0.366%
	Oblique	84.117	0.123	0.146%	73.988	0.232	0.312%
	Synergistic use	84.228	0.012	0.012%	74.208	0.012	0.016%

The processing of nadir-viewing images, along with oblique imagery, enhanced the overall accuracy in all the investigated areas. In more detail, for nine of ten cases (Table 3) the length difference between the GNSS measurements and the measurements on the orthophotos was less than 0.05 m (5 cm). Taking into consideration the terrain steepness and the dense vegetation, we assume that the overall accuracy is excellent.

It is worth mentioning that even in the fifth test area (Patras Port), the combined processing of nadir and oblique images gave results that were more accurate than the single processing of nadir or oblique data. In detail, the length difference of the geodetic lines for the common data set was only 0.012 m (centimeter accuracy), while the respective accuracies for the nadir or oblique data set ranged between 0.12 m and 0.31 m (Table 3).

Another method, utilized for the evaluation of the accuracy of orthophotos resulted from the photogrammetric processing of each acquisition geometry, was based on the comparison of the mean center of the reference geodetic lines and the respective digitized ones. The mean center constitutes the geographical center of a set of features resulting from average x and y coordinates [42,43], which is calculated as:

$$\overline{X} = \frac{\sum_{i=1}^{n} x_i}{n} \text{ and } \overline{Y} = \frac{\sum_{i=1}^{n} y_i}{n} \quad (1)$$

where x_i and y_i are the coordinates for a feature i and n is the total number of features. This approach is widely used for the identification of distribution changes or for the comparison of the distributions of features. The mean centers of the reference geodetic lines of each case study are depicted in Figure 7. The Near tool [44] was applied for the determination of the distance between the reference mean centers and the respective mean center of the digitized lines. Distance variations are displayed in Table 4. As can be noted, the multi-acquisition geometry (nadir and oblique imagery) presented the shortest distances from the reference mean centers in 50% of the UAV campaigns.

Table 4. Near distances of mean centers resulting from the reference mean centers of geodetic lines 1 and 2.

Case Study ID	Acquisition Geometry	Near Distance (m) Mean Center of Line 1	Near Distance (m) Mean Center of Line 2
Moira	Nadir	0.023	0.095
	Oblique	0.015	0.049
	Synergistic use	0.013	0.012
Egkremni	Nadir	0.020	0.022
	Oblique	0.025	0.014
	Synergistic use	0.038	0.021
Zachlorou	Nadir	0.006	0.016
	Oblique	0.070	0.068
	Synergistic use	0.023	0.010
Messarista	Nadir	0.064	0.036
	Oblique	0.477	0.305
	Synergistic use	0.082	0.034
Patras Port	Nadir	0.700	0.812
	Oblique	0.946	0.890
	Synergistic use	0.024	0.009

As a final comparison of the orthophotos derived from different acquisition geometries, we utilized the variations of the dimensions of the aluminum targets. These dimensional variations, regarding the orthophotos covering the landslide of Zachlorou, are presented in Table 5. As can be observed, the synergistic use of nadir and oblique photos increase the accuracy of the orthophoto. The true length of the aluminum target is 0.538 m and the respective measured value from the oblique and nadir imagery is 0.542 m, meaning that the overall difference is only 4 mm. The difference value for the nadir imagery is 8 mm and for the oblique imagery is 10 mm.

Table 5. Comparison of the dimensions of the aluminum target between the different acquisition geometries.

Case Study ID	Acquisition Geometry	Length (m)	Difference (m)	Difference %	Area (m^2)	Difference (m^2)	Difference %
Reference		0.538			0.150		
Zachlorou	Nadir	0.546	−0.008	−1.486%	0.149	0.001	0.666%
	Oblique	0.548	−0.010	−1.858%	0.148	0.002	1.332%
	Synergistic use	0.542	−0.004	−0.743%	0.151	−0.001	−0.666%

3.2. Accuracy Asssessment of DSMs

The evaluation of the accuracy of the extracted DSMs was executed through the computation of RMSE, which estimates the differences between the values of a DMS and the reference high-precision values. The computation is performed by the following equation:

$$RMSE = \sqrt{\frac{1}{n}\sum_{i=n}^{n}\left(h_{ref} - h_i\right)^2} \quad (2)$$

where h_{ref} is the reference elevation, h_i is the DSM elevation at point i, and n is the number of GCPs. A basic condition for the performance of the calculation is that the reference data should be an order better than the data to be evaluated. Thus, GCPs with sub-millimeter accuracy were utilized as reference points in the current study.

The variations in the elevation values between the reference data and the respective photogrammetric DSMs, acquired by different viewing geometries over all case studies, are presented in Table 6. It is obvious that the synergistic use of nadir- and oblique- viewing images during UAV campaigns managed to minimize the elevation errors in most DSMs.

In particular, the RMSE for the DSM arising from the combined use of nadir and oblique imagery in the case study of Egnemni, was calculated at 0.083 m, while the respective errors for the nadir-viewing and oblique-viewing DSMs were estimated at 0.092 m and 0.085 m, respectively. Furthermore, in the case study of Messarista, the RMSE for the DSM that emerged from the simultaneous processing of nadir and oblique imagery was calculated at 0.090 m. The respective RMSE for the nadir-viewing DSM was computed at 0.162 m, while the error of the oblique-viewing DSM was notably larger (i.e., 0.357 m).

Table 6. RMSE values emerging from the generated DSMs.

Case Study ID	Acquisition Geometry	RMSE (m)
Moira	Nadir	0.380
	Oblique	0.260
	Synergistic use	0.140
Egkremni	Nadir	0.092
	Oblique	0.085
	Synergistic use	0.083
Zachlorou	Nadir	0.506
	Oblique	0.498
	Synergistic use	0.475
Messarista	Nadir	0.162
	Oblique	0.357
	Synergistic use	0.090
Patras Port	Nadir	0.159
	Oblique	0.236
	Synergistic use	0.047

Although oblique-viewing images demonstrated an overall good performance in the generation of DSMs, we noticed a correlation between the UAV acquisition geometry and the topographic characteristics of the survey area. Specifically, nadir-viewing geometry was considered more suitable for DSM generation in flat urban (case study 5) or densely built-up (case study 4) areas, since it revealed smaller values in elevation variations than oblique imagery.

4. Discussion

The capability of performing a high accuracy field campaign with excellent repeatability is the basic demand [45–47] for landslide mapping and monitoring for assessing any deformation, usually within a steep and possibly dangerous environment. Sub-centimeter accuracy is a prerequisite in both horizontal and vertical axes in order to assess the activity of a landslide. Aiming to develop a guideline for accurate landslide mapping, we conducted four repeated tests within four active landslides with diverse characteristics. Both the horizontal and the vertical accuracy of the orthophotos and the DSMs were examined.

It was proved that the combination of nadir and oblique imagery produced more accurate results in all the studied cases. This output was in full accordance with the results of a previous study [28] that mentioned that overall accuracy is increased by 50% when nadir and oblique imagery are combined. The statistical comparison of the length of the digitized geodetic lines demonstrated that the integration of nadir and oblique images provides orthophotos with higher accuracy. Oblique imagery seems to provide more accurate results than nadir imagery. The only exception was the Messarista case study, where the processing of the oblique images yielded the worst results, compared to the respective result from the nadir images. This can be justified, as Messarista is a densely built-up village on a steep hill with very narrow backstreets. These narrow side streets cannot be clearly mapped in oblique imagery. Furthermore, the houses are built side-by-side and the front buildings hide the back ones. As a result, nadir imagery provides useful information that cannot be derived from oblique imagery alone.

Similar results were extracted from the measurements of the near distances of the mean centers that were calculated in the ArcMap environment. In almost all the cases, the simultaneous processing of nadir and oblique imagery provided more accurate results. The same combination produced more accurate DSMs in all the study areas. As presented in Table 4, in very steep environments (the four landslide areas) and in a flat area (Patras Port), the vertical accuracy of the oblique-nadir DSM was higher in comparison to the accuracies of either nadir or oblique DSMs.

Summarizing the overall assessment of the results, based on the application of simple statistical calculations, it was proved that the synergistic use of nadir and oblique imagery was considered to be the most appropriate geometry at 80% (i.e., in eight cases), according to the comparison of the length of the two lines of the five study areas (i.e., 10 cases) (Figure 8a). Moreover, comparing only nadir-viewing and oblique-viewing geometry, oblique acquisitions provided more accurate results by 70% (Figure 8b). Figures 9 and 10 display the corresponding evaluations using the comparison of near distances and RMSE calculation.

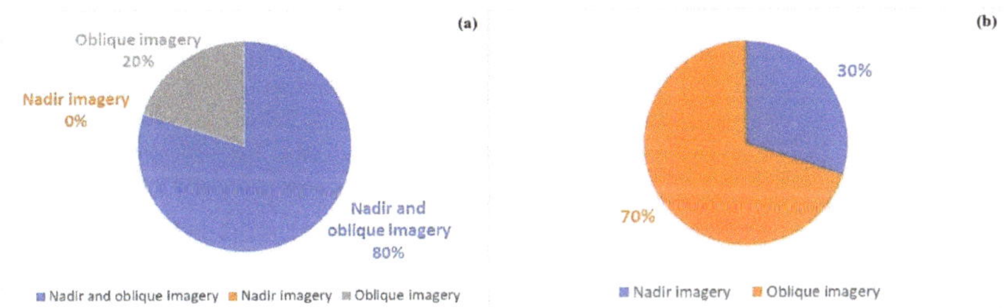

Figure 8. (**a**) Overall assessment of the accuracy of the photogrammetric products in accordance with UAV acquisition geometry using length comparison; (**b**) assessment of the accuracy of the photogrammetric products using length comparison between nadir- and oblique-viewing geometry.

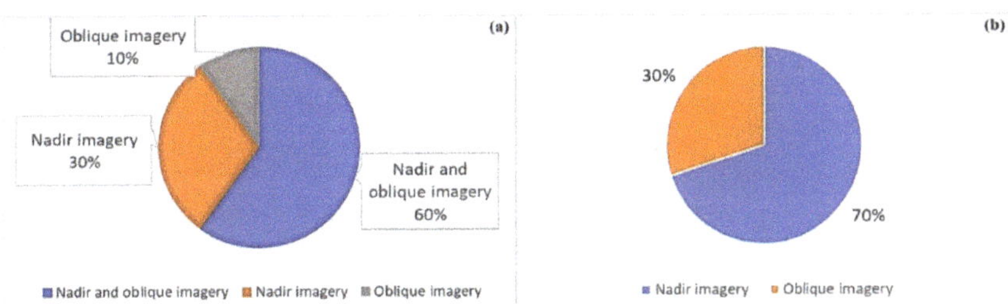

Figure 9. (**a**) Overall assessment of the accuracy of the photogrammetric products in accordance with UAV acquisition geometry using near distances; (**b**) assessment of the accuracy of the photogrammetric products using near distances between nadir- and oblique-viewing geometry.

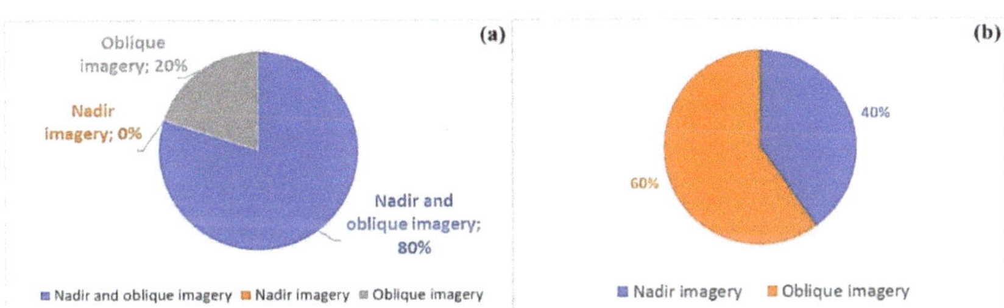

Figure 10. (**a**) Overall assessment of the accuracy of the photogrammetric products in accordance with UAV acquisition geometry using RMSE; (**b**) assessment of the accuracy of the photogrammetric products using RMSE between nadir- and oblique-viewing geometry.

Our findings are in accordance with those of previous studies [20,21,24–26]. Specifically, nadir and oblique images were compared with TLS data for the mapping of a fault plane [20]. The tests proved that both precision and accuracy have been increased when the two sets of images were combined. It was marked [21] that the combined use of nadir and oblique imagery offers a valued methodology for open-pit operations and for the continuous monitoring of the exploitation by managers. UAV data acquired from oblique and facade viewing angles were compared with nadir images in a steep and complex mountainous terrain, and the produced point clouds were compared with TLS data [24]. It was proved that the combination of oblique and façade images, compared to nadir data, increased the geometric accuracy of the derived point cloud data [24]. In a respective test [25], it was demonstrated that the combination of nadir and oblique images ameliorates the accuracy of 3D surface models in an agricultural area when no ground control points or a small number of ground control points are used. Nadir and oblique imageries were compared for 3D building representation [26]. The results proved that oblique data ameliorates the obtainable accuracy of the derived point cloud and of the produced 3D model. As described in [28], supplementing nadir image blocks with oblique images consistently mitigates the systematic error patterns within complex topography. The results from many scenarios proved that the combination of nadir and oblique images increased precision and accuracy, and reduced data gaps. Those previous results are in full accordance with our results. In all the studied areas, it was also proved that the simultaneous processing of the nadir and oblique imagery decreases the RMSE error and increases the geometric precision.

5. Conclusions

The objective of the current study was to develop a guidance for accurate landslide mapping in steep terrains. In this context, we performed four identical tests within four active landslides spreading throughout Western Greece. The main conclusions that emerged from this research included the following:

- It was proved that the acquisition of UAV oblique and nadir imagery and the synergistic processing increase overall centimeter accuracy.
- In general, oblique imagery provides more accurate results in steep terrains compared to nadir imagery. However, in areas that combine high slopes, dense urban settlements, and narrow streets (as in Messarista village), nadir imagery could not be omitted.
- Even in flat areas, such as Patras Port, the combined use of oblique and nadir imagery ameliorates the overall accuracy.
- A UAV flight campaign should be adjusted each time to an investigated area's characteristics and local topography.
- In steep terrains, an average flight altitude between 70 and 110 m above ground level or a ground spatial resolution of around 3 cm are recommended for both nadir and oblique campaigns in order to assess centimeter accuracy.

Author Contributions: Conceptualization, K.G.N.; methodology, K.G.N. and A.K.; software, K.G.N. and A.K.; validation, K.G.N., A.K. and I.K.K.; formal analysis, A.K.; investigation, K.G.N., A.K. and I.K.K.; resources, K.G.N.; data curation, K.G.N., A.K. and I.K.K.; writing—original draft preparation, K.G.N. and A.K.; writing—review and editing, K.G.N., A.K. and I.K.K.; project administration, K.G.N. All authors have read and agreed to the published version of the manuscript.

Funding: This research received no external funding.

Institutional Review Board Statement: Not applicable.

Informed Consent Statement: Not applicable.

Data Availability Statement: The data presented in this study are available on request from the corresponding author. The data are not publicly available, due to privacy considerations.

Conflicts of Interest: The authors declare no conflict of interest.

References

1. Van Westen, C.J. Remote sensing for natural disaster management. *Int. Arch. Photogramm. Remote Sens.* **2000**, *33*, 1609–1617.
2. Garnica-Peña, R.J.; Alcántara-Ayala, I. The use of UAVs for landslide disaster risk research and disaster risk management: A literature review. *J. Mt. Sci.* **2021**, *18*, 482–498. [CrossRef]
3. Salvini, R.; Mastrorocco, G.; Esposito, G.; Di Bartolo, S.; Coggan, J.; Vanneschi, C. Use of a remotely piloted aircraft system for hazard assessment in a rocky mining area (Lucca, Italy). *Nat. Hazards Earth Syst. Sci.* **2018**, *18*, 287–302. [CrossRef]
4. Giordan, D.; Manconi, A.; Remondino, F.; Nex, F. Use of unmanned aerial vehicles in monitoring application and management of natural hazards. *Geomat. Nat. Hazards Risk* **2016**, *8*, 1–4. [CrossRef]
5. Merdaway, A.; Guvenc, I. UAV assisted heterogeneous networks for public safety communications. In Proceedings of the IEEE Wireless Communications and Networking Conference Workshop—2nd International Workshop on Device to Device and Public Safety Communications, New Orleans, LA, USA, 9–12 March 2015; pp. 1–6.
6. Xu, Z.; Yang, J.; Peng, C.; Wu., Y.; Jiang, X.; Li, R.; Zheng, Y.; Gao, Y.; Liu, S.; Tian, B. Development of an UAS for post-earthquake disaster surveying and its application in Ms7.0 Lushan Earthquake, Sichuan, China. *Comput. Geosc.* **2014**, *68*, 22–30. [CrossRef]
7. Feng, Q.; Liu, J.; Gong, J. Urban Flood Mapping Based on Unmanned Aerial Vehicle Remote Sensing and Random Forest Classifier—A Case of Yuyao, China. *Water* **2015**, *7*, 1437–1455. [CrossRef]
8. Liu, C.-C.; Chen, P.-L.; Matsuo, T.; Chen, C.-Y. Rapidly responding to landslides and debris flow events using a low-cost unmanned aerial vehicle. *J. Appl. Remote Sens.* **2015**, *9*, 096016. [CrossRef]
9. Gomez, C.; Purdie, H. UAV-based photogrammetry and geocomputing for hazards and disaster risk monitoring—A Review. *Geoenvironmental Disasters* **2016**, *3*, 23. [CrossRef]
10. Themistocleous, K.; Hadjimitsis, D.G.; Michaelides, S.; Spizzichino, D.J.; Crosta, G.B.; Fernandez Merodo, J.A.; Bee, E. Best practices for monitoring. mitigation and preservation of cultural heritage sites affected by geo-hazards. In Proceedings of the EGU2018, Vienna, Austria, 8–13 April 2018; Volume 20.
11. Agüera-Vega, F.; Carvajal-Ramírez, F.; Martínez-Carricondo, P. Assessment of photogrammetric mapping accuracy based on variation ground control points number using unmanned aerial vehicle. *Measurement* **2017**, *98*, 221–227. [CrossRef]
12. Oniga, V.-E.; Breaban, A.-I.; Pfeifer, N.; Chirila, C. Determining the Suitable Number of Ground Control Points for UAS Images Georeferencing by Varying Number and Spatial Distribution. *Remote Sens.* **2020**, *12*, 876. [CrossRef]
13. Bolkas, D. Assessment of GCP Number and Separation Distance for Small UAS Surveys with and without GNSS-PPK Positioning. *J. Surv. Eng.* **2019**, *145*, 04019007. [CrossRef]
14. Tonkin, T.N.; Midgley, N.G. Ground-Control Networks for Image Based Surface Reconstruction: An Investigation of Optimum Survey Designs Using UAV Derived Imagery and Structure-from-Motion Photogrammetry. *Remote Sens.* **2016**, *8*, 786. [CrossRef]
15. Gindraux, S.; Boesch, R.; Farinotti, D. Accuracy Assessment of Digital Surface Models from Unmanned Aerial Vehicles' Imagery on Glaciers. *Remote Sens.* **2017**, *9*, 186. [CrossRef]
16. Taddia, Y.; González-García, L.; Zambello, E.; Pellegrinelli, A. Quality Assessment of Photogrammetric Models for Façade and Building Reconstruction Using DJI Phantom 4 RTK. *Remote Sens.* **2020**, *12*, 3144. [CrossRef]
17. Žabota, B.; Kobal, M. Accuracy Assessment of UAV-Photogrammetric-Derived Products Using PPK and GCPs in Challenging Terrains: In Search of Optimized Rockfall Mapping. *Remote Sens.* **2021**, *13*, 3812. [CrossRef]
18. Štroner, M.; Urban, R.; Reindl, T.; Seidl, J.; Brouček, J. Evaluation of the Georeferencing Accuracy of a Photogrammetric Model Using a Quadrocopter with Onboard GNSS RTK. *Sensors* **2020**, *20*, 2318. [CrossRef]
19. Štroner, M.; Urban, R.; Seidl, J.; Reindl, T.; Brouček, J. Photogrammetry Using UAV-Mounted GNSS RTK: Georeferencing Strategies without GCPs. *Remote Sens.* **2021**, *13*, 1336. [CrossRef]
20. Amrullah, C.; Suwardhi, D.; Meilano, I. Product accuracy effect of oblique and vertical non-metric digital camera utilization in UAV-photogrammetry to determine fault plane. *ISPRS Ann. Photogramm. Remote Sens. Spat. Inf. Sci.* **2016**, *3*, 41–48. [CrossRef]
21. Rossi, P.; Mancini, F.; Dubbini, M.; Mazzone, F.; Capra, A. Combining nadir and oblique UAV imagery to reconstruct quarry topography: Methodology and feasibility analysis. *Eur. J. Remote Sens.* **2017**, *50*, 211. [CrossRef]
22. Kyriou, A.; Nikolakopoulos, K.; Koukouvelas, I. How image acquisition geometry of UAV campaigns affects the derived products and their accuracy in areas with complex geomorphology. *ISPRS Int. J. Geo-Inf.* **2021**, *10*, 408. [CrossRef]
23. Nesbit, P.R.; Hubbard, S.M.; Hugenholtz, C.H. Direct georeferencing UAV-SfM in high-relief topography: Accuracy assessment and alternative ground control strategies along steep inaccessible rock slopes. *Remote Sens.* **2022**, *14*, 490. [CrossRef]
24. Tu, Y.-H.; Johansen, K.; Aragon, B.; Stutsel, B.M.; Angel, Y.; Camargo, O.A.L.; Al-Mashharawi, S.K.M.; Jiang, J.; Ziliani, M.G.; McCabe, M.F. Combining nadir, oblique, and façade imagery enhances reconstruction of rock formations using unmanned aerial vehicles. *IEEE Trans. Geosci. Remote Sens.* **2021**, *59*, 9987–9999. [CrossRef]
25. Meinen, B.U.; Robinson, D.T. Mapping erosion and deposition in an agricultural landscape: Optimization of UAV imageacquisition schemes for SfM-MVS. *Remote Sens. Environ.* **2020**, *239*, 111666. [CrossRef]
26. Vacca, G.; Dessì, A.; Sacco, A. The use of nadir and oblique UAV images for building knowledge. *ISPRS Int. J. Geo-Inf.* **2017**, *6*, 393. [CrossRef]
27. Lingua, A.; Noardo, F.; Spanò, A.; Sanna, S.; Matrone, F. 3D model generation using oblique images acquired by UAV. *Int. Arch. Photogramm. Remote Sens. Spat. Inf. Sci.* **2019**, *151*, 146–152. [CrossRef]
28. Nesbit, P.R.; Hugenholtz, C.H. Enhancing UAV–SfM 3D Model Accuracy in High-Relief Landscapes by Incorporating Oblique Images. *Remote Sens.* **2019**, *11*, 239. [CrossRef]

29. Anis, Z.; Wissem, G.; Riheb, H.; Biswajeet, P.; Essghaier, G.M. Effects of clay properties in the landslides genesis in flysch massif: Case study of Aïn Draham, North Western Tunisia. *J. Afr. Earth Sci.* **2018**, *151*, 146–152. [CrossRef]
30. Cushing, M. Evolution structurale de la marge nord ouest hellénique dans l' île de Levkas et ses environs (Grèce nord occidentale). In *Thèse 3me Cycle*; Univ. de Paris-Sud: Paris, France, 1985; p. 67.
31. Rondoyianni, T.; Mettos, A.; Paschos, P.; Georgiou, C. *Neotectonic Map of Greece, Scale 1:100.000, Lefkada Sheet*; I.G.M.E.: Athens, Greece, 2007.
32. Tsangaratos, P.; Loupasakis, C.; Nikolakopoulos, K.; Angelitsa, V.; Ilia, I. Developing a landslide susceptibility map based on remote sensing, fuzzy logic and expert knowledge of the Island of Lefkada, Greece. *Environ. Earth Sci.* **2018**, *77*, 363. [CrossRef]
33. Fernández-Blanco, D.; de Gelder, G.; Lacassin, R.; Armijo, R. A new crustal fault formed the modern Corinth Rift. *Earth-Science Rev.* **2019**, *199*, 102919. [CrossRef]
34. Jolivet, L.; Labrousse, L.; Agard, P.; Lacombe, O.; Bailly, V.; Lecomte, E.; Mouthereau, F.; Mehl, C. Rifting and shallow-dipping detachments, clues from the Corinth Rift and the Aegean. *Tectonophysics* **2010**, *483*, 287–304. [CrossRef]
35. Coded Targets and Scale Bars. Available online: https://agisoft.freshdesk.com/support/solutions/articles/31000148855-coded-targets-and-scale-bars (accessed on 1 March 2022).
36. Westoby, M.; Brasington, J.; Glasser, N.; Hambrey, M.; Reynolds, J. 'Structure-from-Motion' photogrammetry: A low-cost. effectivetool for geoscience applications. *Geomorphology* **2012**, *179*, 300–314. [CrossRef]
37. Eltner, A.; Sofia, G. Structure from motion photogrammetric technique. In *Developments in Earth Surface Processes*; Tarolli, P., Mudd, S.M., Eds.; Elsevier: Amsterdam, The Netherlands, 2020; Volume 23, pp. 1–24.
38. Micheletti, N.; Chandler, J.; Lane, S.N. Structure from motion (SFM) photogrammetry. In *Geomorphological Techniques*; British Society for Geomorphology: London, UK, 2015.
39. Clapuyt, F.; Vanacker, V.; Van Oost, K.; Clapuyt, F.; Vanacker, V.; Van Oost, K. Reproducibility of UAV-based earth topography reconstructions based on Structure-from-Motion algorithms. *Geomorphology* **2016**, *260*, 4–15. [CrossRef]
40. Agisoft Metashape User Manual. Available online: https://www.agisoft.com/pdf/metashape-pro_1_7_en.pdf (accessed on 8 March 2022).
41. Tamani, F.; Hadji, R.; Hamad, A.; Hamed, Y. Integrating Remotely Sensed and GIS Data for the Detailed Geological Mapping in Semi-Arid Regions: Case of Youks les Bains Area, Tebessa Province, NE Algeria. *Geotech. Geol. Eng.* **2019**, *37*, 2903–2913. [CrossRef]
42. ArcMap (Mean Center). Available online: https://desktop.arcgis.com/en/arcmap/latest/tools/spatial-statistics-toolbox/mean-center.htm (accessed on 5 March 2022).
43. Mitchell, A. *The ESRI Guide to GIS Analysis*, 1st ed.; ESRI Press: Redlands, CA, USA, 2005; Volume 2.
44. ArcMap (Near Analysis). Available online: https://desktop.arcgis.com/en/arcmap/10.3/tools/analysis-toolbox/near.htm (accessed on 5 March 2022).
45. Lucieer, A.; de Long, S.M.; Turner, D. Mapping landslide displacements using Structure from Motion (SfM) and image correlation of multi-temporal UAV photography. *Prog. Phys. Geogr.* **2014**, *38*, 97–116. [CrossRef]
46. Bhardwaj, A.; Sam, L.; Akanksha; Martín-Torres, F.J.; Kumar, R. UAVs as remote sensing platform in glaciology: Present applications and future prospects. *Remote Sens. Environ.* **2016**, *175*, 196–204. [CrossRef]
47. Eltner, A.; Baumgart, P.; Maas, H.-G.; Faust, D. Multi-temporal UAV data for automatic measurement of rill and interrill erosion on loess soil. *Earth Surf. Process. Landforms* **2014**, *40*, 741–755. [CrossRef]

applied sciences

MDPI

Article

Operational Mapping and Post-Disaster Hazard Assessment by the Development of a Multiparametric Web App Using Geospatial Technologies and Data: Attica Region 2021 Wildfires (Greece)

Triantafyllos Falaras [1,*], Ioanna Tselka [1,2], Ioannis Papadopoulos [1], Maria Nikolidaki [1], Andreas Karavias [1], Despoina Bafi [1], Aliki Petani [1], Pavlos Krassakis [1,3] and Issaak Parcharidis [1]

1 Department of Geography, Harokopio University of Athens, Eleftheriou Venizelou 70, 17676 Athens, Greece; ioannatselka@gmail.com (I.T.); papasmlnros13@gmail.com (I.P.); gp220113@hua.gr (M.N.); karavias@hua.gr (A.K.); gp219313@hua.gr (D.B.); alikipetani@gmail.com (A.P.); pkrassakis@hua.gr (P.K.); parchar@hua.gr (I.P.)

2 School of Rural and Surveying Engineering, Zografou Campus, National Technical University of Athens, Iroon Polytechniou 9, 15780 Athens, Greece

3 Centre for Research & Technology Hellas (CERTH), 15125 Athens, Greece

* Correspondence: falaras@hua.gr

Abstract: The environmental effects of wildfires are a hot issue in current research. This study examines the effects of the 2021 wildfires in the Attica region in Greece based on Earth observation and GIS-based techniques for the development of a web app that includes the derived knowledge. The effects of wildfires were estimated with the use of Sentinel-2 satellite imagery concerning burned area extent and burn severity using a NBR-based method. In addition, the erosion risk was modeled on a pre-fire and post-fire basis with the RUSLE. This study highlights the importance of assessing the effects of wildfires with a holistic approach to produce useful knowledge tools in post-fire impact assessment and restoration.

Keywords: wildfires; geospatial intelligence; web app; operational mapping; burn severity; soil erosion; Attica; RUSLE (revised universal soil loss equation); NBR (normalized burn ratio); remote sensing (rs)

Citation: Falaras, T.; Tselka, I.; Papadopoulos, I.; Nikolidaki, M.; Karavias, A.; Bafi, D.; Petani, A.; Krassakis, P.; Parcharidis, I. Operational Mapping and Post-Disaster Hazard Assessment by the Development of a Multiparametric Web App Using Geospatial Technologies and Data: Attica Region 2021 Wildfires (Greece). *Appl. Sci.* **2022**, *12*, 7256. https://doi.org/10.3390/app12147256

Academic Editor: Edoardo Rotigliano

Received: 19 May 2022
Accepted: 16 July 2022
Published: 19 July 2022

Publisher's Note: MDPI stays neutral with regard to jurisdictional claims in published maps and institutional affiliations.

1. Introduction

Greece in the 2021 fire season suffered from one of the greatest ecological disasters caused by wildfires. Especially in the first days of August, an intense prolonged heatwave struck Greece making it the worst that the country faced in almost 34 years. During this period of extreme conditions (27 July to 16 August 2021) in the Greek territory, many wildfires broke out due to the drought that increased wildfire vulnerability. All the means available, from the fire brigade to the local authorities and support from other countries, were used to fight these fires. In addition, evacuations of settlements were performed alongside the use of 112 to send alert messages to citizens, while the Copernicus Emergency Management Service (EMS) was also activated for rapid mapping of the wildfires in most cases [1].

Over the last couple of years, climate change has dramatically affected the Mediterranean region. The vulnerability of this region to drought and rising temperatures has led to severe ecological disasters annually. Specifically, Greece has been dealing with great environmental catastrophic events due to wildfire incidents and remarkably strong heatwaves [2]. The assessment of the wildfires' effect spatially provides a leading role in the documentation of vulnerable areas in different timescale evaluations [3,4].

Fire incidents constitute a dynamic process in planet Earth alteration affecting the global climate system [5]. Lately, wildfires show an increased occurrence, especially during the summer months, causing a rise in the overall period of the fire season [6]. The Mediterranean countries have been damaged over the last couple of years, due to extensive fire incidents causing irreversible damage to the environment [7]. Especially, high mountain regions characterized by Mediterranean climates seem to be more vulnerable to wildfire incidents due to the occurrence of high rainfall intensity in sparse vegetation cover areas [8]. Fire impacts on vegetation, soil, atmosphere, and society are dependent on fire characteristics such as severity and size [9]. Particularly, in Mediterranean ecosystems where fires occur in summer, the increased temperatures and evapotranspiration succeeded by the increased precipitation in autumn cause alterations to the ecosystems including increased soil loss [10].

Geospatial intelligence (GEOINT) has as a basic principle, the collection, analysis, and combination of all the available data, both geospatial and satellite imagery for Earth's geographical area, so as to utilize them in creating useful products in planning, decision making, and emergency response [11]. The integrated use of remote sensing and GIS proved very important in disaster management, with satellite imagery providing a synoptic view, spatiotemporal changes monitoring, and of course sufficient spatial coverage alongside GIS, which enables the utilization of all the geospatial information available. The capabilities of remote sensing for this purpose can be found in their operational use and their damage assessment after a wildfire event [12]. The use of satellite-based remote sensing in wildfires is well-known in the literature. For the mapping of burned areas, plenty of methods have been developed [13] with the use of multispectral data through the decades, including spectral indices among them such as the burn area index (BAI) [14] and normalized burn ratio (NBR) [15]. Additionally, with the use of satellite imagery, not only the burned area but also further information regarding the impact can be retrieved, as in the case of the dNBR [16], where burn severity can also be assessed. In addition, the combination of remote sensing with GIS in the case of wildfires provides a wide range of geospatial information regarding the post-fire period including the mapping of burned areas, soil erosion estimation, and recovery of the ecosystems [17].

Soil erosion caused by land degradation affects both natural and human resources, resulting from insufficient agricultural productivity, and is characterized as one of the most important threats worldwide [18]. Forest fires constitute one of the most critical causes of soil erosion due to their ability to burn large amounts of vegetation cover leading to an increase in runoff and sediment transfer [19,20]. For this purpose, several indices have been developed so as to assess the damage caused by wildfires on soil properties, such as the dNBR (differential NBR), which [8] is calculated based on pre- and post-fire satellite images [21].

Research on soil erosion has been extensively carried out over the past few decades since it is inextricably linked to the world's most serious environmental problems [8]. There are a wide variety of models that have been developed throughout the years considering soil loss assessment in several regions around the world [22]. The most commonly known models are the Universal Soil Loss Equation [23], its revised version (RUSLE) [24], and the Water Erosion Prediction Project (WEPP) [25]. Those models are used for the estimation of long-term average soil erosion values based on information extracted from rainfall data, soil properties, topography, and land cover management.

A recent study [2] assessed post-fire effects by mapping burned areas and their burn severity and also the soil erosion for an area within the Lokroi Municipality in Central Greece. This study made use of NBR to map the burned areas and to assess the burn severity with the use of Sentinel-2 and Landsat-8 imagery. In addition, to this affected area, the RUSLE was applied to estimate soil erosion. The combined use of these Earth observation and GIS-based methodologies was important before and after the wildfire and led to the identification of the vulnerable areas within the affected region. Another study was conducted [26] regarding the 2021 wildfires in Greece in Attica, Evia, and Peloponnese. In that study, the erosion vulnerability to these areas was evaluated after the fire events with the use of a GIS Boolean logic-based model. The results of the study led to the erosion risk assessment of these areas, making evident the increased susceptibility to surface runoff erosion. A similar study [27] studied southern Italy's Basilicata region, following an integrated approach with GIS and remote sensing regarding post-fire erosion risk. The soil fire severity was estimated by the use of Sentinel-2 images while the RUSLE was applied for the modeling of pre-fire and post-fire erosion for the first year after the events. Results of the study revealed the complex relations between fire severity and soil erosion factors while helping with soil loss estimation after the fire events.

Based on all the above-mentioned, this study aims at the assessment of the impact that the 2021 fire season wildfires had in the region of Attica, Greece. More analytically, four major wildfires of Schinos, Varympompi, Lavreotiki, and Vilia were mapped in an operational way with the use of Sentinel-2 imagery with the normalized burn ratio spectral index-based approach, thus extracting the burned area extent and assessing the burn severity while Corine Land Cover data was also used to analyze the consequences each fire caused. Furthermore, the soil erosion in the affected areas via the implementation of the revised universal soil loss equation (RUSLE) was assessed to estimate the soil loss and potential high-risk areas. In this way, to meet the purpose of this study effectively to gain geospatial intelligence on the effects of these wildfires, the results of all the applied methodologies were used to develop a web app that constitutes an invaluable tool in post-disaster hazard assessment. The results of the study showed the spatial damage that the wildfires caused in Attica in 2021 and they also demonstrated a significant increase in soil erosion in the affected areas.

2. Wildfires in Attica Region 2021

In the region of Attica, which is located in Central Greece where the capital of Greece, Athens, is, during the 2021 fire season four major wildfires occurred with three of them happening in the above-mentioned August period resulting in extensive damages. To all the four wildfires the Copernicus EMS (Emergency Management Service) was activated with the activations EMSR51O, EMSR527, EMSR540, and EMSR542. Spatially, according to Figure 1, these wildfires were located in western, north-eastern, and south-eastern Attica with the closest to the metropolitan area of Athens being the fire in Varympompi [1,28,29].

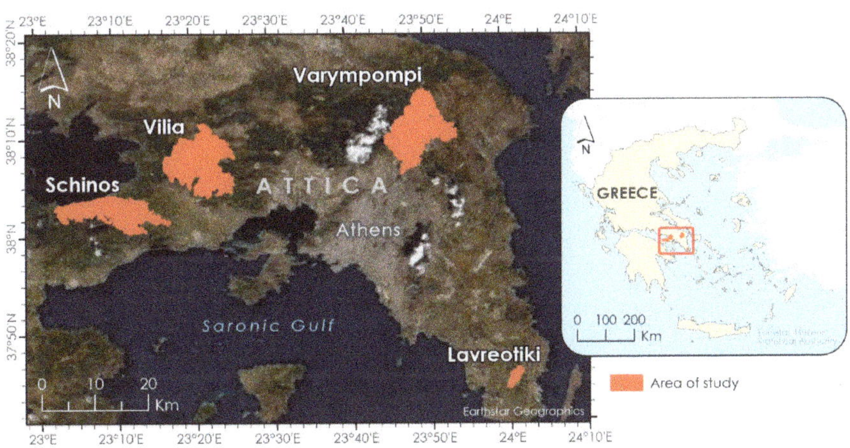

Figure 1. Areas of study.

Starting with the May 19th, 2021 wildfire in Schinos that broke out in the evening of that day close to the Schinos settlement in Korinthia. The fire moved from the east to western Attica, reaching to the north of the Kineta settlement during almost five days of activity. Damages were limited to the build-up environment and extended to the forests of the area. The affected area includes the northern slopes of the Geraneia Mountains, which includes a Natura 2000 site, and the geomorphological status of the area with high slopes aids the development of floods and other post-fire hazards to the region [28].

The next fire is that of Varympompi, which broke out on the midday of August 3rd, 2021, in the Ano Varympompi area, and lasted for the next few days. More precisely, the northern part of the Attica region was affected from Adames in Kifisia to Ippokrateios, Politeia, mainly within the base of Parnitha Mountain. Although the impact included the forest vegetation of the area, an important part of the urban areas of settlements was impacted due to the complexity of the forest and building mixture [1,26].

The wildfire in Lavreotiki started on the morning of August 16th in Markati in the northwest of Lavreotiki. Close-to-settlement areas such as coniferous forests, natural vegetation, and agricultural land were affected by the wildfire. The meteorological conditions made it difficult for the fire brigade to put the fire under control despite the smaller extent than the previously mentioned wildfires [1,26].

Lastly, the wildfire in Vilia also broke out on August 16th at midday a few hours after this in Lavreotiki in Pateras Mountain close to the Vilia settlement. The wildfire was burning an area predominantly consisting of sclerophyllous vegetation and coniferous forests for more than five days with a rapid spread day by day. The above-mentioned conditions in that period led to the devastation of a large area between Vilia and Nea Peramos despite the huge effort and the increased ground and aerial support [1,29].

3. Materials and Methods

3.1. Data and Software

In Table 1, the used datasets are briefly presented. Free and open accessible datasets were selected to be used for each methodological part described in the following subsections.

Table 1. Utilized datasets.

Datasets	Format	Resolution Spatial ǀ Temporal	Source	Purpose/Use
Sentinel-2 imagery [1]	Optical Level-2A	10 m (Used) ǀ 2021	Copernicus Open Access Hub	Wildfire Mapping and *C* Factor
Corine Land Cover 2018	Vector (Polygon)	- ǀ 2018	Copernicus Land Monitoring Service	Wildfire Mapping and *P* Factor
EU-DEM v1.0	Raster	25 m ǀ 2000 (2016)	Copernicus Land Monitoring Service	*LS* Factor
Meteorological data	Text	- ǀ 2021	Meteo.gr of the National Observatory of Athens (NOA)	*R* Factor
Topsoil physical properties for Europe (based on LUCAS topsoil data)	Raster	500 m ǀ 2015	European Soil Data Centre (ESDAC) of Joint Research Centre (JRC)	*K* Factor
Topsoil organic carbon (LUCAS) for EU25	Raster	500 m ǀ 2014	European Soil Data Centre (ESDAC) of Joint Research Centre (JRC)	*K* Factor

[1] The used Sentinel-2 acquisitions are presented in detail in Table 2.

For the operational wildfire mapping, optical/multispectral Sentinel-2 mission satellite images of the ESA Copernicus program were utilized which are open and accessible from the Copernicus Open Access Hub [30]. The products used were the atmospherically corrected and scene classified Level-2A Bottom-of-Atmosphere (BOA) [31]. A key characteristic of Sentinel-2 images important in operational purposes is the 5-day interval between satellite acquisitions over an area and their availability on the Copernicus Open Access

Hub a few hours after their acquisition. In Table 2, the full list of Sentinel-2 acquisitions is given for each purpose of this study regarding the mapping of wildfires and the *C* factor of the RUSLE. The image selection was redundant to the cloud-free scene availability, taking into consideration the temporal coherence for each purpose.

Table 2. List of used Sentinel-2 acquisitions for each area and purpose.

Area	Wildfire Start Date	Sentinel-2 Acquisition Date	Relation with the Wildfire Event	Purpose
Schinos	19 May 2021	18 May 2021 23 May 2021	Pre-event Post-event	Wildfire Mapping
Varympompi	3 August 2021	29 July 2021 8 August 2021	Pre-event Post-event	
Lavreotiki	16 August 2021	13 August 2021 18 August 2021	Pre-event Post-event	
Vilia	16 August 2021	29 July 2021 26 August 2021	Pre-event Post-event	
All the areas	-	10 May 2021 17 September 021	Pre-event Post-event	*C* Factor

Land cover information for the affected study areas was retrieved with the use of the Corine Land Cover 2018 (CLC 2018) in vector polygon format from vector geodatabase open available from Copernicus Land Monitoring Service [32]. The CLC 2018 is highly accurate (≥85% accuracy) with a minimum mapping unit (MMU) of 25 hectares and a mapping minimum width (MMW) of 100 m, while it is detailed enough by having 44 land cover classes at its most analytical level-3 [33]. In this study, it was used to map the affected area of the wildfire land cover and for the RUSLE's *P* factor.

Regarding the meteorological data required for the estimation of the *R* factor of the RUSLE, these were retrieved from the Meteo.gr of the National Observatory of Athens (NOA) [34]. Data regarding the precipitation for the examined period were obtained from the respective meteorological stations of the Meteo.gr network within and around of each area of study.

The EU-DEM v1.0 is a digital elevation model (DEM) that was obtained freely from the Copernicus Land Monitoring Service for the application of the RUSLE's *LS* factor. With a resolution of 25 m, it was produced with the fusion of SRTM and ASTER GDEM datasets [35].

The last dataset used refers to the topsoil properties necessary for the *K* factor calculation of the RUSLE. This dataset consists of 500 m rasters acquired from the European Soil Data Centre (ESDAC) of the Joint Research Centre (JRC). More specifically, the dataset was derived from the topsoil data collected during the Land Use and Cover Area frame Statistical survey (LUCAS). It includes the content percentage (%) of clay, sand, silt, and the predicted topsoil soil organic carbon content in g C kg^{-1} [36,37].

The software used includes the free and open ESA STEP SNAP v8.0 remote sensing software for the processing of the satellite images and the commercial GIS suite of ESRI that includes ArcGIS Desktop v.10, ArcGIS Pro v2.8–2.9, and ArcGIS Online with the Web AppBuilder.

3.2. Methodology

The methods applied for the purposes of this study are described in the following subsections. In Figure 2, the flowchart presents all the steps followed for each part of the methodology with the mapping of the wildfires presented on the left and the implementation of the RUSLE on the right part. In the middle, the different products are presented with the following addition of them to the web app.

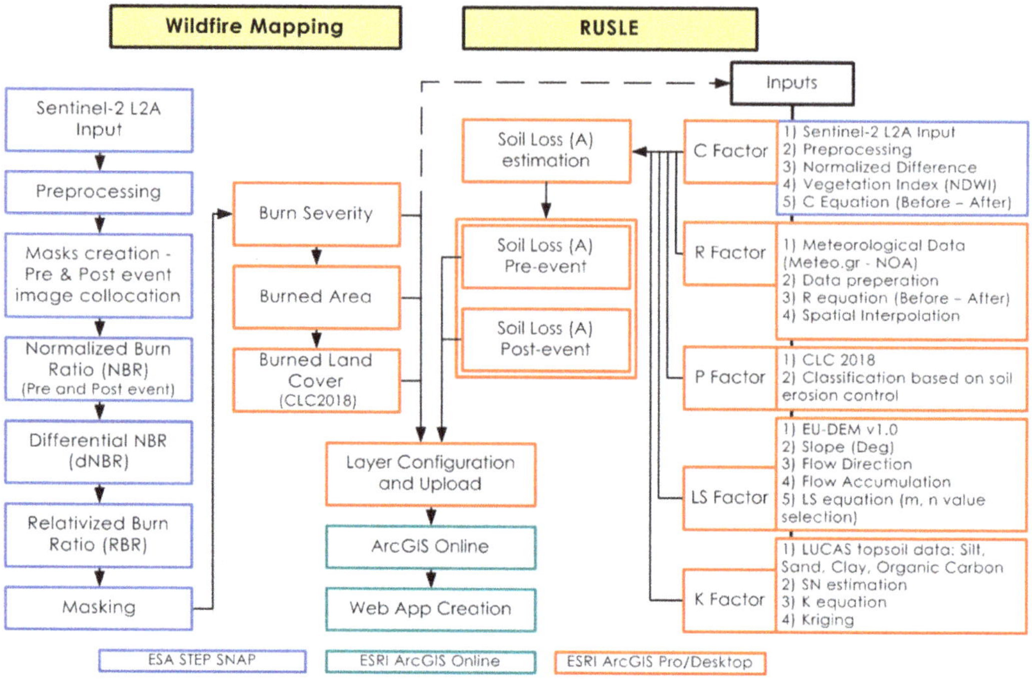

Figure 2. Methodology flowchart.

3.2.1. Operational Wildfire Mapping

The operational mapping of the previously mentioned wildfires was performed with the use of the Sentinel-2 L2A imagery listed in Table 2. In order to have the most accurate and timely available results, cloud-free images acquired immediately before and after each wildfire event were carefully selected. These images were imported in SNAP and then preprocessed with nearest neighbor resampling of the spectral bands to 10 m resolution and then subset to the area of interest (AOI) extent to enhance processing performance. The next steps included the creation of a cloud and water mask with band maths operations based on the available scene classification from the L2A product and the estimation of the water mask with the normalized difference water index (NDWI) [38] and the collocation of the pre-fire and post-fire event imagery followed to merge them into one product.

For this study, for the wildfire mapping, the related spectral indices were applied. More analytically, the normalized burn ratio (NBR) is the used index that utilizes the spectral bands of near-infrared (NIR) in which burned areas present low reflectance and shortwave-infrared (SWIR) in which burned areas show high reflectance for the mapping of burned areas and the assessment of burn severity based on Equation (1). The index has high values where healthy vegetation exists and low values in burned areas and areas with low or no vegetation presence [15,16,39,40].

$$\text{NBR} = \frac{\text{NIR} - \text{SWIR}}{\text{NIR} + \text{SWIR}} \quad (1)$$

The NBR index was calculated before (pre-fire) and after (post-fire) the fire event and then the difference between the two, the dNBR (Equation (2)), results in a better distinction of the burned areas while the burn severity is also assessed. Burn severity is a term used to express the degree of a wildfire's impact on an area's ecosystem. The assessment of burn severity contributes to quantifying the impact of a wildfire and aids restoration attempts and natural disaster management. Based on the dNBR values, the burn severity is classified into categories as is shown in Table 3 with the higher values corresponding to higher severity and thus the more severe impact of the wildfire on an area [40–44].

$$dNBR = PrefireNBR - PostfireNBR \quad (2)$$

After the dNBR estimation, the relativized burn ratio (RBR) was calculated based on Equation (3). This index aids the assessment of burn severity in a diversity of ecosystems and regions by enhancing accuracy. Additionally, it aids the identification of the changes after a fire in areas with low vegetation [45].

$$RBR = \frac{dNBR}{prefireNBR + 1.001} \quad (3)$$

Table 3. Burn severity classification according to dNBR values [16,46].

dNBR Value *	Burn Severity
0.100–0.269	Low severity
0.270–0.439	Moderate-low severity
0.440–0.659	Moderate-high severity
0.660–1.300	High severity

* Values $\leq$ 0.099 represent unburned areas.

The final RBR raster was masked using the above-mentioned cloud and water mask in order to be not only clear from any possible clouds, but also to eliminate existing water surfaces due to their spectral characteristics which make them attributed falsely as burned areas. Results production followed with the masked RBR raster import to the GIS software and its reclassification according to the Table 3 burn severity classes. Conversion to vector polygons, burned area extraction, and clipping of CLC 2018 to the burned area extent were the last steps including the area estimations.

3.2.2. Revised Universal Soil Loss Equation (RUSLE)

The Revised Universal Soil Loss Equation (RUSLE) model provides a widely known and useful tool for soil erosion assessment [2,22,47,48]. Developed in the late 1970s [23], it was formulated so as to conclude a better estimation of the initial parameters defined by the Universal Soil Loss Equation (USLE) [49]. The derived methodology represents the influence of topography, soil properties, meteorological parameters, and land cover on surface and rill erosion [2,47,48]. The average soil loss assessment is enabled by the RUSLE according to the specific study area characteristics [2] and it is based on an empirical equation (Equation (4)), constituting of five factors [24], which can be easily implemented through a GIS framework:

$$A = R \times K \times LS \times C \times P \quad (4)$$

where A is the estimated mean seasonal soil loss (ton h^{-1} $Season^{-1}$), R represents the rainfall erosivity factor (MJ mm ha^{-1} h^{-1} $Season^{-1}$), K is the soil erodibility factor (ton h MJ^{-1} mm^{-1}), LS is the slope length and steepness factor (dimensionless), C is the cover management factor (dimensionless), and P represents the conservation practice factor (dimensionless). These factors are of vital meaning in soil erosion assessment, and they should be very carefully estimated in order to lead to highly accurate results (source).

R factor plays a crucial role in RUSLE modeling since it defines the possibility of erosion risk [49] and it is estimated based on meteorological data from meteorological stations surrounding the study areas as previously mentioned [50]. Using the empirical formulation (Equation (5)) developed by Loureiro and Couthino [50], monthly rainfall data concerning seasonal time periods of 2021, were used to calculate the R factor:

$$R = \frac{\sum_{i=1}^{12}(7.05 \times r_{10} - 88.92 \times d_{10})}{N} \tag{5}$$

where R represents the rainfall erosivity factor (MJ mm ha^{-1} h^{-1} season^{-1}), N is the total months calculated annually, r_{10} is the monthly rainfall exceeding 10 mm, and d_{10} is the number of days when daily rainfall exceeded 10 mm per month. Specifically, rainfall data acquired from each weather station were implemented using a spatial interpolation technique, inverse distance weighting (IDW), so as to define unknown meteorological values within the study areas.

K factor refers to the rainfall impact on soil properties resulting in soil erosion due to sediment penetration, detachment, and transport [23,24,27]. The soil erodibility factor is affected by soil properties such as structure, organic matter, permeability, and soil texture [48]. In this study, the estimation of the K factor was based on Equation (6) developed by Williams and Renard [51]:

$$= 0.2 + 0.3\exp\left(0.0256 \times Sa \times \left(1 - \tfrac{Si}{100}\right)\right) \times \left(\tfrac{Si}{Cl+Si}\right)^{0.3} \times \left(1.0 - \tfrac{0.25\times C}{C+\exp(3.72-2.95C)}\right) \times \left(1 - \tfrac{0.7\times SN}{SN+\exp(-5.51+22.9SN)}\right) \tag{6}$$

where K is the soil erodibility factor (ton h MJ^{-1} mm^{-1}), Sa represents the percentage of salt, Si is the percentage of silt, Cl is the percentage of clay, C is the percentage of organic carbon, and SN stands for SN = 1 − (Sa/100). Soil data acquisition was based on datasets provided by the ESDAC database [37]. The resulted K values were implemented on the spatial interpolation method Kriging to produce values covering the total surface of the study areas.

The topographic effect on soil erosion is determined by the impact of the slope length and steepness factor (LS factor) [27]. LS factor constitutes the result of slope length (L) and slope steepness (S) multiplication. Increased slope steepness values determine increased soil erosion risk due to the increase in the velocity and erosivity of the accumulated runoff [2]. The calculation of the LS factor was based on Equation (7) of Moore and Burch [52]:

$$LS = \left(\frac{U}{L_0}\right)^m \times \left\{[\sin\left(\frac{\beta \times 0.01745}{S_0}\right)]^n\right\} \times (m+1) \tag{7}$$

where LS is the topographic factor, U is the flow accumulation multiplied with the pixel size, L_0 is the slope length (22.13 m), β is the slope in degrees, S_0 is the slope percentage (9%), m is sheet erosion ranging from 0.4 to 0.6, and n is rill erosion ranging from 1 to 1.3. The LS factor was created based on data derived from a digital elevation model (DEM) into a GIS setting. The rill erosion values corresponded to n = 1.1 while the sheet erosion values ranged according to the examination area each time. Specifically, for the regions of Schinos, Vilia, and Varympompi, the sheet erosion value was set as m = 0.45, whereas for the Keratea region the corresponding value was m = 0.5.

C factor demonstrates the cover management factor providing a measure of soil erosion rate as controlled by the cropping and management practices within a region [23,24]. The cover management factor was calculated based on satellite data acquisition corresponding to specific time periods. Specifically, the satellite data consisted of Sentinel-2 images for the regions of Schinos, Vilia, Varympompi, and Keratea, corresponding to time periods before and after the fire. Particularly, through the acquired data, normalized derived vegetation index (NDVI) images were generated to produce the factor's values based on the formulation (Equation (8)) developed by Durigon et al. [53]:

$$C = \frac{1 - NDVI}{2} \tag{8}$$

The support practice *P* factor represents the practices on agricultural land that affect the soil erosion processes [54] *P* factor was estimated based on the Corine Land Cover 2018 dataset. According to Yang et al. [55], all Corine Land Cover classes were assigned the value of 1, except in agricultural regions where the *P* factor was given the value of 0.5.

3.2.3. Web App Development

The final processing stage includes the preparation of the results layers to be suitable to be published as web layers on ArcGIS Online. ArcGIS Online is a cloud-based web GIS Software as a Service Software (SaaS) that is accessible from any device with internet access and which is interactive and enables web map creation among other capabilities. After the publication of the web layers, the web map needed for the web app was created. Finally, the web app in the Web AppBuilder was then developed with the addition of the previously mentioned map, the setting of the user interface (UI), parameters, and widgets [56,57].

4. Results

4.1. Burned Areas and Burn Severity

The total burned area of the four wildfires in Attica during the 2021 fire season reached 243.98 km^2 making evident the large extent of damage caused to the region. As Figure 3 presents, the wildfire in Vilia burned the largest area reaching almost 96 km^2, followed by Varympompi with 78.95 km^2, Schinos with 64.05 km^2, and Lavreotiki that has a burned an area of 5 km^2.

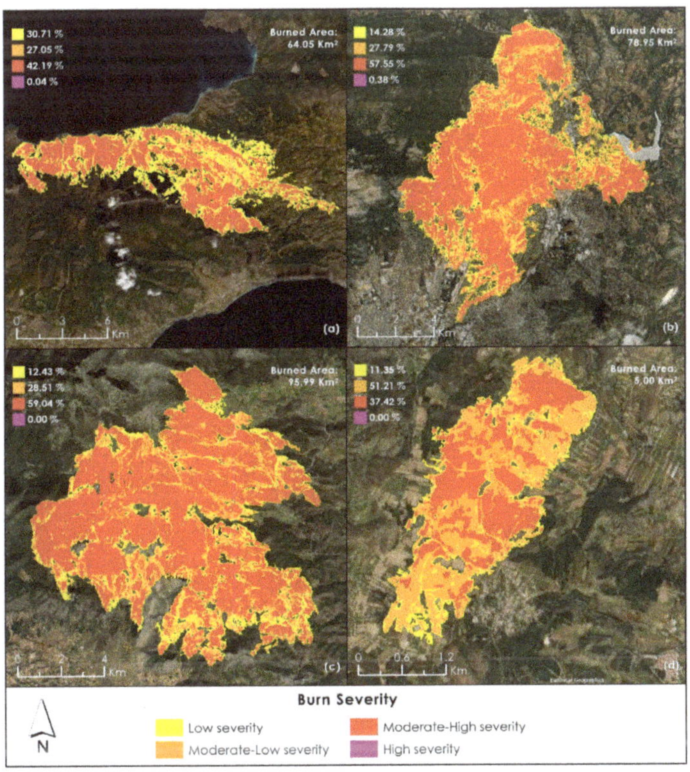

Figure 3. Burn Severity of 2021 Attica region wildfires: (**a**) Schinos; (**b**) Varympompi; (**c**) Vilia; (**d**) Lavreotiki. The burned area classification based on burn severity is illustrated including the percentage of the burned area total of each area per class.

Analyzing the burn severity of these wildfires, generally the impact was severe enough with the moderate-high severity accounting for 53.69% or 131 km^2 out of the total burned area with the moderate-low (28.36%) and low (17.8%) following, while high severity covers 0.14%. More specifically, considering each area based on Figure 3, it is evident that the moderate-high severity characterizes all the burned areas except Lavreotiki, where lower severity levels prevail, whereas in Schinos a mostly close-to-equal distribution is observed among burn severity levels. It should be noted that in the cases of Varympompi and Vilia, the percentages of burn severity levels coverage in each case follow an almost identical pattern in the moderate severity levels percentage.

Regarding the affected land cover based on CLC 2018 as presented in Figure 4 and Table 4, the total numbers show that the third CLC category which includes forests, shrubs, and/or herbaceous vegetation is heavily impacted by the wildfires covering 82.71% or 201.8 km^2 of the total burned area, thus highlighting the ecological disaster in the region. To put it another way, coniferous forests (72.48 km^2), sclerophyllous vegetation (52.47 km^2), transitional woodland-shrub (46.13 km^2), and mixed forests (24.73 km^2) comprised the most affected land cover. In Schinos, the burned area consisted of 62.80% of coniferous forests, in Varympompi transitional woodland-shrubs and mixed forests took up 46.60%, although it should be mentioned that a significant part of the artificial surfaces category was impacted in the region reaching 9.56% of the total burned area, including discontinuous urban fabric. Continuing with Lavreotiki, coniferous forests, natural grasslands, and sclerophyllous vegetation accounted for 64.44% of the total burned area, and lastly, in Vilia Sclerophyllous vegetation was 49.21% of the burned area.

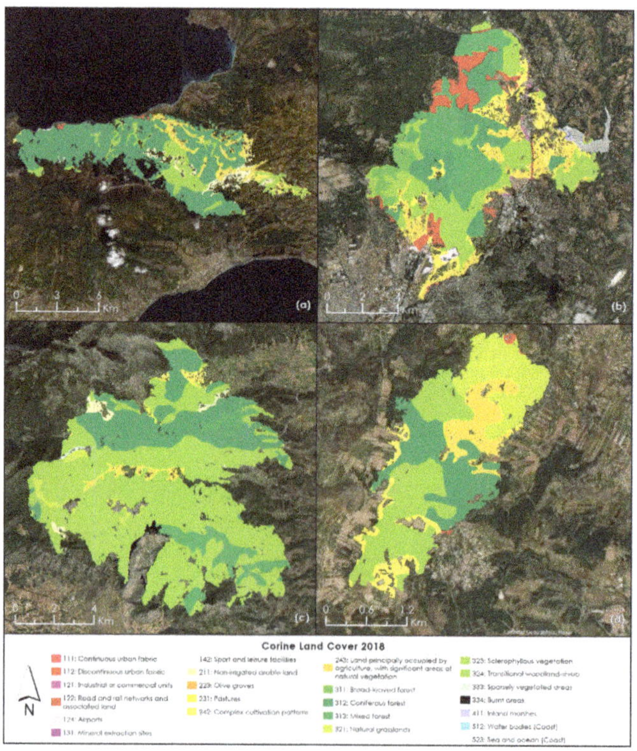

Figure 4. Burned land cover of Attica region wildfires based on Corine Land Cover 2018: (**a**) Schinos; (**b**) Varympompi; (**c**) Vilia; (**d**) Lavreotiki. The third level of Corine Land Cover 2018 was utilized and clipped to each burned area to assess the affected land cover.

Table 4. Burned land cover of Attica region wildfires based on Corine Land Cover 2018 (% of the burned area total of each area per land cover category).

CLC 2018	Schinos	Varympompi	Lavreotiki	Vilia
111: Continuous urban fabric		0.02%		
112: Discontinuous urban fabric	0.42%	7.96%	0.47%	
121: Industrial or commercial units		0.29%		
122: Road and rail networks and associated land		0.36%		
124: Airports		0.31%		
131: Mineral extraction sites		0.00%		
142: Sport and leisure facilities		0.63%		
211: Non-irrigated arable land	5.47%			1.29%
223: Olive groves	0.16%			
231: Pastures		0.23%		0.19%
242: Complex cultivation patterns	2.42%	10.76%	8.21%	0.76%
243: Land principally occupied by agriculture, with significant areas of natural vegetation	10.58%	9.87%	13.06%	1.45%
311: Broad-leaved forest		4.33%		
312: Coniferous forest	62.79%	15.41%	23.18%	19.73%
313: Mixed forest	0.01%	22.11%		7.58%
321: Natural grasslands			20.92%	1.57%
323: Sclerophyllous vegetation	3.04%	2.88%	20.35%	49.21%
324: Transitional woodland-shrub	14.77%	24.49%	13.81%	17.34%
333: Sparsely vegetated areas	0.16%			0.21%
334: Burnt areas (Previous wildfire)				0.65%
411: Inland marshes		0.32%		
512: Water bodies (Coast)		0.02%		
523: Sea and ocean (Coast)	0.18%			

The correlation between burn severity and land cover led to useful findings. More analytically, transitional woodland-shrubs were the main land cover category characterized by high severity along with broad-leaved forests. Moderate-high severity existed mostly in coniferous forests (41 km^2), schlerophyllous vegetation (30.96 km^2), and transitional woodland-shrub (28.65 km^2). In Vilia, 28.49 km^2 of schlerophyllous vegetation belonged to this burn severity level as well as 19.3 km^2 of coniferous forests in Schinos, 12.31 km^2 of mixed forest in Varympompi, and 0.79 in Lavreotiki accordingly. Regarding moderate-low severity, this was found in coniferous forests (18 km^2), schlerophyllous vegetation (15.57 km^2), and transitional woodland-shrubs (11.5 km^2). In each area, the following prevailed: the schlerophyllous vegetation of Vilia (13.5 km^2), coniferous forests of Schinos (10.27 km^2), transitional woodland-shrubs of Varympompi (4.26 km^2), and schlerophyllous vegetation of Lavreotiki (0.6 km^2). Lastly, low severity is primarily met in coniferous forests (13.4 km^2) and it mainly characterizes coniferous forests in Schinos (10.9 km^2), sclerophyllous vegetation in Vilia (5.14 km^2), and complex cultivation patterns in Varympompi (2.99 km^2) and Lavreotiki (0.17 km^2).

4.2. Soil Erosion Risk

Regarding the soil erosion risk derived from the use of the RUSLE concerning the time periods before and after the fire incidents in the Attica region in 2021, the spatial distribution of soil loss is presented in the following Figures 5–8. Due to the assessed areas' heterogeneity of characteristics, the soil loss results in ton/ha/season were converted into a universal classification based on each area as presented in Table 5. More analytically, the 5 classes (from very low to very high erosion risk) were specified on an equal interval for each case estimated from the standard deviation average of soil loss before and after the fire events. In this way, a comparable classification, taking into consideration each area's soil loss results, was constructed which enables the identification of the change concerning erosion risk in pre-fire and post-fire results.

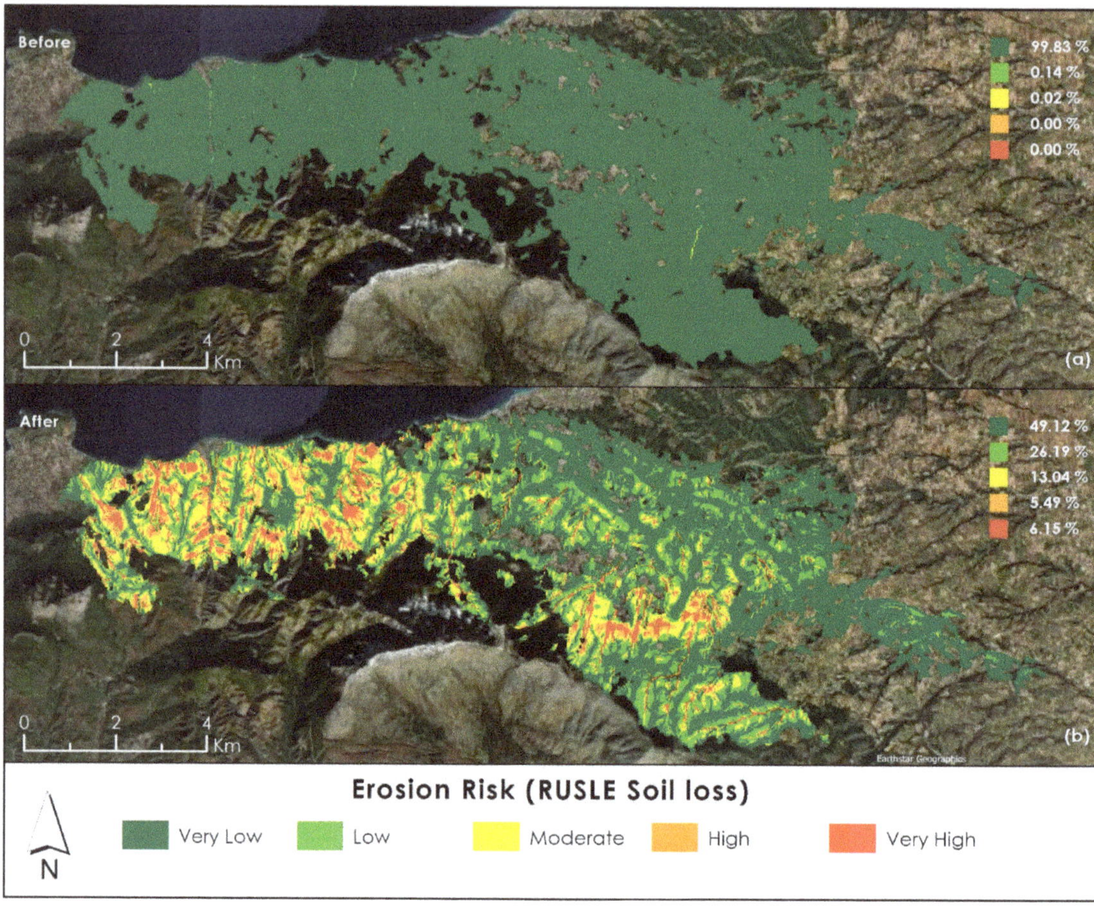

Figure 5. Erosion Risk according to RUSLE soil loss for Schinos burned area: (**a**) before the fire; (**b**) after the fire. The erosion risk as classified is presented including the percentage of the burned area total of each area per class.

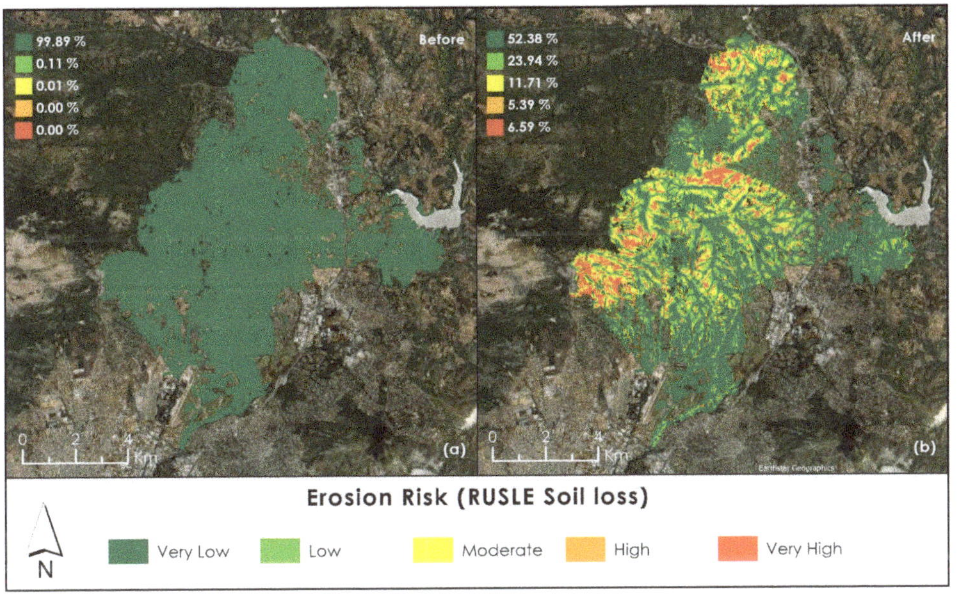

Figure 6. Erosion Risk according to RUSLE soil loss for Varympompi burned area: (**a**) before the fire; (**b**) after the fire. The erosion risk as classified is presented including the percentage of the burned area total of each area per class.

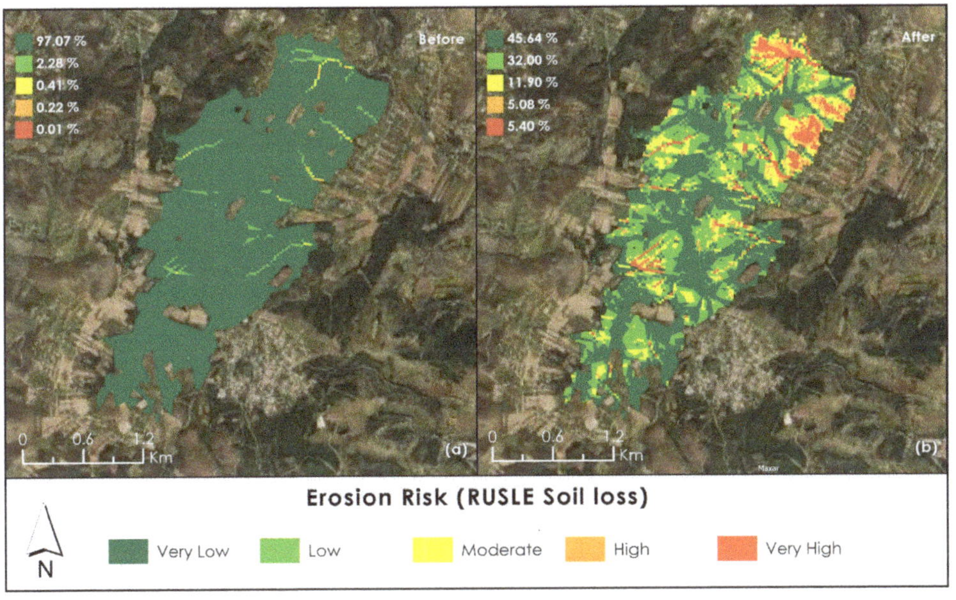

Figure 7. Erosion risk according to RUSLE soil loss for Lavreotiki burned area: (**a**) before the fire; (**b**) after the fire. The erosion risk as classified is presented including the percentage of the burned area total of each area per class.

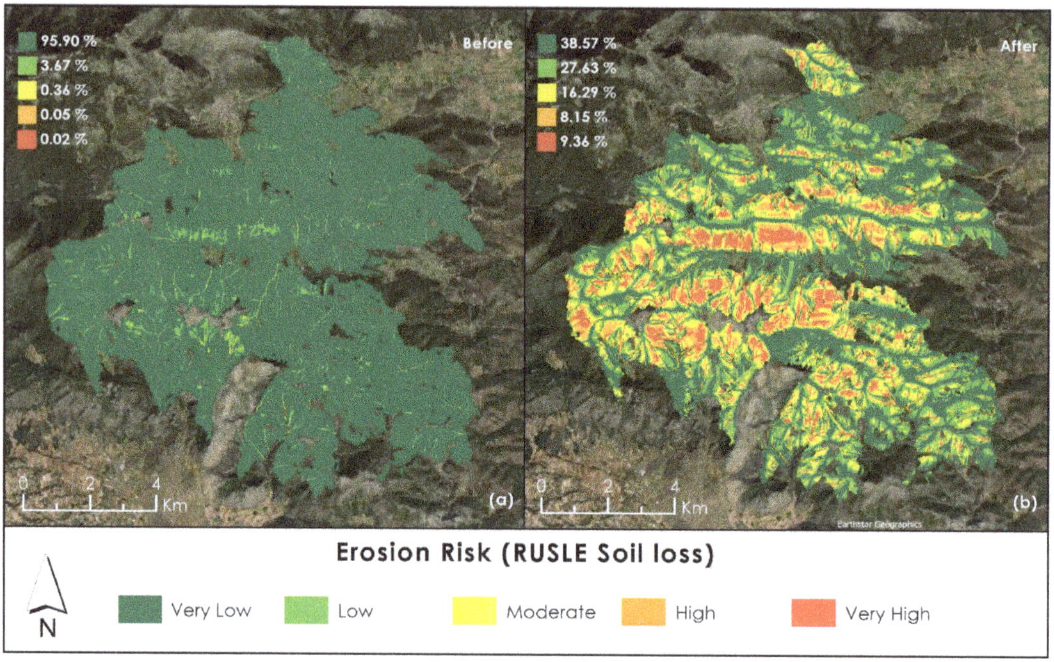

Figure 8. Erosion risk according to RUSLE soil loss for Vilia burned area: (**a**) before the fire; (**b**) after the fire. The erosion risk as classified is presented including the percentage of the burned area total of each area per class.

Table 5. Erosion risk classification for RUSLE soil loss (ton/ha/season), presenting each class for each area.

Erosion Risk	Schinos	Varympompi	Lavreotiki	Vilia
Very Low	0.00–122.44	0.00–235.67	0.00–42.60	0.00–1957.77
Low	122.44–244.88	235.67–471.34	42.60–85.20	1957.77–3915.54
Moderate	244.88–367.32	471.341–707.01	85.20–127.80	3915.54–5873.31
High	367.32–489.76	707.01–942.68	127.80–170.40	5873.31–7831.08
Very High	>489.76	>942.681	>170.401	>7831.08

Soil loss results, in general, show a clear differentiation between the pre-fire and post-fire states in all four areas. More specifically, the pre-fire state is characterized by very low soil erosion risk, which is predominant in all cases. Regarding the post-fire soil erosion risk, it was clearly increased in the four areas of interest. Very low soil erosion risk was reduced to 45.76% whereas the low and moderate soil erosion risk constitutes a combined 40.12% of the burned areas. Moreover, the rest areas are characterized by high and very high soil erosion risk values of almost 7% each. It has to be mentioned that the results of the RUSLE cover a slightly reduced area than the original burned areas due to the slope estimation in the *LS* factor.

To begin with, the analysis of Schinos burned areas (Figure 5). Figure 5a illustrates that the area has a very low soil erosion risk. In the post-fire state (Figure 5b) the increased soil erosion risk is evident in the area with a reduction of very low to almost 50% area coverage with a main concentration in an axis from the north to the southeastern parts of the region. To analyze more, the western part of the area is characterized by concentrations of moderate to very high erosion risk as well as the southern part with an important concentration on the east–west axis. Moderate (13.04% or 7.43 km^2) and low (26.19% or 14.92 km^2) are distributed over the area covering a total large extent. A further analysis associated with the burn severity shows that moderate-high severity prevails in all erosion risk classes while considering that land cover coniferous forests are the most affected.

Regarding the burned areas of Varympompi in Figure 6a, the area in the pre-event soil erosion risk was very low. Analyzing the post-fire soil erosion risk as of Figure 6b, very low erosion risk covered 52.37% of the area, especially in the southern and eastern parts. Low erosion risk was distributed over the area characterizing 23.94% of it alongside moderate (11.71% or 8.14 km^2). The higher erosion risk categories took up the rest, almost 12% of the area, with very high erosion risk covering important parts to the west in the Partnitha Mountain base and around the Afidnes area in the north-central to northern parts. The distribution of erosion risk based on burn severity followed the same trends proportionally as the predominant moderate-high severity category. Assessing erosion risk with land cover, very low erosion risk characterized mostly transitional woodland-shrubs and low mixed forests alongside moderate, and high while very high was mainly spread within mixed and coniferous forests and transitional woodland-shrubs.

Proceeding to Lavreotiki, the pre-fire state as in previous cases showed a very low soil erosion risk (Figure 7a). In Figure 7b, the post-fire erosion risk presented an increase with very low and low erosion risk taking up 77.63% or 3.38 km^2 of the area. In the rest of the area, the moderate risk was sparsely distributed (11.90%) while the almost equal coverage of high and very high erosion risk was mainly located in the north-northeastern part of the region. Concerning the burn severity, the erosion risk followed a distribution based on moderate-low severity which covered the larger area. From the CLC 2018 perspective, the very low and low erosion risk was found primarily in coniferous forests and schlerophylous vegetation while moderate, high, and very high was found in natural grasslands.

In the last area, Vilia, the soil erosion risk is presented in Figure 8. During the pre-fire period, the area was at very low erosion risk (Figure 8a). The post-fire erosion risk was higher in all the previously mentioned areas as of Figure 8b. Starting with the very low erosion risk it only covered 38.57% of the Vilia burned area followed by the low with 27.63%. To continue with the moderate severity which was dispersed all over the area it encompassed 16.29% or 14.73 km^2 of it. An important part covered the high (9.36%) together with very high (8.14%) soil erosion risk with the area's topography playing a key role in their spatial distribution located mainly in large parts in the central part of it. As already seen in the previous areas, the same pattern regarding burn severity was followed with the predominant moderate-high severity in this case covering the larger part of each risk category. Lastly, sclerophyllous vegetation, as of land cover, in every soil erosion risk class, was the major affected land cover.

4.3. Web App

The web app is presented in Figure 9 and it includes all the above-mentioned results layers. Within the user interface, the various widgets are visible including map navigation ones and information, legend, layer list, and basemap selection. Others include location retrieval, search, drawing, sharing, measuring, and swiping between layers while coordinates can also be retrieved and converted. The web app is easily accessible from any device through the link: https://learn-students.maps.arcgis.com/apps/webappviewer/index.html?id=b55b196c8f464d28b18182809e590a33, accessed on 2 May 2022.

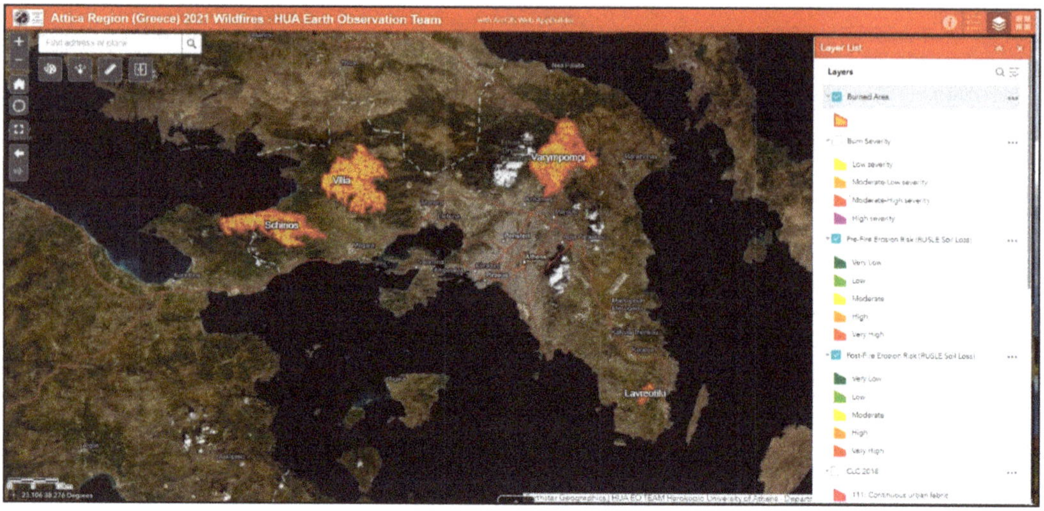

Figure 9. Screenshot of the developed web app. The user interface is shown with the web map of the affected areas and the widgets.

5. Discussion

This study achieved the target of assessing the impact of the wildfires that struck the Attica region in 2021 by combining remote sensing and GIS techniques and free open available datasets. The methods used led to the production of results which helped in understanding the effects of the four examined wildfires. The operational wildfire mapping with the use of Sentinel-2 optical imagery with the application of NBR was effective in mapping the burned areas and assessing their burn severity on a near-real-time basis and with a very good spatial resolution. The later soil erosion risk modeling derived from the RUSLE was sufficient enough to trace the soil loss before and after the wildfire events and understand the risk that each area faces after them. The highlight of this study is the addition of all these products to the web app developed for this purpose thus enabling the getting of an inclusive geospatial intelligence of the post-fire situation.

As mentioned above, the wildfires of 2021 constitute one of the greatest disasters Greece has faced. The analysis that has been implemented in this manuscript was related to the wildfires that broke out in the Attica region. Specifically, the study areas consisted of the wildfire incidents that broke out in Varympompi, Schinos, Vilia, and Lavreotiki. Those regions had been extensively affected by the fires with the burned areas varying from 5 km^2 to 96 km^2. The burn severity demonstrated some very interesting patterns spatially, predominated by moderate-high burn severity values in all regions, except Lavreotiki, where moderate-low severity values presented high frequency as well. Based on the results, the fires affected mostly forestry area consisting of broad-leaved, sclerophyllous, and mixed vegetation.

Wildfire incidents usually lead to post-catastrophic events such as the soil erosion phenomenon. In this study, advanced erosion modeling was applied to the above-mentioned regions in Attica in order to estimate the erosion risk resulting from the wildfire incidents. Particularly, RUSLE modeling was implemented so as to calculate the soil loss rates considering the study area. According to the results, erosion mapping after the wildfire incidents seemed to present high soil loss values in comparison to their pre-fire state. Post-fire mapping displayed very high erosion values according to RUSLE modeling in all four areas of interest.

Analyzing the results of the burned areas, moderate-high severity characterized the burned areas which were also found in large concentrations in all of them except Lavreotiki.

31. *Sentinel-2 User Handbook*; European Space Agency: Paris, France, 2015; Available online: https://sentinel.esa.int/documents/247904/685211/Sentinel-2_User_Handbook (accessed on 27 December 2021).
32. CLC 2018. Copernicus Land Monitoring Service. Available online: https://land.copernicus.eu/pan-european/corine-land-cover/clc2018 (accessed on 21 December 2021).
33. Büttner, G.; Kosztra, B.; Soukup, T.; Sousa, A.; Langanke, T. *CLC2018 Technical Guidelines*; European Environment Agency: Copenhagen, Denmark, 2017; Available online: https://land.copernicus.eu/user-corner/technicallibrary/clc2018technicalguidelines_final.pdf (accessed on 27 December 2021).
34. Meteo.gr Meteo Search. Available online: https://meteosearch.meteo.gr/ (accessed on 14 January 2022).
35. EU-DEM v1.0. Copernicus Land Monitoring Service. Available online: https://land.copernicus.eu/imagery-in-situ/eu-dem/eu-dem_v1_0_and_derived_products/eu_dem_v1.0 (accessed on 23 December 2021).
36. Panagos, P.; Van Liedekerke, M.; Jones, A.; Montanarella, L. European Soil Data Centre: Response to European policy support and public data requirements. *Land Use Policy* **2012**, *29*, 329–338. [CrossRef]
37. European Soil Data Centre (ESDAC). Available online: https://esdac.jrc.ec.europa.eu (accessed on 25 January 2022).
38. McFeeters, S.K. The use of the Normalized Difference Water Index (NDWI) in the delineation of open water features. *Int. J. Remote Sens.* **1996**, *17*, 1425–1432. [CrossRef]
39. Miller, J.D.; Thode, A.E. Quantifying burn severity in a heterogeneous landscape with a relative version of the delta Normalized Burn Ratio (dNBR). *Remote Sens. Environ.* **2007**, *109*, 66–80. [CrossRef]
40. Key, C.H.; Benson, N. Measuring and remote sensing of burn severity: The CBI and NBR. In Proceedings of the Joint Fire Science Conference and Workshop, Boise, ID, USA, 15–17 June 1999.
41. Keeley, J.E. Fire Intensity, Fire Severity and Burn Severity: A Brief Review and Suggested Usage. *Int. J. Wildland Fire* **2009**, *18*, 116–126. [CrossRef]
42. Casady, G.M.; van Leeuwen, W.J.D.; Marsh, S.E. Evaluating Post-Wildfire Vegetation Regeneration as a Response to Multiple Environmental Determinants. *Environ. Model. Assess.* **2010**, *15*, 295–307. [CrossRef]
43. Eidenshink, J.; Schwind, B.; Brewer, K.; Zhu, Z.-L.; Quayle, B.; Howard, S. A Project for Monitoring Trends in Burn Severity. *Fire Ecol.* **2007**, *3*, 3–21. [CrossRef]
44. Veraverbeke, S.; Lhermitte, S.; Verstraeten, W.; Goossens, R. The temporal dimension of differenced Normalized Burn Ratio (dNBR) fire/burn severity studies: The case of the large 2007 Peloponnese wildfires in Greece. *Remote Sens. Environ.* **2010**, *114*, 2548–2563. [CrossRef]
45. Parks, S.A.; Dillon, G.K.; Miller, C. A New Metric for Quantifying Burn Severity: The Relativized Burn Ratio. *Remote Sens.* **2014**, *6*, 1827–1844. [CrossRef]
46. UN-SPIDER Knowledge Portal. Normalized Burn Ratio (NBR). Available online: https://un-spider.org/advisory-support/recommended-practices/recommended-practice-burn-severity/in-detail/normalized-burn-ratio (accessed on 2 April 2022).
47. Polykretis, C.; Alexakis, D.; Grillakis, M.G.; Manoudakis, S. Assessment of intra-annual and inter-annual variabilities of soil erosion in Crete Island (Greece) by incorporating the dynamic "nature" of R and C-factors in RUSLE modeling. *Remote Sens.* **2020**, *12*, 2439. [CrossRef]
48. Ouma, Y.O.; Lottering, L.; Tateishi, R. Soil Erosion Susceptibility Prediction in Railway Corridors Using RUSLE, Soil Degradation Index and the New Normalized Difference Railway Erosivity Index (NDReLI). *Remote Sens.* **2022**, *14*, 348. [CrossRef]
49. Renard, K.G.; Freimund, J.R. Using Monthly Precipitation Data to Estimate the R-Factor in the Revised USLE. *J. Hydrol.* **1994**, *157*, 287–306. [CrossRef]
50. de Santos Loureiro, N.; de Azevedo Coutinho, M. A New Procedure to Estimate the RUSLE EI30 Index, Based on Monthly Rainfall Data and Applied to the Algarve Region, Portugal. *J. Hydrol* **2001**, *250*, 12–18. [CrossRef]
51. Williams, J.R.; Renard, K.G.; Dyke, P.T. EPIC: A New Method for Assessing Erosion's Effect on Soil Productivity. *J. Soil Water Conserv.* **1983**, *38*, 381.
52. Moore, I.D.; Burch, G.J. Physical Basis of the Length-Slope Factor in the Universal Soil Loss Equation. *Soil Sci. Soc. Am. J.* **1986**, *50*, 1294–1298. [CrossRef]
53. Durigon, V.L.; Carvalho, D.F.; Antunes, M.A.H.; Oliveira, P.T.S.; Fernandes, M.M. NDVI Time Series for Monitoring RUSLE Cover Management Factor in a Tropical Watershed. *Int. J. Remote Sens.* **2014**, *35*, 441–453. [CrossRef]
54. Vidali, M. Estimation of Soil Erosion Model in the Reservoir of the Pinios Dam of Ilia Prefecture. Master's Thesis, University of Patras, Patras, Greece, 2013.
55. Yang, D.; Kanae, S.; Oki, T.; Koike, T.; Musiake, K. Global Potential Soil Erosion with Reference to Land Use and Climate Changes. *Hydrol. Process.* **2003**, *17*, 2913–2928. [CrossRef]
56. ArcGIS Online. Available online: https://www.esri.com/en-us/arcgis/products/arcgis-online/overview (accessed on 20 March 2022).
57. ArcGIS Web AppBuilder. Available online: https://www.esri.com/en-us/arcgis/products/arcgis-web-appbuilder/overview (accessed on 20 March 2022).

58. Panagos, P.; Borrelli, P.; Meusburger, K.; van der Zanden, E.H.; Poesen, J.; Alewell, C. Modelling the Effect of Support Practices (P-Factor) on the Reduction of Soil Erosion by Water at European Scale. *Environ. Sci. Policy* **2015**, *51*, 23–34. [CrossRef]
59. Zevgoli, E.; Xanthopoulos, G.; Psomiadis, E.; Papanikolaou, I. Forest Fire Severity Assessment Through Field Sampling and Satellite Remote Sensing Methods and Correlation of Their Results. In Proceedings of the 20th Congress of Hellenic Forestry Society, Trikala, Greece, 3–6 October 2021; pp. 333–344.

applied sciences

MDPI

Article

Modeling the Ignition Risk: Analysis before and after Megafire on Maule Region, Chile

Gabriela Azócar de la Cruz [1,2,3,*], Gabriela Alfaro [3,4], Claudia Alonso [2,3], Rubén Calvo [3,5] and Paz Orellana [3]

1 Department of Social Work, University of Chile, Av. Ignacio Carrera Pinto 1045, Ñuñoa, Santiago 7800284, Chile
2 Center for Climate and Resilience Research (CR), Blanco Encalada 2002, Floor 4, Santiago 8370449, Chile
3 Nucleus of Transdisciplinary Systemic Studies, University of Chile, Santiago 7820436, Chile
4 Industrial Engineering Department, University of Chile, Av. Víctor Jara 3769, Estación Central, Santiago 9170124, Chile
5 Institute of Geography, Pontificia Universidad Católica de Chile, Campus San Joaquín—Avda. Vicuña Mackenna 4860, Macul, Santiago 7820436, Chile
* Correspondence: gazocarde@uchile.cl

Abstract: Wildland fires are a phenomenon of broad interest due to their relationship with climate change. The impacts of climate change are related to a greater frequency and intensity of wildland fires. In this context, megafires have become a phenomenon of particular concern. In this study, we develop a model of ignition risk. We use factors such as human activity, geographic, topographic, and land cover variables to develop a bagged decision tree model. The study area corresponds to the Maule region in Chile, a large zone with a Mediterranean climate. This area was affected by a megafire in 2017. After generating the model, we compared three interface zones, analyzing the scar and the occurrences of ignition during and after the megafire. For the construction of georeferenced data, we used the geographic information system QGIS. The results show a model with high fit goodness that can be replicated in other areas. Fewer ignitions are observed after the megafire, a high recovery of urban infrastructure, and a slow recovery of forest plantations. It is feasible to interpret that the lower number of ignitions observed in the 2019–2020 season is a consequence of the megafire scar. It is crucial to remember that the risk of ignition will increase as forest crops recover. Wildland fire management requires integrating this information into decision-making processes if we consider that the impacts of climate change persist in the area.

Keywords: wildfire; ignition risk; model; megafire; climate change; bagged decision tree; wildland urban interface

Citation: Azócar de la Cruz, G.; Alfaro, G.; Alonso, C.; Calvo, R.; Orellana, P. Modeling the Ignition Risk: Analysis before and after Megafire on Maule Region, Chile. *Appl. Sci.* **2022**, *12*, 9353. https://doi.org/10.3390/app12189353

Academic Editors: Efthymios Lekkas and Spyridon Mavroulis

Received: 15 August 2022
Accepted: 14 September 2022
Published: 18 September 2022

1. Introduction

Wildfires are an extreme phenomenon of great global concern [1]. The frequency and intensity of these events have progressively increased in different areas of the world [2]. Studying the causes and risks of forest fires has become a field of broad interest. Various investigations and technical reports on wildfires agree that at least 90% of these have an anthropogenic origin [3–7]. It is not entirely clear, however, how many of the fires caused by human actions are due to negligence and how many are intentional. Identifying and punishing the people responsible for wildfires is a complex problem, given that the evidence is not recordable or disappears due to the fire [8]. A series of environmental factors favor the spread of wildfires, such as the combination of temperature, humidity, and wind [4,9–11]. Added to this are research results on the influence of productive activities and the characteristics of wildland–urban interface zones [6,12,13]. All the above indicates the need to deepen the analysis of the interaction between human action and its ecological environment to investigate the conditions that affect the fire origin and spread. In addition, there is a need to connect this with the impacts of socio-environmental disasters.

Climate change is a highly complex phenomenon that has become the focus of interdisciplinary studies on the relationship between society and the environment, particularly disaster risks. Its link to wildfires exemplifies this evident relationship [14–18]. Due to climate change, different areas of the world have been affected by the increase in heat waves, the increase in the magnitude of periods of drought, the decrease and absence of rainfall, and soil degradation [19–21]. All these phenomena are not direct causes of wildfires, but are factors that influence their recurrence and magnitude. Heat, drought, lack of rain, and land degradation affect the availability of fuel material, which favors the spread of wildland fires [4,22]. These factors also affect the intensity of fires and the damage they cause [13,23–25]. An interesting example of the interaction between climate change and wildfires is the increase in ignition points in mountainous areas that are difficult for people to access [26]. Lightning generated in dry electrical storms (without rain) can cause fires in native forests in these areas, where difficult access is an obstacle to their control [27–29].

The relationship between climate change and wildland fires increases concern about its social and environmental impacts. Wildland fires impact people's health due to burns or the large amount of CO_2 they generate, produce irreparable damage to infrastructure and homes located in interface areas, and affect productive activities [3,30,31]. On the other hand, wildland fires destroy flora and fauna, generate desertification, and deteriorate biodiversity [20,31]. The interaction with climate change increases the damage capacity of these impacts.

In this scenario, it is necessary to generate tools that allow technical organizations and communities to have information that enables better wildland fire risk management [20,32]. For this, it is essential to advance the study of the behavior of risk factors in particular territorial contexts. This will allow the delivery of valuable and valid information to territorial planners, risk managers, and community leaders about the prevention, preparation, firefighting, and mitigation actions they must promote in their environments. Forest fire risk analysis is, therefore, contextual, since it depends on the characteristics of the socio-ecological systems in which it is carried out. We understand socio-ecological systems as those that emerge from the relationship between biophysical and social factors, sustaining a set of human needs and environmental conditions in interaction [33–36].

From disaster risk management, the concept of risk integrates the following two main components: damage and future projection. Risk has been defined as the probability of occurrence of a future event that can cause possible damage [37,38]. Therefore, the risk of wildland fires refers to the negative impacts that these can have on communities and ecological environments [20]. The probability of wildland fire occurrence depends on the interaction between different geographic, topographic, climatic, land cover, and human action variables [11,39–41]. The interaction between these factors determines the magnitude of the projected damages [11,42].

The social theory of risk expands the analysis of this concept and its application in disaster risk management. This integrates the decision as a relevant component of the risk. The probability of damage is the consequence of an unwanted decision that wants to be avoided. For decisions to manage the risk, it is necessary to know what and how this damage is produced [43]. In the context of wildland fire risk, this means integrating knowledge about what makes wildland fires so that the decisions adopted for their management are those that prevent their occurrence or minimize their damage. In this sense, risk analysis implies anticipating possible negative results of a decision [44,45].

The study of fire risk can be divided between research on propagation and the ignition of forest fires. This work addresses the risk of wildland fire ignition due to its relevance in developing prevention and preparedness strategies. A fire occurs when an ignition source (human or natural) meets the available combustible material [3]. Ignition risk is the probability that a fire will start at a given point in the territory [46]. Depending on the conditions and characteristics of the space where the ignition point is located, the flames can increase in intensity and propagate, developing into a forest fire [5,47]. Ignition depends on a wide range of variables associated with the point where it occurs. These can be classified into (a) natural conditions, such as plant species and plantations height, humidity, temperature, topography, and local climate, [3,4,11,47,48] and (b) anthropogenic conditions, such as land use and cover, distance from roads, distance from urban or inhabited areas and urban infrastructure [6,20,23,26,47,49]. The ignition risk increases when the associated variables interact at a certain point.

Wildland urban interfaces (WUI) are areas in which these factors interact and are enhanced. Research indicates that the risk of wildland fires is greater when human settlements mix with vegetation [40,50]. This acquires relevance in the context of megafires. The most significant damage caused by wildland fires is in the ecological environment in which they occur. The damage on urban surfaces is usually negligible if we compare it with vegetated or forested surfaces. In megafires, however, the risk of damage in inhabited urban areas increases [51–53]. Their high intensity characterizes megafires compared to the general pattern of wildland fires. Other characteristics are their broad ecological and social impact, the obstacles they impose on their management, and the danger of reactivation [19,54]. All these factors imply that these events tend to cross WUIs, affecting people and urban infrastructure more significantly than forest fires. Along with this, a megafire can change the land cover, reducing the ecosystem services that the land provides to the surrounding populated areas [51,55].

In this study, we analyze the risk of wildland fire ignition by modeling and mapping these events. Our case study is the Maule region in Chile, which was affected by the megafire of 2017. Considering the high vulnerability that this area presents to the impacts of climate change and the uncertainty about its future effects, we developed a model of ignition risk from historical fire data in the region. First, we identify the variables that best explain ignition. Based on these variable values, the model assigned a probability of fire occurrence to points on the map. We then validated the model with data from fires during the megafire and later years. Finally, we identify four interface areas affected by the megafire of 2017 and compare the state of the territory before and after this event. This last exercise aims to analyze the current fire risk conditions in a region that, due to its high exposure to climate change, may once again be affected by events of significant magnitudes, such as the 2017 megafire.

2. Case Background

Between January and February 2017, the south-central zone of Chile experienced one of the largest megafires in history [56]. The amount of heat energy released during the months that this event occurred exceeded the scales used internationally until then [57]. Official figures indicate that the fire destroyed 529,974 hectares. Although the greatest damage occurred in forest areas, a set of WUIs was affected, with 3000 homes lost and 11 deceased persons [16,51].

Megafires are usually generated in Mediterranean landscapes, such as the affected area in Chile. The climatic conditions and homogeneous land cover, given the vast extension of forest plantations, were favorable conditions for this event [58,59]. These conditions are characteristic of the Maule region, one of the most affected by the 2017 event. In this region, the fire destroyed 252,556 hectares, equivalent to a third of the forest area of the region [51]. The mega-drought that has affected this region since 2010 led to the fire spread and a high level of damage [57,60,61].

Due to this disaster, the resources provided for firefighting were increased in the country, which positively increased the response capacity of specialized agencies [61]. Little has been addressed, however, in the prevention of wildland fires in science and public policy. We believe studying the risk of wildland fire ignition will contribute to the characterization of areas highly exposed to these fires, information that can be used in prevention policies and actions. On the other hand, the set of conditions that increased the chance of the development of the megafire in the Maule region has not changed in recent years, making it necessary to analyze the risk of ignition of wildland fires and the associated factors.

2.1. Study Area

Region of Maule

The study area is located in the Maule Region, in central Chile, between 34°41′ and 36°33′ south latitude. The surface area is 30,296 km^2, equivalent to 10% of the national territory [62]. Its population is 1,044,950 inhabitants, with a density of 34.5 inhabitants per square km, with 73% of the population living in urban areas [63]. Its topography integrates mountains of 4000 m.a.s.l. in the Andes Mountains, an intermediate depression, and the Coastal Mountains with mountain ranges with moderate to steep slopes. The climate is a temperate Mediterranean climate, with a dry season of six months in the north and four months in the south. The primary use of the land is grassland and scrubland (25%), agricultural land (22%), and forest plantations (20%) [64]. It has an area of native forest of 581,515 hectares and 634,893 forest hectares [65]. Given the topographical and climatic characteristics, the forest crops are mainly located in the Cordillera de la Costa [64]. For the analysis of the changes in the landscape after the 2017 megafire, we selected the following three communes in the region: Constitución, Empedrado, and Cauquenes (Figure 1). The selected communes correspond to the populations most affected by the 2017 fires [66].

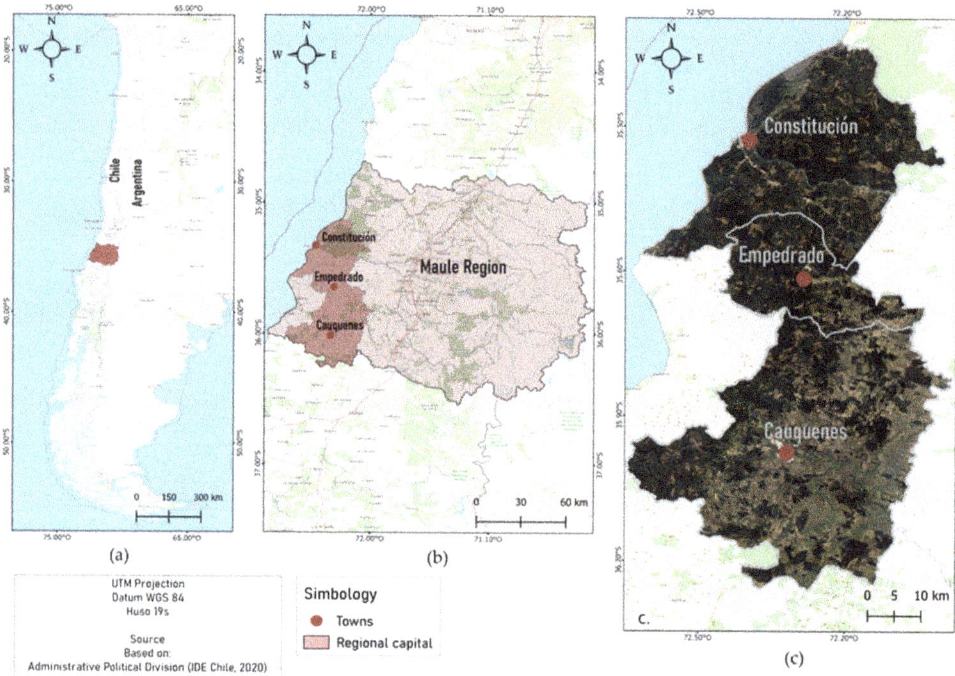

Figure 1. (**a**) Location of the study area in Chile. (**b**) Location of the study area on a regional scale. (**c**) Communal scale study area.

It should be noted that in Empedrado, the town of Santa Olga was entirely consumed by the fire, becoming an emblematic case of megafires' impact on a population in a WUI [15,67]. The incident negatively impacted the water and electricity supply and generated contamination of water sources for human and animal consumption. In these communes, the labor sources associated with agricultural and livestock production and the forestry industry were also affected [67].

Constitución, located between latitude 35°19′ south and longitude 72°24′ west, is a coastal city with a total area of 1344 km^2 [68]. It is located on the south bank of the mouth of the Maule River in the Pacific Ocean. The maritime influence keeps its daily temperatures moderate, with an annual average of around 14 °C. It is characterized by hot summers and mild winters, with an average annual rainfall of 662 mm. Its population is 46,068, of which 19% reside in rural areas [63]. The main economic activity is forestry and agriculture. The use of the land is native forest (5%), plantations (41%), thickets (34%), crops, and grasslands (19%) [68].

In the south of the region is the commune of Empedrado, between the coordinates 35°36′ south latitude and 72°16′ west. Its surface is 565 km^2 and is located in the coastal mountain range. Its annual temperatures range between 10 °C and 33 °C [69]. Its total population is 4142 inhabitants, of which 27% live in rural areas [63]. The main economic activity is agricultural and forestry production, including agriculture, livestock, dairy production, wines, and liquors. Land use is divided into the native forest (17%), plantations (17%), bushes (17%), agricultural (17%), grasslands (17%), meadows (17%), grassland crop rotation (17%) and others (17%) [69].

Between 35°58′ south latitude and 72°18′ west longitude is the commune of Cauquenes, with an area of 2216 km^2 [70]. Its population is 40,441 inhabitants, with a rural population of 18% [63]. It has a main body of water, the Cauquenes River, whose flow has decreased due to the drought affecting the region. Its climate is a Mediterranean climate, with average temperatures of around 25 °C in January and 7 °C in July. The economy of the commune is diverse and includes manufacturing activities (20.5%), forestry (17.3%), and electricity production (13.2%), among others. Land use is divided into the native forest (32%), forest plantations (32%), and thickets (31%), which together cover 95% of the surface [70].

3. Materials and Methods

We developed a model of the risk of ignition of forest fires in the Maule region through a machine learning model. For this, we defined the ignition of forest fires as a dependent variable and selected a set of independent variables as possible predictors of ignition. The construction of the database was in two stages. First, spatially represented points of the ignition variable (binary variable) were generated. Then, the values of the set of independent variables were estimated for each of these points.

The dependent variable was generated from the Corporation Nacional Forestal (CONAF) data, published as official data on its website (CONAF https://www.conaf.cl/incendios-forestales/incendios-forestales-en-chile/estadisticas-historicas/ (accessed on 20 May 2022)). From this, we obtained information on the coordinates, start date, control, cause, and magnitude, of the fires that occurred in the Maule region between 2013 and 2015. The period defined for collecting information was due to the need to have a rich source of available data on fire ignition and independent variables before the megafire of the 2016–2017 season. On the other hand, we decided to generate the model with data before the megafire to contrast it with what happened during that event to validate its results. With this, we looked for a model that allowed us to explain the ex-post distribution of the ignition points of forest fires, while predicting the risk of ignition in the future.

We assigned the value 1 (one) to the points where a fire occurred in the indicated years and a value 0 (zero) to the points where there was no ignition. The assigning points that represented areas with no ignition of wildland fires were carried out randomly. Points with the value 0 were assigned to located areas more than 500 m from an ignition point. For the construction of georeferenced data, we used the geographic information system QGIS (version 3.20.3. for Windows/Copyright © 2000, 2001, 2002, 2007, 2008 Free Software Foundation, Inc. <http://fsf.org/>).

We obtained 3784 points, of which 1892 indicate the ignition of forest fires between 2013 and 2015 in the Maule region; these correspond to the red points in Figure 2. The 1892 blue points generated by the model represent areas with no ignition in the same period.

The independent variables selected for the construction of the model were based on the work of Miranda et al. [40]. According to the authors, the scientific evidence indicates that this set of variables is the one that best represents the territorial characteristics of those places where wildland fires occur in interface areas. Based on the results of this group of researchers, we worked with 14 independent variables organized into the following 3 classes: (a) human activity, (b) geography and topography, and (c) land cover (Table 1). Each variable was spatially represented in a 30-m resolution raster. To assign the values of the independent variables at each point, we defined a zone of influence of 500 m (centered at the point of ignition). These buffers were built by taking a circumference of a radius of 500 m around each point of the raster. Each one was assigned the value (percentages, averages) of each explanatory variable to avoid bias.

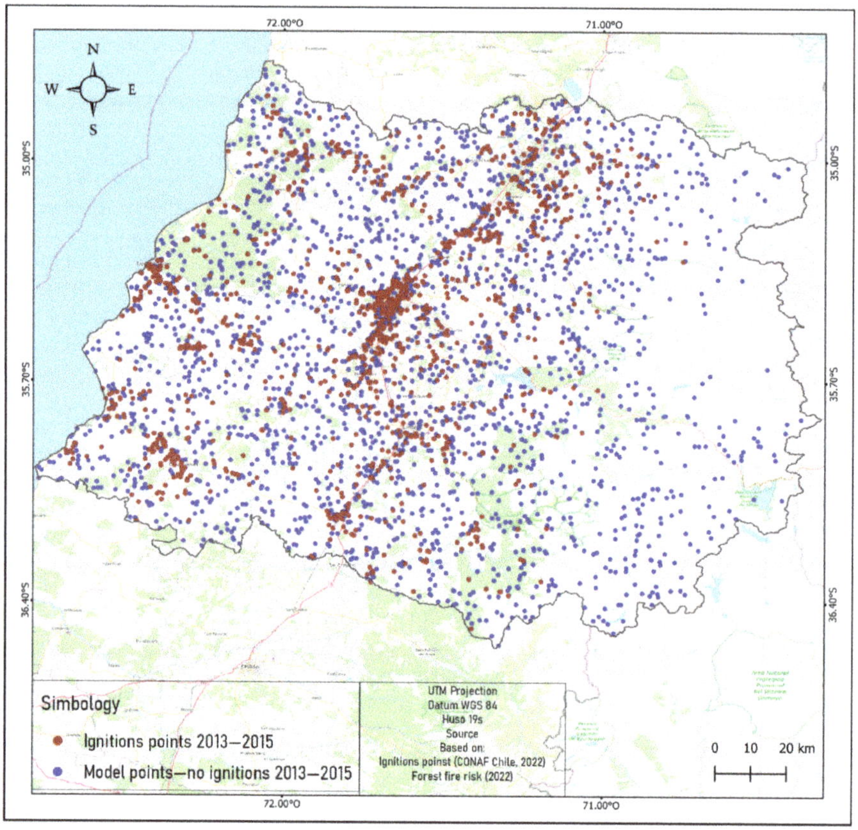

Figure 2. Binary dependent variable, ignition of wildland fires in the Maule region between the years 2013 to 2015.

Each of the 3784 points was assigned values for these 14 variables. Thus, the model was trained with the values obtained between 2013 and 2015.

We seek to identify the variables that most influenced the fire's start in the past. With this, we estimate what factors generate a greater probability of a fire. Based on the methodology used by Miranda [40], we use the bagged decision tree, BDT model [75]. This ensemble machine learning method combines different "weak" classification sub-models to obtain a "strong" one. The processing of the model was carried out using the MATLAB R2020a (see: https://www.mathworks.com/products/matlab.html?s_tid=hp_products_matlab (accessed on 10 April 2022)).

Bagging (which is short for bootstrap aggregating) consists of building different submodels using random samples, with replacement, and then assembling the results. Various subsets of the training set data are created. The model has that name, since it trains the submodels using bagging. A model is trained with each subset, and the final predictions are averaged, making it more robust. An example of a bagging model is shown in Figure 3.

Table 1. Independent variables, names, and sources.

Variables Classes	Name	Label	Source
Human activity	DensPop	Population density of the point (inhabitants/census district)	Instituto Nacional de Estadísticas: https://www.ine.cl/herramientas/portal-de-mapas/geodatos-abiertos (accessed on 2 April 2022) [71]
	DistCit	Distance from point to nearest city (m)	
	DistRoad_buff	Average distance from the 500 m radius buffer to the nearest road (m)	Ministerio de Obras Públicas: https://ide.mop.gob.cl/geomop/ (accessed on 2 April 2022) [72]
Geography and topography	Exposition	Point exposition (indices)	Earth Resources Observation and Science Center (EROS), https://www.usgs.gov/centers/eros/data-tools (accessed on April 2022) [73]
	Slope	Point slope (grades)	
	Elev	Point elevation (m.a.s.l.)	
Land cover	Crop_buff	Proportion of crops in a 500 m radius buffer (proportion)	Zhao Y et al. 2016 [74]
	Nat_buff	Proportion of native forest in a 500 m radius buffer (proportion)	
	ForPlan_buff	Proportion of forests plantation in a 500 m radius buffer (proportion)	
	Grass_buff	Proportion of grassland in a 500 m radius buffer (proportion)	
	Scrub_buff	Proportion of scrubs in a 500 m radius buffer (proportion)	
	Imper_buff	Proportion of impermeable land in a 500 m radius buffer (proportion)	
	BareSoil_buff	Proportion of bare soil in a 500 m radius buffer (proportion)	
	LC	Type of land cover predominant in the point (categorical)	

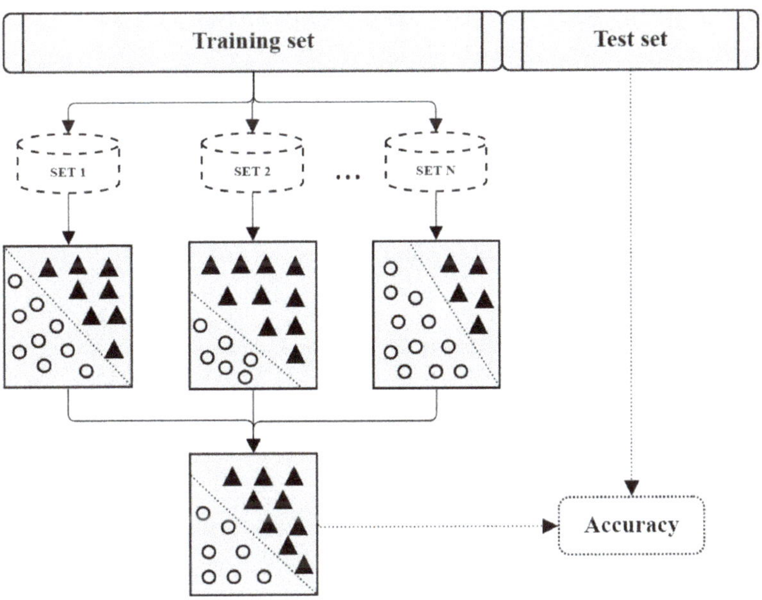

Figure 3. Diagram of the bagged model.

For this work, we use the bagged decision tree, where each model is a decision tree. The training set was built with 80% of the database, and the remaining 20% was used to test and verify the accuracy of the results. Once the model was trained, it generated the fire ignition susceptibility classification for each point on the 30 m × 30 m quadrant map.

Finally, a comparative temporal matrix of four WUI was built to analyze changes in the distribution of wildland fire ignition. The matrix was elaborated through the QGIS software. The satellite images come from the Sentinel-2 mission. These images present an atmospheric correction at ground level, providing spectral radiance levels similar to reality. The vector layer of the fire scar was extracted from the Landscape Fire Scars database for Chile [76]. Previous layers corresponding to the ignition points of the 2016–2017 and 2019–2020 seasons were used on the satellite images. It should be noted that the 2020–2021 season was not considered in the analysis. We decided to exclude this season, since the lower number of wildland fires it registers is associated with fewer people in transit in the interface areas, due to the confinement measures adopted in the context of the COVID-19 pandemic.

4. Results

4.1. Model Fit

The performance of the model was estimated through the global adjustment indicator AUC. This corresponds to the area under the curve ROC (relative operating characteristic), which represents the ratio between the true positives (TPR, true positive ratio) and the false positives (FPR, false positive ratio), as the model predicts values (Figure 4). When the AUC takes a value of 1, the model has perfect prediction. When it takes a value of 0.5, it is a model without explanatory power that does not discriminate between categories. For the model, an AUC value of 0.85 is obtained, which means that it has a high level of fit.

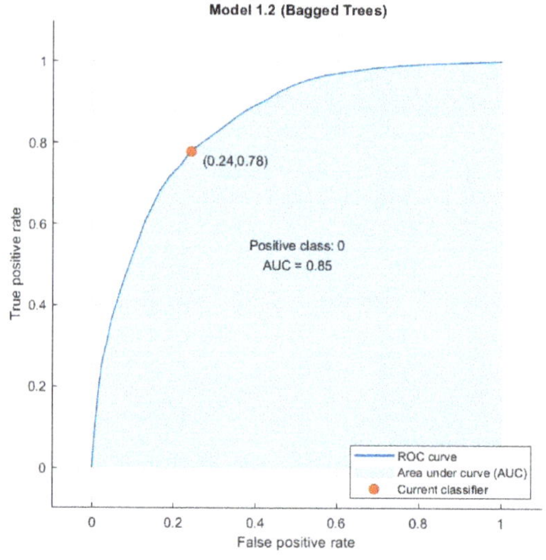

Figure 4. ROC curve of the fit of the ignition model.

Another measure of the model's effectiveness is the confusion matrix, which indicates the level of correct and incorrect classifications. In Figure 5, we can observe that 75.8% of values 1 (ignition) and 77.9% of values 0 (not ignition) were correctly classified.

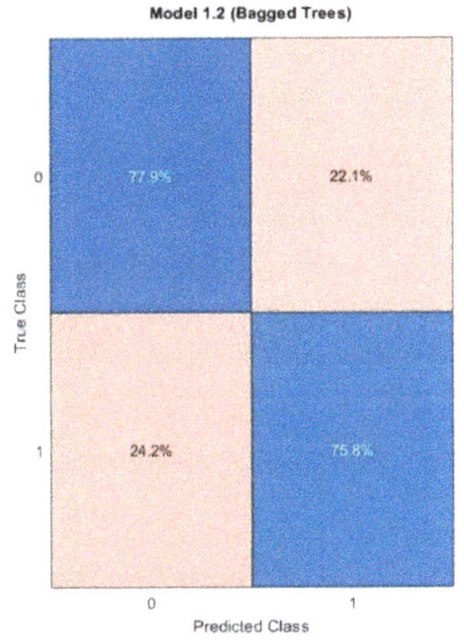

Figure 5. Confusion matrix.

One of the advantages of decision tree models is that their results show the explicative importance of the variables. This importance represents the prediction model's error increase after the variable's value has been permuted (separated in a branch of the tree). This is the normalized average of how much this variable changes the final classification result. In this model, the importance of the variables is as follows, as we can observe in Table 2.

Table 2. Variables' importance.

Variable Importance	Variable Label	Variable Name
18.9599929	Proportion of crops in a 500 m radius buffer (proportion)	Crop_buff
15.2302175	Average distance from the 500 m radius buffer to the nearest road (m)	DistRoad_buff
15.046572	Distance from point to the nearest city (m)	DistCit
8.80795483	Proportion of forest plantation in a 500 m radius buffer (proportion)	ForPlan_buff
7.63480897	Proportion of grassland in a 500 m radius buffer (proportion)	Grass_buff
6.82820899	Proportion of scrubs in a 500 m radius buffer (proportion)	Scrub_buff
6.35721698	Point exposition (indices)	Exposition
5.09869124	Population density of the point (inhabitants/census district)	DensPop
4.0326676	Point elevation (m.a.s.l.)	Elev
4.01163185	Point slope (grades)	Slope
2.78965119	Proportion of impermeable land in a 500 m radius buffer (proportion)	Imper_buff
2.31063774	Proportion of native forest in a 500 m radius buffer (proportion)	Nat_buff
1.84278487	Type of land cover predominant in the point (categorical)	LC
1.04896344	Proportion of bare soil in a 500 m radius buffer (proportion)	BareSoil_buff

In order of importance, the variables that best explain the ignition of wildland fire in the Maule region are the proportion of crops, the distance to the nearest road, and the distance to the nearest city. The second group of variables with a medium explanatory capacity integrates the proportion of forest plantations, the proportion of grassland, the proportion of scrub, exposure, and population density.

4.2. Ignition Risk Model Map

One of the model results is the assignment of ignition risk levels to the different zones of the study area map. This is achieved due to the georeferencing of each point. Each pixel is classified according to its level of susceptibility to the ignition of a forest fire.

In the map of the Maule region, we classify 33,696,273 pixels of a 30 m resolution raster. Each of these pixels is assigned the values of the independent variables of the trained model. The result of the classification delivers values between 0 and 1. It is a continuous variable that represents levels of probability of ignition of forest fires. This graphic representation is built on the map of the region. Figure 6 shows the result of the classification; the blue zones have a lower susceptibility to fire ignition, the yellow zones have a medium probability, and the red zones have a high probability.

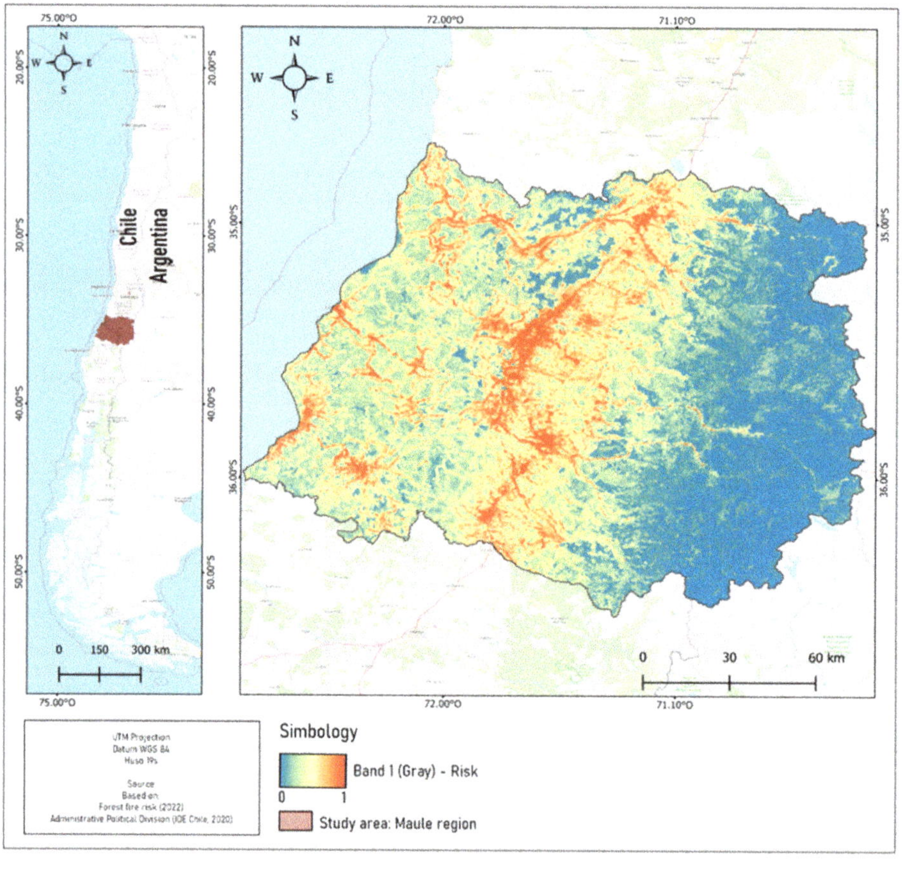

Figure 6. Ignition risk map in the Maule region, Chile.

4.3. Model Validation

To verify the explanatory capacity of the model, we superimposed on the map that shows the results of the ignition risk model the ignition points recorded in the 2016–2017 season. These account for the distribution of the ignition points of the megafire in the Maule region (red points in Figure 7). In this comparison, it is important to consider that the modeling of the ignition risk was carried out based on data from 2013 and 2015. Figure 7 clearly shows that most of the points where a fire started are in areas the model classified as having a high probability of ignition. These results contribute to validating the model and provide useful information to risk management in the future.

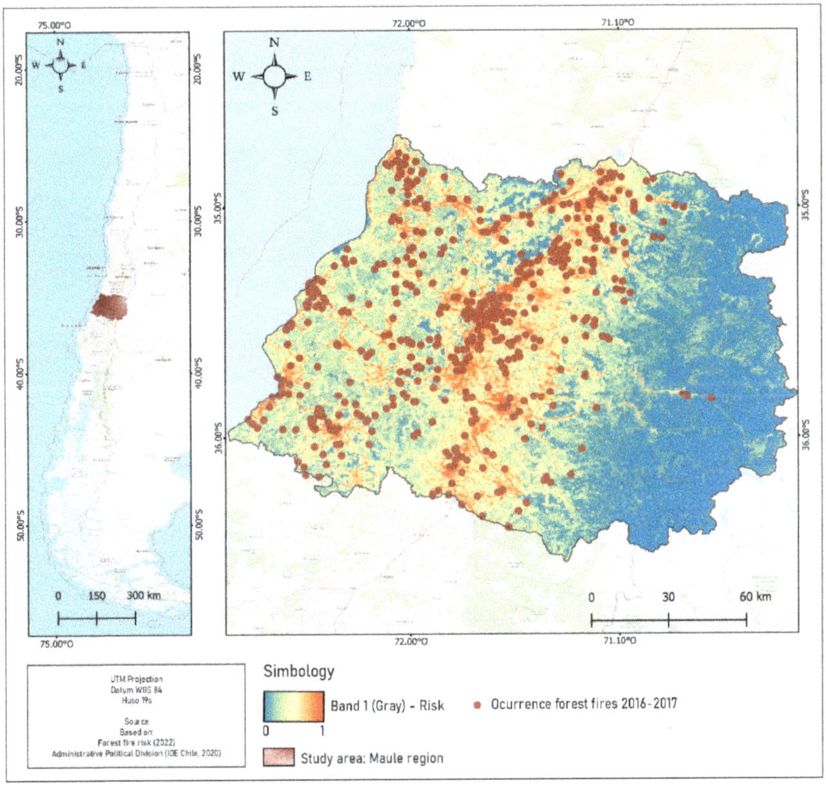

Figure 7. Ignition risk map and ignition points during 2016–2017 season.

In addition, we compare the model results with the real ignitions between 2016 and 2020. Table 3 shows that 85% of fires in that period occurred in points classified as medium, high, and very high risk by the model.

Table 3. Percentage of fire ignition during 2016–2020 according to model probability levels.

Ignition Probability	Risk Level	Points Frequency	Ignitions 2016–2020 (%)
0–20	Very low	113	3.19%
20–40	Low	424	11.99%
40–60	Medium	728	20.58%
60–80	High	941	26.60%
80–100	Very high	1331	37.63%
	TOTAL	3537	100.00%

4.4. Temporal Comparative Matrix of Interface Zones

To deepen the analysis of our model and generate guidance for decision-makers, professionals in charge of forest fire prevention and preparedness, and communities, we analyzed three interface zones severely damaged by the 2017 megafire (Figure 8). The matrix shows the results of the ignition risk modeling (a), the ignition points of the 2016–2017 season (b), the scar of the 2017 megafire (c), and the ignition points of the 2019–2020 season (d).

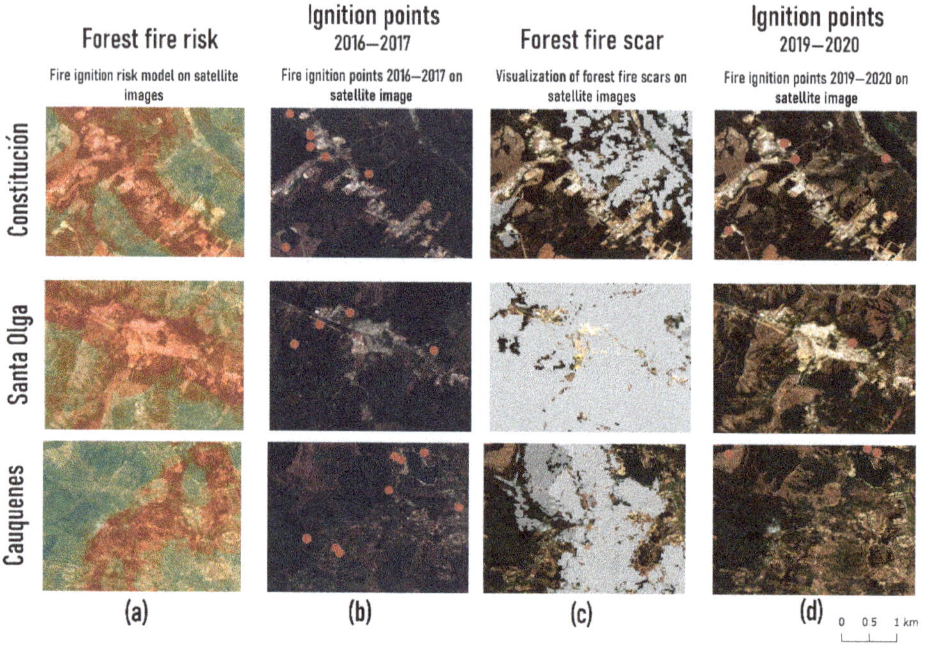

Figure 8. Matrix comparative of interface zones.

The results show that fire ignition in the two seasons under analysis occurred in areas the model classifies with a high probability (see columns a, b, and d). In these areas, crop areas of diverse composition can be observed—especially forestry—near the ignition points by roads and cities. These characteristics correspond to the variables with the most significant weight in the probability of ignition of our model.

Comparative analysis of satellite images indicates that the number of ignitions in each WUI is lower in the 2019–2020 season. Changes in the territory's characteristics can also be observed, including a notable decrease in the density of forest crops in each of the analyzed areas (column d). This change in the territory's composition can be attributed to the 2017 megafire scar (column c). A large part of the forest plantations, grassland, and scrubs in these areas was destroyed by the fires of 2016–2017. Our model classifies these variables with a medium influence on the probability of ignition.

It should be noted that the town of Santa Olga was destroyed by the fires of 2017. As a reparative measure, it was quickly rebuilt in the following years. The reconstruction of the town meant an improvement in the conditions and quality of life of its population, which received better quality housing and urban infrastructure that did not exist before the megafire [67]. These improvements, however, do not mean a reduction in the ignition risk in the area adjacent to this town, since it continues to be surrounded by extensive and dense forest plantations.

5. Discussion and Conclusions

The ignition risk model generated in this work makes it possible to identify those sectors of the wildland urban interface zones with a higher probability of forest fire ignition. In turn, it tells us what characteristics of the study area make it more prone to ignition. Depending on their predictive capacity, two groups of variables allow us to understand why ignition is more probable in these sectors. The group of variables with the most significant predictive capacity includes the crop proportion, the distance to the nearest road, and the distance to the nearest city. These are variables of anthropogenic origin if we consider that they are associated with productive activities (crops) and urban characteristics (roads and cities). The second group of variables has a medium ability to predict ignition. Among these, we find variables of anthropogenic origin, such as the proportion of forest plantations and population density, and variables of natural origin, such as grasslands, scrubs, and exposure.

The results coincide with previous studies that have verified the coincidence between forest fires' risk and high proportions of land use destined for crops [26]. In this regard, particular importance has been assigned to forest crops in the ignition of fires [22]. It has been shown that the uncontrolled growth of forest species increases the combustible material and the ignition risk [16,22,77–79]. On the other hand, it has been pointed out that replacing the native forest with homogeneous pine and eucalyptus plantations in the central-southern zone of Chile has led to a greater production of large-scale wildland fires [56]. Our results help confirm the strong influence of human activities on forest fire ignition [4,10,20,24,40]. As in our work, the proximity of ignition points to roads and human settlements are factors that previous studies highlight as risk factors [3].

The analysis of the WUI satellite images indicates that the territory's characteristics changed due to the scar from the 2017 megafire. The density of forest plantations is the most evident change. The images show that forest plantations are in the process of recovery. On the other hand, the urban infrastructure has recovered rapidly. This maintains the high levels of risk associated with the distance from roads and cities and population density. In particular, Santa Olga town reconstruction shows that the repair process focuses on restoring the urban infrastructure and improving the living conditions of the people affected by the fires. However, integrating the ignition risk prevalent in the Maule region into decisions is not evident.

It is feasible to interpret that the lower number of ignitions observed in the 2019–2020 season is a consequence of the megafire scar. Variables associated with crops, scrubs, and grasslands decrease their influence on ignition, given that such combustible materials are less dense. However, it is crucial to remember that the risk of ignition will grow as forest crops recover. This information must be incorporated into the decisions adopted for the prevention and preparation of forest fires. Along with this, the authorities and communities must integrate into their decisions how climate change contributes to ignition probability.

The decrease in rainfall, soil degradation, and heat waves are impacts of climate change related to the variables that the ignition risk model classifies with a strong and medium influence. Previous studies have confirmed that these impacts of climate change affect soil humidity, creating conditions for high flammability of combustible materials [4,10,11,56]. The exposure of the land to sunlight is another climate change variable that our model associates with the ignition risk. According to Maniatis (2021), in areas more exposed to sunlight, the vegetation loses humidity and becomes more flammable. The Maule region has been highly impacted by climate change. This condition is not expected to vary, at least in the short and medium term [60,61,80]. For this reason, it is imperative to consider the variables above in wildland fire risk management, especially given the uncertainty that a new megafire might occur in the region.

The creation of models that make it possible to understand and make visible the risk of wildland fire ignition is necessary but not sufficient. If disaster risk is still understood only as the future projection of damage, these models only serve to identify the risk, not to manage it. Forest fire management requires integrating risk as a decision. This means improving the policies that regulate the density of forest plantations, the maintenance and expansion of firebreaks in interface zones, land use, and post-fire cleaning of areas where tree species' seeds are irregularly disseminated. It also means integrating the ignition risk into the measures to repair the damage caused by wildfires.

Developing ignition models, such as the one in this work, is undoubtedly necessary to manage wildland fires. An example is the verification of sectors with a greater probability of ignition that, according to the model, are strongly associated with the areas where multiple fires occurred in 2017. This information will allow fire governance actors to implement better strategies to prevent wildland fire ignition. Nevertheless, we must be aware that, even so, forest fires will continue to occur. The risk of ignition is higher in regions such as the study area, where climate change has strongly impacted territory characteristics and conditions. For this reason, integrating the ignition risk into decisions also means implementing campaigns and actions to improve the ability to react to this kind of phenomenon.

It should be noted that the model developed in this work accounts for the particularities of the study area. The model was trained with historical data from the Maule region in Chile. Therefore, the results obtained are valid only for the case analyzed. However, working with the bagged decision tree model has the following significant advantage: it is possible to replicate it in other areas with different characteristics. It is possible because the model learns from the data provided to it. In addition, variables such as human activity, geography, topography, and land cover can be obtained for other areas of interest. The only limitation is that these layers of information, particularly land cover, are scarce in many countries.

One of the challenges that arise from this work is to continue deepening the characterization of the ignition risk. We believe that obtaining better views of the interface areas and the factors associated with risk is necessary. This will allow professionals and people to integrate this information. Visually identifying the components and characteristics of the territories where the ignition risk is higher can contribute to a better understanding of how to prevent and prepare for wildland fires.

Author Contributions: Conceptualization, G.A.d.l.C.; methodology, G.A.d.l.C., G.A., R.C. and C.A.; software, G.A., R.C. and C.A.; validation, G.A and R.C.; formal analysis, G.A.d.l.C., G.A. and R.C; investigation, G.A.d.l.C., C.A. and P.O.; resources, G.A.d.l.C., G.A., C.A. and P.O.; data curation, G.A and C.A.; writing—original draft preparation, G.A.d.l.C.; writing—review and editing, G.A.d.l.C.; visualization, C.A.; supervision, G.A.d.l.C.; project administration, G.A.d.l.C.; funding acquisition, G.A.d.l.C. All authors have read and agreed to the published version of the manuscript.

Funding: This research was funded by the National Agency of Research and Development of Chile, ANID, Fondecyt N° 11190483.

Institutional Review Board Statement: Not applicable.

Informed Consent Statement: Not applicable.

Data Availability Statement: The dependent variable was generated from the National Forestry Corporation of Chile (CONAF) https://www.conaf.cl/incendios-forestales/incendios-forestales-en-chile/historical-statistics/ (accessed on 2 April 2022)). The independent variables can be found at the Instituto Nacional de Estadísticas, https://www.ine.cl/herramientas/portal-de-mapas/geodatos-abiertos (accessed on 2 April 2022); Ministerio de Obras Públicas, https://ide.mop.gob.cl/geomop/ (accessed on 2 April 2022); Earth Resources Observation and Science Center (EROS), https://www.usgs.gov/centers/eros/data-tools (accessed on 22 May 2022). Sentinel-2 images for mapping-based use are located on the Copernicus Open Access Hub (https://scihub.copernicus.eu/dhus/#/home, accessed 22 May 2022). The vector layers used for the administrative political division can be downloaded from the Geospatial Data Infrastructure (IDE, Chile) (https://www.ide.cl/index.php/limites-y-fronteras/item/1528-division-administrative-policy-2020, accessed on 11 April 2022).

Conflicts of Interest: The authors declare no conflict of interest.

References

1. CEPAL. *Recomendaciones Institucionales y Metodológicas para la Medición de Indicadores ODS Relacionados con Desastres y con el Marco de Sendai para la Reducción del Riesgo de Desastres. Grupo de Trabajo Sobre la Medición y Registro de Indicadores Relativos a la Reducción del Riesgo de Desastres en América Latina y el Caribe 2020–2021*; Report; CEPAL: Santiago, Chile, 2021.
2. Seidl, R.; Thom, D.; Kautz, M.; Martin-Benito, D.; Peltoniemi, M.; Vacchiano, G.; Wild, J.; Ascoli, D.; Petr, M.; Honkaniemi, J.; et al. Forest disturbances under climate change. *Nat. Publ. Group* **2017**, *7*, 395–402. [CrossRef] [PubMed]
3. Maniatis, Y.; Doganis, A.; Chatzigeorgiadis, M. Fire Risk Probability Mapping Using Machine Learning Tools and Multi-Criteria Decision Analysis in the GIS Environment: A Case Study in the National Park Forest Dadia-Lefkimi-Soufli, Greece. *Appl. Sci.* **2022**, *12*, 2938. [CrossRef]
4. Kang, Y.; Jang, E.; Im, J.; Kwon, C.; Kim, S. Developing a new hourly forest fire risk index based on catboost in South Korea. *Appl. Sci.* **2020**, *10*, 8213. [CrossRef]
5. Xofis, P.; Konstantinidis, P.; Papadopoulos, I.; Tsiourlis, G. Integrating remote sensing methods and fire simulation models to estimate fire hazard in a south-east mediterranea protected area. *Fire* **2020**, *3*, 31. [CrossRef]
6. Ricotta, C.; Bajocco, S.; Guglietta, D.; Conedera, M. Assessing the influence of roads on fire ignition: Does land cover matter? *Fire* **2018**, *1*, 24. [CrossRef]
7. Ganteaume, A.; Camia, A.; Jappiot, M.; San-Miguel-Ayanz, J.; Long-Fournel, M.; Lampin, C. A Review of the Main Driving Factors of Forest Fire Ignition Over Europe. *Environ. Manag.* **2013**, *51*, 651–662. [CrossRef] [PubMed]
8. Quiroz, N.; Walls, R.; Cicione, A. Towards understanding fire causes in informal settlements based on inhabitant risk perception. *Fire* **2021**, *4*, 39. [CrossRef]
9. Özbayoğlu, A.; Bozer, R. Estimation of the burned area in forest fires using computational intelligence techniques. *Procedia Comput. Sci.* **2012**, *12*, 282–287. [CrossRef]
10. Schaefer, A.J.; Magi, B.I. Land-cover dependent relationships between fire and soil moisture. *Fire* **2019**, *2*, 55. [CrossRef]
11. Van Hoang, T.; Chou, T.Y.; Fang, Y.M.; Nguyen, N.T.; Nguyen, Q.H.; Canh, P.X.; Toan, D.N.B.; Nguyen, X.L.; Meadows, M.E. Mapping forest fire risk and development of early warning system for NW Vietnam using AHP and MCA/GIS methods. *Appl. Sci.* **2020**, *10*, 4348. [CrossRef]
12. Hassan, Q.K.; Rahaman, K.R.; Ahmed, M.R.; Hossain, S.M. Examining post-fire perceptions of selected mitigation strategies after the 2016 horse river wildland fire in alberta, Canada. *Appl. Sci.* **2021**, *11*, 10155. [CrossRef]
13. Jolly, W.M.; Freeborn, P.H.; Page, W.G.; Butler, B.W. Severe fire danger index: A forecastable metric to inform firefighter and community wildfire risk management. *Fire* **2019**, *2*, 47. [CrossRef]
14. Connors, S.; Dionne, M.; Hanák, G.; Musulin, R.; Aellen, N.; Amjad, M.; Bowen, S.; Carrascal, D.R.; Coppola, E.; Dal Moro, E.; et al. Climate Science: A Summary for Actuaries—What the IPCC Climate Change Report 2021 Means for the Actuarial Profession. 2022. Available online: www.actuaries.orgwww.actuaries.org\T1\textbar{}secretariat@actuaries.org (accessed on 3 June 2022).
15. Bowman, D.M.J.S.; Moreira-Muñoz, A.; Kolden, C.A.; Chávez, R.O.; Muñoz, A.A.; Salinas, F. Human-environmental drivers and impacts of the globally extreme 2017 Chilean fires. *Tapuya Lat. Am. Sci. Technol. Soc.* **2021**, *4*, 350–362. [CrossRef] [PubMed]
16. Bowman, D.M.J.S.; Williamson, G.J. River flows are a reliable index of forest fire risk in the temperate Tasmanian wilderness world heritage area, Australia. *Fire* **2021**, *4*, 22. [CrossRef]
17. Parisien, M.A.; Parks, S.A.; Krawchuk, M.A.; Little, J.M.; Flannigan, M.D.; Gowman, L.M.; Moritz, M.A. An analysis of controls on fire activity in boreal Canada: Comparing models built with different temporal resolutions. *Ecol. Appl.* **2014**, *24*, 1341–1356. [CrossRef]
18. Westerling, A.L.; Hidalgo, H.G.; Cayan, D.R.; Swetnam, T.W. Warming and earlier spring increase Western U.S. forest wildfire activity. *Science* **2006**, *313*, 940–943. [CrossRef]
19. Evers, C.; Holz, A.; Busby, S.; Nielsen-Pincus, M. Extreme Winds Alter Influence of Fuels and Topography on Megafire Burn Severity in Seasonal Temperate Rainforests under Record Fuel Aridity. *Fire* **2022**, *5*, 41. [CrossRef]
20. Ghorbanzadeh, O.; Blaschke, T.; Gholamnia, K.; Aryal, J. Forest fire susceptibility and risk mapping using social/infrastructural vulnerability and environmental variables. *Fire* **2019**, *2*, 50. [CrossRef]
21. Liu, Q.; Shan, Y.; Shu, L.; Sun, P.; Du, S. Spatial and temporal distribution of forest fire frequency and forest area burnt in Jilin Province, Northeast China. *J. For. Res.* **2018**, *29*, 1233–1239. [CrossRef]
22. Meneses, B.M. Vegetation recovery patterns in burned areas assessed with landsat 8 OLI imagery and environmental biophysical data. *Fire* **2021**, *4*, 76. [CrossRef]
23. Caggiano, M.D.; Hawbaker, T.J.; Gannon, B.M.; Hoffman, C.M. Building loss in WUI disasters: Evaluating the core components of the wildland–urban interface definition. *Fire* **2020**, *3*, 73. [CrossRef]
24. Jolly, W.M.; Cochrane, M.A.; Freeborn, P.H.; Holden, Z.A.; Brown, T.J.; Williamson, G.J.M.; Bowman, D.M. Climate-induced variations in global wildfire danger from 1979 to 2013. *Nat. Commun.* **2015**, *6*, 1–11. [CrossRef] [PubMed]
25. Flannigan, M.; Stocks, B.; Turetsky, M.; Wotton, M. Impacts of climate change on fire activity and fire management in the circumboreal forest. *Glob. Change Biol.* **2009**, *15*, 549–560. [CrossRef]
26. Montiel-Molina, C.; Vilar, L.; Romão-Sequeira, C.; Karlsson, O.; Galiana-Martín, L.; de Lomana, G.M.G.; Palacios-Estremera, M.T. Have historical land use/land cover changes triggered a fire regime shift in central Spain? *Fire* **2019**, *2*, 44. [CrossRef]
27. Couto, F.T.; Iakunin, M.; Salgado, R.; Pinto, P.; Viegas, T.; Pinty, J.P. Lightning modelling for the research of forest fire ignition in Portugal. *Atmos. Res.* **2020**, *242*, 104993. [CrossRef]

28. Dowdy, A.J. Climatology of thunderstorms, convective rainfall and dry lightning environments in Australia. *Clim. Dyn.* **2020**, *54*, 3041–3052. [CrossRef]
29. Castedo-Dorado, F.; Rodriguez-Perez, J.R.; Marcos-Menendez, J.L.; Alvarez-Taboada, M.F. Modelling the probability of lightning-induced forest fire occurrence in the province of León (NW Spain). *For. Syst.* **2001**, *20*, 95–107. [CrossRef]
30. Podschwit, H.; Miller, C.; Alvarado, E. Spatiotemporal prescribed fire patterns in Washington state, USA. *Fire* **2021**, *4*, 19. [CrossRef]
31. Ahn, Y.S.; Ryu, S.-R.; Lim, J.; Choong, H.L.; Shin, J.H.; Choi, W.I.L.; Lee, B.; Jeong, J.-H.; An, K.W.; Jung, I.S. Effects of forest fires on forest ecosystems in eastern coastal areas of Korea and an overview of restoration projects. *Landsc. Ecol.* **2014**, *10*, 229–237. [CrossRef]
32. Herbert, C. Assessing the Effectiveness of Green Landscape Buffers to Reduce Fire Severity and Limit Fire Spread in California: Case Study of Golf Courses. *Fire* **2022**, *5*, 44. [CrossRef]
33. Fischer, J.; Gardner, T.A.; Bennett, E.M.; Balvanera, P.; Biggs, R.; Carpenter, S.; Daw, T.; Folke, C.; Hill, R.; Hughes, T.P.; et al. Advancing sustainability through mainstreaming a social-ecological systems perspective. In *Current Opinion in Environmental Sustainability*; Elsevier: Amsterdam, The Netherlands, 2015; Volume 14, pp. 144–149.
34. Folke, C. Resilience: The emergence of a perspective for social-ecological systems analyses. *Glob. Environ. Change* **2006**, *16*, 253–267. [CrossRef]
35. Olsson, P.; Folke, C.; Hahn, T. Social-Ecological Transformation for Ecosystem Management: The Development of Adaptive Co-management of a Wetland Landscape in Southern Sweden. *Ecol. Soc.* **2004**, *9*, 2. [CrossRef]
36. Olsson, P.; Folke, C.; Berkes, F. Adaptive comanagement for building resilience in social-ecological systems. *Environ. Manag.* **2004**, *34*, 75–90. [CrossRef] [PubMed]
37. Scheer, D.; Benighaus, C.; Benighaus, L.; Renn, O.; Gold, S.; Röder, B.; Böl, G.F. The Distinction Between Risk and Hazard: Understanding and Use in Stakeholder Communication. *Risk Anal.* **2014**, *34*, 1270–1285. [CrossRef]
38. Dong, X.; Shao, G.; Limin, D.; Zhanqing, H.; Lei, T.; Hui, W. Mapping forest fire risk zones with spatial data and principal component analysis. *Sci. China Ser. E Technol. Sci.* **2006**, *49*, 140–149. [CrossRef]
39. Aragoneses, E.; Chuvieco, E. Generation and mapping of fuel types for fire risk assessment. *Fire* **2021**, *4*, 59. [CrossRef]
40. Miranda, A.; Carrasco, J.; González, M.; Pais, C.; Lara, A.; Altamirano, A.; Weintraub, A.; Syphard, A.D. Evidence-based mapping of the wildland-urban interface to better identify human communities threatened by wildfires. *Environ. Res. Lett.* **2020**, *15*, 094069. [CrossRef]
41. Chuvieco, E.; Aguado, I.; Yebra, M.; Nieto, H.; Salas, J.; Martín, M.P.; Vilar, L.; Martínez, J.; Martín, S.; Ibarra, P.; et al. Development of a framework for fire risk assessment using remote sensing and geographic information system technologies. *Ecol. Model.* **2010**, *221*, 46–58. [CrossRef]
42. Keane, R.E.; Drury, S.A.; Karau, E.C.; Hessburg, P.F.; Reynolds, K.M. A method for mapping fire hazard and risk across multiple scales and its application in fire management. *Ecol. Model.* **2010**, *221*, 2–18. [CrossRef]
43. Luhmann, N. *Organización y decisión. Autopoiesis, Acción y Entendimiento Comunicativo*; Universidad Iberoamericana: Barcelona, España, 1995.
44. Giddens, A. *The Sociology of Risk and Gambling Reader*; Routledge: London, UK, 2006.
45. Beck, U. *A Critical Introduction to Risk Society*; Pluto Press: London, UK, 2004.
46. Abram, N.J.; Henley, B.J.; Gupta, A.S.; Lippmann, T.J.R.; Clarke, H.; Dowdy, A.J.; Sharples, J.J.; Nolan, R.H.; Zhang, T.; Wooster, M.J.; et al. Connections of climate change and variability to large and extreme forest fires. In *Communications Earth and Environment*; Nature Publishing Group: Berlin, Germany, 2021. [CrossRef]
47. Martin, J.; Hillen, T. The spotting distribution of wildfires. *Appl. Sci.* **2016**, *6*, 177. [CrossRef]
48. Pérez-Sánchez, J.; Jimeno-Sáez, P.; Senent-Aparicio, J.; Díaz-Palmero, J.M.; de Cabezas-Cerezo, J.D. Evolution of burned area in forest fires under climate change conditions in southern Spain using ANN. *Appl. Sci.* **2019**, *9*, 4155. [CrossRef]
49. Minsavage-Davis, C.D. Evaluating the Performance of Fire Rate of Spread Models in Northern-European Calluna vulgaris Heathlands. *Fire* **2022**, *5*, 46. [CrossRef]
50. Syphard, A.D.; Rustigian-Romsos, H.; Keeley, J.E. Multiple-scale relationships between vegetation, the wildland–urban interface, and structure loss to wildfire in California. *Fire* **2021**, *4*, 12. [CrossRef]
51. De la Barrera, F.; Ruiz, V.; Quense, J. *Evaluación del Impacto de los Incendios de Chile Centro-Sur en el Verano del año 2017*; Primera entrega; International Association of Landscape Ecology (IALE)-Chile: Santiago, Chile, 2017.
52. Hendrychová, M.; Kabrna, M. An analysis of 200-year-long changes in a landscape affected by large-scale surface coal mining: History, present and future. *Appl. Geogr.* **2016**, *1*, 151–159. [CrossRef]
53. Lee, C.; Schlemme, C.; Murray, J.; Unsworth, R. The cost of climate change: Ecosystem services and wildland fires. *Ecol. Econ.* **2015**, *116*, 261–269. [CrossRef]
54. San-Miguel-Ayanz, J.; Moreno, J.M.; Camia, A. Analysis of large fires in European Mediterranean landscapes: Lessons learned and perspectives. *For. Ecol. Manag.* **2013**, *294*, 11–22. [CrossRef]
55. Moritz, M.A.; Batllori, E.; Bradstock, R.A.; Gill, A.M.; Handmer, J.; Hessburg, P.F.; Leonard, J.; McCaffrey, S.; Odion, D.C.; Schoennagel, T.; et al. Learning to coexist with wildfire. *Nature* **2014**, *515*, 58–66. [CrossRef]
56. McWethy, D.B.; Pauchard, A.; García, R.A.; Holz, A.; González, M.E.; Veblen, T.T.; Stahl, J.; Currey, B. Landscape drivers of recent fire activity (2001–2017) in south-central Chile. *PLoS ONE* **2018**, *13*, e0205287. [CrossRef]

57. CONAF. *Análisis de la Afectación y Severidad de los Incendios Forestales*; Report; Corporación Nacional Forestal (CONAF): Santiago, Chile, 2017.
58. Levin, N.; Tessler, N.; Smith, A.; Mcalpine, C. The Human and Physical Determinants of Wildfires and Burnt Areas in Israel. *Environ. Manag.* **2016**, *58*, 549–562. [CrossRef] [PubMed]
59. Salis, M.; Ager, A.A.; Finney, M.A.; Arca, B.; Finney, M.A. Analyzing spatiotemporal changes in wildfire regime and exposure across a Mediterranean fire-prone area. *Nat. Hazards* **2014**, *71*, 1389–1418. [CrossRef]
60. Santelices-Moya, R.; Gibson-Carpintero, S.; Cabrera-Ariza, A.; Santini-Junior, L.; Venegas-González, A. Reduced Rainfall Variability Reduces Growth of Nothofagus alessandrii Espinosa (Nothofagaceae) in the Maule Region, Chile. *Forests* **2022**, *13*, 1184. [CrossRef]
61. González, M.E.; Gómez-González, S.; Lara, A.; Garreaud, R.; Díaz-Hormazábal, I. The 2010–2015 Megadrought and its influence on the fire regime in central and south-central Chile. *Ecosphere* **2018**, *9*, e02300. [CrossRef]
62. ODEPA. *Región del Maule: Información Regional*; Oficina de Estudios y Políticas Agrariasm: Santiago, Chile, 2018.
63. INE. *División Político-Administrativa y Censal Región del Maule*; Instituto Nacional de Estadísticas: Santiago, Chile, 2019.
64. Díaz-Hormazábal, I.; González, M.E. Spatio-temporal analyses of wildfires in the region of Maule, Chile. *Bosque* **2016**, *37*, 147–158. [CrossRef]
65. CONAF. *Monitoreo de Cambios, Corrección Cartográfica y Actualización del Catastro de los Recursos Vegetacionales Nativos de la Región Del Maule*; Corporación Nacional Forestal: Santiago, Chile, 2018. Available online: http://www.biblioteca.digital.gob.cl/bitstream/handle/123456789/2341/Resumen%20Ejecutivo%20Catastro%20Maule_2016.pdf (accessed on 2 June 2022).
66. Valencia, D.; Saavedra, J.; Brull, J.; Santelices, R. Fire severity damages caused on Nothofagus alessandrii forest on the Maule Region of Chile. *Gayana Botánica* **2018**, *75*, 531–534. [CrossRef]
67. Gálvez, M.G.; Gallegos, F.; Turén, V. Unfinished extinction and the velocities of capitalist sacrifices in the woodlands of central Chile. *Tapuya Lat. Am. Sci. Technol. Soc.* **2021**, *4*, 1939491. [CrossRef]
68. Municipalidad de Constitución. *Plan de Desarrollo Comunal Ilustre Municipalidad de Constitución Informe Etapa 1*; Municipalidad de Constitución: Maule, Chile, 2016.
69. CIREN. *Comuna de Empedrado*; Recursos Naturales: Empedrado, Chile, 2020.
70. Municipalidad de Cauquenes. *Plan de Desarrollo Comunal 2014–2018*; Ilustre Municipalidad de Cauquenes: Cauquenes, Chile, 2018.
71. Instituto Nacional de Estadísticas (INE). 2022. Available online: https://www.ine.cl/herramientas/portal-de-mapas/geodatos-abiertos (accessed on 5 June 2022).
72. Ministerio de Obras Públicas (MOP). 2022. Available online: https://ide.mop.gob.cl/geomop/ (accessed on 5 June 2022).
73. Earth Resources Observation and Science Center (EROS). Available online: https://www.usgs.gov/centers/eros/data-tools (accessed on 5 June 2022).
74. Zhao, Y.; Feng, D.; Yu, L.; Wang, X.; Chen, Y.; Bai, Y.; Hernández, H.J.; Galleguillos, M.; Estades, C.; Biging, G.S.; et al. Detailed dynamic land cover mapping of Chile: Accuracy improvement by integrating multi-temporal data. *Remote Sens. Environ.* **2016**, *183*, 170–185. [CrossRef]
75. Breiman, L. Bagging Predictors. *Mach. Learn.* **1996**, *24*, 123–140. [CrossRef]
76. Miranda, A.; Mentler, R.; Moleto-Lobos, I.; Alfaro, G.; Aliaga, L.; Balbontín, D.; Barraza, M.; Baumbach, S.; Calderón, P.; Urrutia, V.; et al. The Landscape Fire Scars Database: Mapping historical burned area and fire severity in Chile. *Earth Syst. Sci. Data Discuss.* **2022**, *14*, 3599–3613. [CrossRef]
77. Martinez, J.L.; Lucas-borja, M.E.; Plaza-alvarez, P.A.; Denisi, P.; Moreno, M.A.; Hernández, D.; González-Romero, J.; Zema, D.A. Comparison of satellite and drone-based images at two spatial scales to evaluate vegetation regeneration after post-fire treatments in a mediterranean forest. *Appl. Sci.* **2021**, *11*, 5423. [CrossRef]
78. Nunes, L.J.R.; Raposo, M.A.M.; Pinto Gomes, C.J. A historical perspective of landscape and human population dynamics in guimarães (Northern Portugal): Possible implications of rural fire risk in a changing environment. *Fire* **2021**, *4*, 49. [CrossRef]
79. Berry, L.E.; Sitters, H.; Abrams, R.W.; Heneberg, P.; Arsenault, A. Regional Case Studies: Southeast Australia, Sub-Saharan Africa, Central Europe, and Boreal Canada: Case Study: The Ecology of Mixed-Severity Fire in Mountain Ash Forests. In *The Ecological Importance of Mixed-Severity Fires: Nature's Phoenix*; Elsevier Inc.: Amsterdam, The Netherlands, 2015; pp. 210–222. [CrossRef]
80. Henriquez, E.M.; González, P.S.N. Análisis del Déficit Hídrico de la Región del Maule, Chile. *Rev. Interam. De Ambiente Y Tur. RIAT* **2012**, *7*, 25–32. [CrossRef]

applied sciences

MDPI

Article

Landslides Triggered by Medicane Ianos in Greece, September 2020: Rapid Satellite Mapping and Field Survey

Sotiris Valkaniotis [1,*], George Papathanassiou [2], Vassilis Marinos [3], Charalampos Saroglou [3], Dimitrios Zekkos [4], Vasileios Kallimogiannis [3], Efstratios Karantanellis [5], Ioannis Farmakis [6], Georgios Zalachoris [7], John Manousakis [8] and Olga-Joan Ktenidou [9]

1 Department of Civil Engineering, Democritus University of Thrace, 67100 Xanthi, Greece
2 Department of Geology, Aristotle University of Thessaloniki, 54124 Thessaloniki, Greece
3 School of Civil Engineering, National Technical University of Athens, 10682 Athens, Greece
4 Department of Civil and Environmental Engineering, University of California at Berkeley, Berkeley, CA 94720, USA
5 Department of Earth and Environmental Sciences, University of Michigan, Ann Arbor, MI 48109, USA
6 Department of Geological Sciences and Engineering, Queen's University, Kingston, ON K7L 3N6, Canada
7 Omikron Kappa Consulting S.A., 15122 Athens, Greece
8 Elxis Group, 10682 Athens, Greece
9 Geodynamic Institute, National Observatory of Athens, 11810 Athens, Greece
* Correspondence: svalkani@civil.duth.gr; Tel.: +30-2541079688

Citation: Valkaniotis, S.; Papathanassiou, G.; Marinos, V.; Saroglou, C.; Zekkos, D.; Kallimogiannis, V.; Karantanellis, E.; Farmakis, I.; Zalachoris, G.; Manousakis, J.; et al. Landslides Triggered by Medicane Ianos in Greece, September 2020: Rapid Satellite Mapping and Field Survey. *Appl. Sci.* **2022**, *12*, 12443. https://doi.org/10.3390/app122312443

Academic Editors: Spyridon Mavroulis, Efthymios Lekkas and Valerio Comerci

Received: 19 July 2022
Accepted: 2 December 2022
Published: 5 December 2022

Abstract: Medicanes, a type of strong hurricanes/cyclones occurring in the Mediterranean, can be the source of major geohazard events in Mediterranean coastal and inland areas. Medicane Ianos that hit Greece during 17–19 September 2020 caused widespread damage, with numerous landsides and floods being the most prominent. Following the landfall of Medicane Ianos, a series of field surveys were launched together with rapid response through satellite imagery. We focused on two of the areas most affected by Medicane Ianos, Cephalonia island and Karditsa, Thessaly, both in Greece. A rapid landslide inventory for the Karditsa region was prepared using Copernicus Sentinel-2 satellite imagery, the first of its kind for a severe weather event in Greece. The mountainous area of Karditsa region in western Thessaly experienced the unprecedented number of 1696 landslides, mapped through satellite imagery and examined in the field. Cephalonia Island experienced a smaller number of landsides but damaging debris flows and severe structural damages. The rapid landside inventory was then compared to new methods of automated landslide mapping through change detection of satellite imagery.

Keywords: landslides; landslide inventory; rapid mapping; remote sensing; Sentinel-2; Ianos; Medicane; Greece

1. Introduction

As a distinctive part of the geomorphic evolution of active mountain belts, landslides play an important role in gradually changing the landscape. The triggering of landslides is related to earthquake, meteorological and human-induced factors. The former cases are reported in active tectonic zones and are characterized by the sudden occurrence of co-seismic landslides covering large areas close to the earthquake fault rupture. As it has been demonstrated by [1], fault rupture geometry and kinematics plays a significant role to the spatial distribution and density of the coseismic landslides. On the other hand, most of the landslide-related phenomena are generated by intense rainfalls. It is well known that long periods of low intensity rainfall can trigger deep-seated landslides while short duration heavy intensity rainfall is mostly related to shallow mass movements [2]. Climate change, caused by global warming in the recent years, is expected to lead to an increase of the rate of landslide phenomena in the near future [3].

One of the outcomes of climate change is a more frequent occurrence of tropical-like cyclones in areas of dry climate and relatively shallow seas. The Mediterranean Basin is one of the most cyclogenetic regions worldwide as a result of its characteristic morphology [4]. Medicanes, a concatenation of Mediterranean Sea with hurricanes, resemble tropical cyclones but present certain differences. The formation of Medicanes is highly controlled by the air–sea interaction [5]. Generally, high sea surface temperatures favor their formation especially when the air temperature is relatively low [6]. Tropical cyclones emerge when the temperature of the sea surface exceeds 26 °C. Nevertheless, the correspondent temperatures in the Mediterranean range between 18 to 23 °C [7,8], much lower than the threshold required for the tropical cyclone formation. Moreover, the expected lifespan of Medicanes is shorter than that of tropical storms.

The return period of Medicanes is higher in the central and western Mediterranean Sea [9] despite the fact that the eastern basin is warmer (a favorable factor for the development of tropical-like cyclones). This phenomenon is probably related to the fact that the central and western Mediterranean is prone to cold upper-air intrusions from north and central Europe [6]. A Medicane usually carries enough energy to travel large distances (sometimes hundreds of kms) causing torrential rains and strong winds until its deterioration. Its environmental and socioeconomic impact can be devastating.

One of these meteorological events is Medicane "Ianos" that occurred in September 2020 and formed as a result of a cluster of convection off the Libyan coast on 14 September 2020 [10–13]. Over the following days, it moved to the north and intensified before making landfall over Greece on 17 September 2020. After impacting western and central Greece, it changed its course and reached south Crete island by 20 September 2020. Medicane Ianos triggered intense rainfall at the central and southwestern part of Greece (Figure 1), including the two areas focused in this study; western Thessaly and Ionian Islands. Accumulated rainfall peaked at 769 mm for 17–18 September in Cephalonia island in the Ionian Sea, while Pertouli and Mouzaki stations in western Thessaly peaked at 317 mm and 268 mm [12]. This natural phenomenon induced thousands of mostly shallow landslides, debris flows and floods mainly at Central Greece and particularly at the area of Karditsa, Thessaly, Central Greece [14–16]. Landslides blocked most of the narrow mountainous roads and damaged numerous bridges [15,17]. This resulted in the isolation of communities located in higher elevations and delayed emergency response and recovery [14,17].

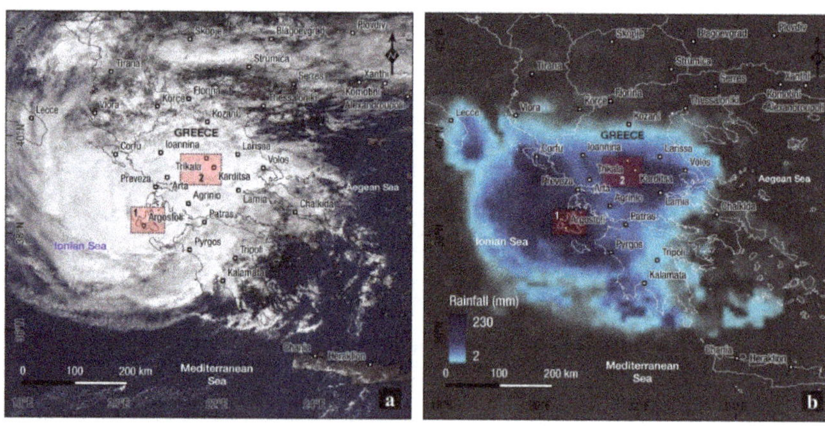

Figure 1. (**a**) Medicane Ianos over Greece, Sentinel-3 OLCI true color image acquired on 18 September 2020, one day after Medicane landing. (**b**) Satellite precipitation (NASA IMERG) of 18 September 2020. Overview boxes show the location of the two studied areas; 1—Cephalonia island, 2—western Thessaly.

2. Medicane Ianos-Induced Landslides

The first step for understanding the mechanism of the cyclones-induced landslides and to mitigate in near future similar types of catastrophes is to rapidly document them. This can be achieved by immediately organizing a field survey after the event to acquire perishable field data before it disappears. Landslide debris and rockfalls are commonly cleaned out of the road network as part of the recovery operations, but their removal obscures critical information used to analyze the triggers for slope failures. Furthermore, physical processes such as weathering and erosion could additionally alter the landslide-formed landscape after the landslide event, especially concerning small size shallow landslides. Data recovered from landslide studies are essential to understanding landslide mechanisms and the associated risks, which greatly aids in identifying susceptible zones, augmenting emergency response, and mitigating potential property damage and loss of life. This research applies a mixed-method approach, primarily consisting of a desk study followed by a detailed post-event field survey aiming to report the failures in order to compile a landslide-event inventory map.

In this study, the landslide phenomena triggered in the area of Thessaly and Ionian Islands are presented and preliminary analyzed with the aim to relate their spatial distribution with some basic geological and geomorphological parameters, i.e., geology, aspect, slope and elevation. Although large parts of central and western Greece were affected by Medicane Ianos, Karditsa and Cephalonia island are presented here in detail, as they experienced the most dense and severe landslide phenomena. In order to achieve this, we used information obtained during extensive field work and data provided by rapid satellite imagery mapping captured a few hours and days after the event.

Satellite-derived precipitation (e.g., Global Precipitation Measurement–GPM) enables monitoring of rainfall amount and patterns in near real-time (Figure 1b), and precipitation maps can be derived for forecasting or study of landslide events such as hurricanes and Medicanes [18,19]. However, results from Medicane Ianos satellite-derived precipitation showed an inconsistency with ground station measurements, with significantly lower values measured from the satellite sensors [12]. This, in addition to the poor coverage of the affected area (mountainous western Thessaly) by ground stations does not permit us to recreate an accurate map of precipitation during Medicane Ianos.

2.1. Landslides in Thessaly

2.1.1. Geologic Setting

Thessaly is the largest plain of central Greece surrounded by mountain ranges, most notably Pindos mountains in the west. The study area is part of the western Karditsa Prefecture and a smaller part of Evritania to the southwest. It includes Plastira dam lake at the center, and the high-elevation Agrafa Mountains (part of Pindos range) in the western part.

The oldest post-alpine sediments in the area are the Molassic formations of the Mesohellenic trench [20,21]. The majority of the area is covered by the flysch formation of the Pindos geotectonic unit [21–23]. Flysch formation consists of alternations of shales, sandstones, and limestones that are highly susceptible to slope failures. Bedrock formations of the Alpine units also include limestones and cherts of Pindos, Koziakas and Sub-Pelagionian zones, while ophiolites can be found in the eastern and southwestern border of the area [21,23,24].

2.1.2. Remote Sensing-Based Landslide Inventory

To investigate and map the distribution of landslides in the Karditsa, Thessaly area, we used post-event acquisitions of Copernicus Sentinel-2 optical satellite imagery. Sentinel-2A/B multispectral imagery has a ground resolution of 10–60 m and covers wide swaths, thus enabling the rapid imaging of the affected area. Multiple Sentinel-2 frames were selected for the post-event period, and acquired 20, 25 and 30 September 2020, as large parts of the area were covered by clouds in the 20 and 25 September frames. Images acquired before the event, on 5, 10 and 15 September 2020, were used as references in order to visually identify and map landslides (Figures 2 and 3). The minimum size of landslides mapped was limited by the satellite imagery resolution (10 m for Sentinel-2), and consequently smaller-sized slides or rockfalls were not possible to identify and were not included. In addition, it was not feasible to classify most of the landslides into different types based on this remote sensing procedure. A small portion of slope failures was classified by the field surveys as presented in following section, validating the outcome of the statistical analysis presented in Figure 4.

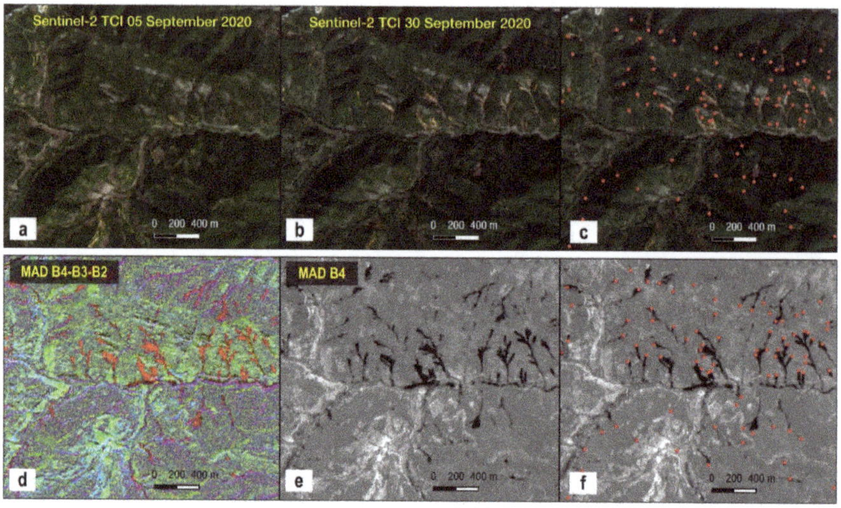

Figure 2. Example of manually mapping landslides within the study area using Sentinel-2 imagery. Pre- and post-event true color Sentinel-2 images (**a**,**b**) and an overlay (**c**) of mapped landslide points with red color (mostly headscarp points). Change detection MAD processing (**d**,**e**) assisted visual picking of headscarp points (**f**).

Due to the small size of the majority of the identified landslides (most having a size of 2–4 pixels in Sentinel-2 imagery), landslide polygons were not digitized in this preliminary inventory. Digitizing was focused on marking the initiation point/headscarp (long avalanches or earthflows) or the approximate center of the landslide feature when the former was not possible due to small size and image resolution. For assistance during manual picking of landslides, we used additional analysis products, such as change analysis rasters (MAD-Multivariate Alteration Detection Transformation [25]) and multi-temporal single band RGB composites (Figure 2). Co-registration issues present between Sentinel-2 frames were resolved using GeFolki registration algorithm [26].

This preliminary inventory (Figure 3) includes 1697 landslides, limited to a narrow mountainous area of Karditsa and part of Evritania regions. Examination of Sentinel-2 imagery over the Pindos mountain region showed no or few sparse landslides, outside this area. The largest concentration of landslides was found around Amarantos village, southeast of Plastira lake, with up to 15 landslides per square kilometer (Figure 3).

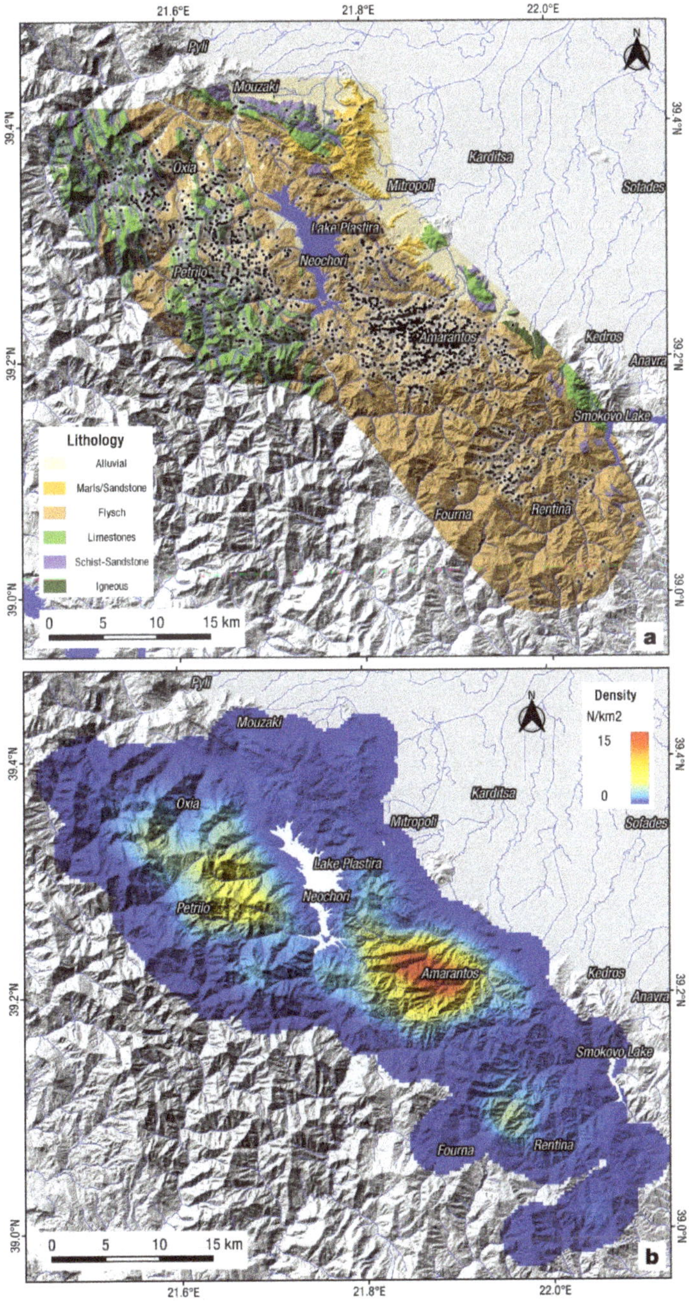

Figure 3. (**a**) Mapped landslides (black dots) triggered by Medicane Ianos during 17–19 September 2020, and simplified lithology of the area (see text for references). (**b**) Density map of Ianos landslides (using a radius of 1000 m); two major concentrations of landslides west of Lake Plastira and around Amarantos village.

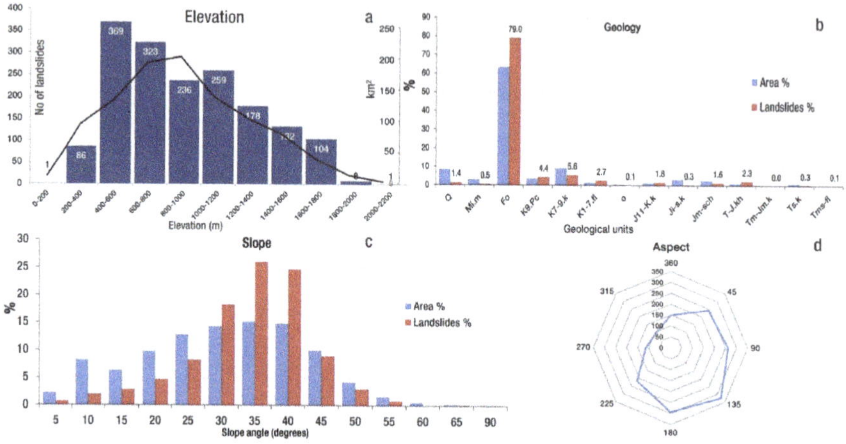

Figure 4. Statistical correlation of the spatial distribution of Ianos slope failures with: (**a**) elevation (histogram with area covered by each class in km^2 with a black line), (**b**) geology, (**c**) slope angles and (**d**) histogram of slope aspect values.

2.1.3. Statistical Analysis of Slope Failures

The spatial distribution of the slope failures was correlated with topographic parameters e.g., slope angle, aspect and elevation and the geologic characteristics of the study area, as provided by the relevant 1:50,000 scale geological maps. In particular, information related to the spatial distribution of the geological formations acquired by the official geological map sheets in 1:50,000 scale available from the Institute of Geological and Mineral Exploration [27–32]. Afterwards, we combined those formations into a smaller number of generalized groups based on similar lithological features (Table 1). The topographic parameters were extracted from a digital elevation model (DEM) of 5 m resolution, provided by the Hellenic Cadastre.

As shown in Figure 4, about 40% of the generated landslides were mapped on elevation ranging from 400 to 800 m; approximately 21% of the total number of landslides was documented from 400 to 600 m and 19% was reported on areas located between 600 and 800 m elevation. On higher elevation areas, the percent of slope failures was 13.91% between 800 and 1000 m, 15.26% between 1000 and 1200 m, and 10.49% for areas at elevation from 1200 and 1400 m. A lower percentage of slope failures has been identified at an elevation higher than 1400 m while a sharp decrease of landslides is related to areas located higher than 1800 m and lower than 400 m elevation. These statistics make sense as the areas located at high elevation are limited, while the latter outcome is due to the spatial distribution of geological units. More specifically, the lower parts of the study areas are mostly flat and covered by Quaternary sediments (loose clays, gravel and sand) and not by the flysch formation, which as it is demonstrated next is the predominant unit for landslide triggering among the geological units, along with slope gradient and relief.

The largest part of the study area is covered by the flysch formation (Fo), a generally weak, complex, and of variable rock mass quality unit. Flysch is composed by different rhythmic alternations of competent/strong sandstone layers and is incompetent, of generally low-strength siltstone/clayey schist beds, while it is associated with intensive folding and fracturing. At the mountainous area, Cretaceous carbonates are mapped (K7-9.k), while Quaternary deposits (Q) are concentrated mostly in the basin of Thessaly. These three geological units cover almost 81% of the study area. In particular, the flysch and K7-9.k cover 63% and 9.41%, respectively, and the quaternary deposits 8.5%. As it was expected, most of the landslides, approximately 79%, were reported in the area that is geologically covered by the flysch formation, followed by the geological units of K7-9.k and K9.Pc with

5.6% and 4.37%, respectively. To facilitate the reader, only the geological affected by slope failures are plotted in the relevant diagram.

Table 1. Description of geological formations (based on [26–31]).

Unit Name	Symbol	Area km²	Landslides %	Age	Description
Quaternary deposits	Q	111.38	1.4	Quaternary	Alluvial and fluvial sediments, scree, debris and terrace deposits
Molassic formations	Mi.m	40.81	0.5	Oligocene-Miocene	Conglomerates, marls, sandstones, Limestone (Mi.m and Mi.k)
Flysch	Fo	829.64	79		The formation consists of alternations of sandstone, shales, siltstones (fo.st) and more seldom conglomerates (fb)
Transition Beds	K9-Pc	48.08	4.4	Cretaceous-Paleocene	Alternations of limestones, sandstones, shales and marls (K9-Pc, K7-Pc)
Cretaceous carbonates	K7-9.k	123.21	5.6	Cretaceous-Paleocene	Limestones
Pindos "First" Flysch	K1-7.fl	20.05	2.7	Cretaceous	Older flysch formation; red cherts, shales, conglomerates and sandstones
Ophiolites	O	5.54	0.1	Jurassic-Cretaceous	Mafic and ultra-mafic igneous rocks; peridotites, serpentinites, dunites, basalts. Also contain syn-sedimentary shales, limestones and conglomerates
Jurassic carbonates	J11-K.k	17.61	1.8	Jurassic-Cretaceous	Thin bedded limestones and cherts
Jurassic carbonates	Ji-s.k	42.9	0.3	Jurassic	Limestones
Jurassic carbonates	Jm.sch	35.68	1.6	Jurassic	Alternations of cherts, limestones and shales
Triassic carbonates	T-J.kh	14.41	2.3	Triassic-Jurassic	Limestones, cherts, sandstones
Triassic carbonates	Ts.k	14.20	0.3	Upper Triassic	Limestones
Triassic carbonates	Tm-Jm.k	2.15	0	Middle Triassic-Jurassic	Limestones, cherts, sandstones
Triassic Basal Beds	Tms.fl	2.33	0.1	Middle Triassic	Flysch formation, thin beds of cherts, shales and limestones

A strong correlation with landslide occurrence exists for the areas where the slope angle ranges between 25° and 40° (Figure 4c). In these areas, the identified cases of slope failures are 1167, which is approximately 69% of the total failures. On lower (20–25°) and higher (40–45°) slope angle areas, the reported landslides are 139 and 151, respectively and rapidly decrease for slope angles higher than 45° and lower than 20°. The median slope angle of the landslides is approximately 32°, which is in agreement with the one documented by [32] for the slope failures triggered by Hurricane Maria, Puerto Rico 2017.

Analyzing the spatial distribution of slope failures with respect to the aspect of the slopes, a correlation between east-faced slopes and triggering of slope failures is observed (Figure 4d). In particular, considering the slopes with an aspect between 45° and 135°, it was found that 587 (34.5%) cases are identified. A high number of failures (294) was additionally reported on slopes with aspects ranging from 135° to 180°, while the lowest percentage of landslides (5.6%) was identified on areas oriented from 270° to 315°.

2.1.4. Types of Landslides and Induced Phenomena

Few days after the landfall and occurrence of Medicane Ianos, several teams of engineering geologists and engineers participated in field surveys aiming to map and document as many as possible slope failures, and validate the preliminary inventory produced based on remote sensing techniques [17]. This combination of desktop studies, remote sensing and field survey was also applied for documenting earthquake-induced secondary effects few kilometers to the east of the here study area, triggered by a seismic sequence, which occurred in March 2021 [33].

A total number of 73 landslides were documented during these field reconnaissance surveys in Karditsa Prefecture, which mainly caused partial or total damage to the roads and in some cases caused failures of riverbanks. This remote area is characterized by a combination of weak sedimentary sequences, high elevation deviations and steep slopes, as described in the previous section. The majority of remotely mapped landslides were found away from the local road network or in inaccessible areas, and thus the relative small number of field-surveyed landslides.

The documented landslides were mostly found in flysch formations, validating the outcome of the remote sensing approach, as it was earlier presented. Following the information obtained by these surveys, the failure mechanisms are strongly related to the heterogeneity degree of the flysch formation (siltstone-sandstone participation), the intensity of tectonic disturbance and the weathering degree.

In particular, the main types of landslides that were detected were:

1. **Rotational and translational landslides** formed mainly in siltstone flysch and siltstone and sandstone flysch in alternations. Those landslides are characterized by a complex geometry with their lower part frequently consisting of a slump that follows the geometry of a stronger underlying bed. In many cases, the accumulated material covered the road without resulting in a complete failure.
2. **Debris flows** in fractured limestones, scree, sandstone/conglomerate flysch. These were encountered in areas with steep morphology with deep gullies where mass transport of rock pieces and boulders was favored. These failures, i.e., debris flows, heavily impacted the road network and some villages located at higher elevations.
3. **Rockfalls and structural failures** (planar and wedges) mainly in limestone formations. Structurally controlled slides were recorded, under favorable kinematic conditions of the persistent bedding planes and their low shear strength due to the clayey nature. Such slides and rock falls were not very large in volume (few m^3) but resulted in road closure and in some cases minor to moderate structural failures.

Based on the information obtained, the majority of landslides are classified as rotational (Figures 5 and 6), causing the most severe damage to the human-made environment in the wider investigated area. As it has been previously mentioned, these slides occurred mainly in the weathered flysch mantle and the siltstone flysch formation, due to their overall low rock mass strength.

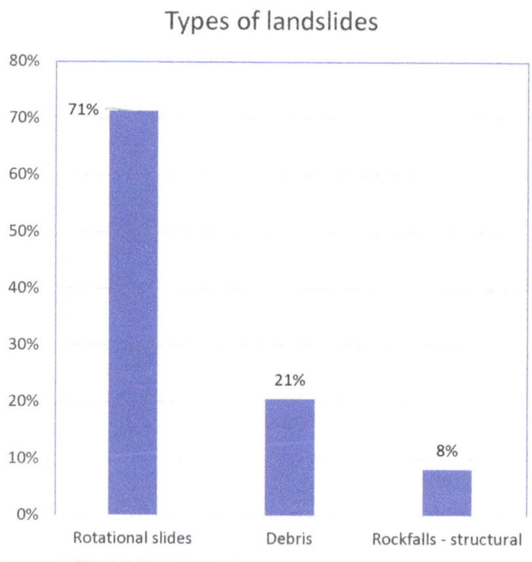

Figure 5. Types of landslides identified in the field reconnaissance surveys.

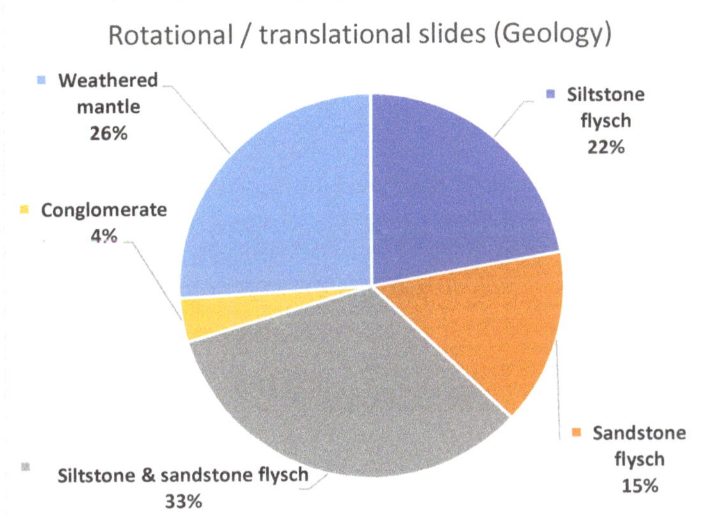

Figure 6. Different lithology types affected by rotational and translational slides.

Regarding the rotational landslides, most of them had a depth ranging between 10 and 15 m while a significant portion of failures was relatively shallow (<5 m). Their width ranges between 20 to 80 m, while the smaller slides have a width between 5 and 20 m. The width of the largest landslides was up to 200 m. Debris flows in most cases affected road sections with limited width, between 10 and 20 m, except a few cases where the affected width is almost 300 m.

The most severe landslide-induced failures in Karditsa Prefecture were around Amarantos village, located to the south-east of Plastira lake. In Figure 7a, a landslide that occurred in siltstone flysch causing the road failure is shown. In particular, the road was totally covered by slipped material across a length of 200 m, while its depth was estimated at about 15 m. Its height is approximately 70 m. The slide failure plane was probably constrained by the presence of a deeper stronger flysch layer downslope. In Figure 7b, a rotational slide in the weathered mantle of siltstone flysch is also shown. This landslide resulted in the collapse of the road pavement for a length of 10 m.

Figure 7. Characteristic rotational failures around Amarantos area. (**a**) large failure in siltstone flysch west of Itamos village and (**b**) failure in weathered flysch mantle material causing damage to the road.

Extensive debris flows were mapped in Livadia and between the communities of Pefkofito and Vlassi (Figure 8). Specifically, the thickness of the debris at the road level is

5 m while the impacted area is 300 m long in the former case. The debris flow material mainly consisted of limestone fragments and clay materials (Figure 8a).

Figure 8. (**a**) Characteristic debris flow at Livadia village; (**b**) extensive debris flows in gullies in the area between Pefkofito and Vlassi.

Regarding the second case, located in the area between Pefkofito and Vlassi (Figure 8b), it should be pointed out that this area is characterized by steep morphology while the extensive amounts of debris have been transported via deep gullies from higher elevations.

Moreover, the erosion and flooding of the Karitsiotis River damaged and blocked the road crossing south of Belokomiti village, Plastira Lake, as shown in Figure 9. Although the bridge itself did not collapse, road embankments on both sides were partially or completely washed out. The intersection of the roads west of the bridge was the most damaged part, blocking road traffic towards communities to the south and west of Lake Plastira. Riverbank erosion also caused several slope failures, while the flooding of the river caused slope undercutting that resulted in road failures (Figure 9).

Figure 9. Damaged bridge crossing of Karitsiotis River near Belokomiti. Ground (**a**) and vertical aerial (**b**) images of the damaged bridge undergoing repairs on 1 October 2020. UAS orthophoto (**c**) and digital surface model (**d**) of the Karitsiotis River bridge area (surveys 1 October 2020).

3. Landslides in Cephalonia Island

3.1. Geologic Setting

Cephalonia is part of the Ionian Islands and is located at the westernmost part of Greece. Geologically, the island is within the outermost edge of the ongoing subduction of the African plate under the Eurasian plate [34]. The bedrock of the island consists of two main formations: (i) the Pre-Apulian unit, which covers most of the island and consists mainly of a thick sequence of limestone and dolomite, overlain by a much thinner sequence of marl and pelite, and (ii) the Ionian unit, which covers part of the southeastern coastal areas and consists of limestone, shale, and breccia [35–37]. The basement is overlain by extensive sedimentary deposits in the southern and eastern part of the Paliki peninsula and the south part of the main island. These deposits consist of alluvial fans, mainly deposited along the stream channels (Figure 10). The dominance of carbonate rocks in combination with favorable climate conditions have facilitated the formation of karst units especially in the north and central part of the island. In particular, the Erissos Peninsula is dominated by an extensive, and partially karstified, planation surface. Cephalonia island is also characterized by steep slopes, especially along its western shoreline. Ridges are arranged in a NNW–SSE direction. The principal water divide has a NW–SE direction, with the main flow directions towards Sami Bay in the NE, and Poros in the SW.

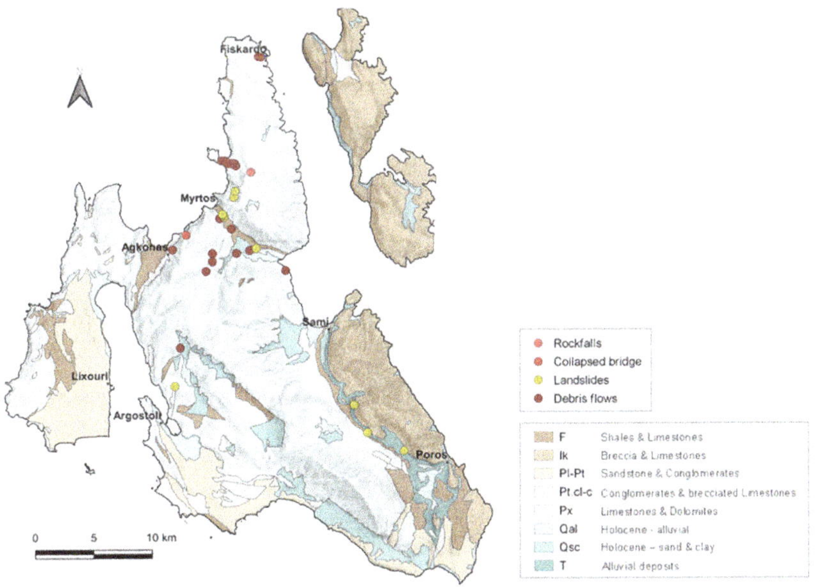

Figure 10. Lithological map of Cephalonia Island with observed landslides and related phenomena following Medicane Ianos. Modified from [35,37].

3.2. Types of Landslides and Induced Phenomena

Cephalonia Island was heavily impacted by Medicane Ianos. Ianos was the most damaging natural hazard event for Cephalonia island, since the earthquakes of 2014 that caused building damages, widespread landslides, rockfalls and liquefaction over the island [38–41]. Severe rainfall at Cephalonia peaked at 759 mm (17–18 September 2020) during the passing of Medicane Ianos from the Ionian Sea islands [12]. In particular, the highest severity of the Medicane-induced phenomena was reported at the northern-central part of the island towards Erissos peninsula, between Sami and Fiskardo. Debris flows and landslides occurred in several locations across the island (Figures 10 and 11). Often the debris flows covered wide areas (~40–70 m in width), while typically the average size of debris blocks reached 20–50 cm (Figure 11d). Moreover, the earth flows were often associated with road embankment failures (Figure 11c). The latter was also observed in cases of severe riverbank erosion/scour. The road network of the island was damaged and/or obstructed in several locations due to landslides, while a critical simple-span reinforced concrete bridge near the village of Agkonas collapsed due to scour of the west pier foundation by extensive gully flooding and debris (Figure 11e), heavily impacting the transportation network of the island.

Figure 11. Characteristic damages in Cephalonia Island following Medicane Ianos: (**a**) Riverbank erosion, (**b**) debris flow within a rural community, (**c**) road embankment failure, (**d**) extensive debris flow, and (**e**) bridge collapse.

Debris flows appeared to be the most common feature of the effect of Medicane Ianos across Cephalonia. Debris, collected or still in place, was present along gullies and riverbeds, across the island. Most notably, Assos village, situated in the north-west of the island, was severely impacted by a major debris flow, as it was covered by approximately 1.5 m of earth/debris material (Figure 11b). The source of the debris was traced to the hills east of the village, where significant surface erosion and ground cracking were observed. At places, eroded zones reached several meters in width. Eroded limestone boulders and residual soil material were present along the entire path of the flow. UAS-enabled mapping of the entire area indicates the change along the sea-shore due to material deposition (Figure 12). An initial, rough, estimation of the debris flow volume is 20,000 m^3.

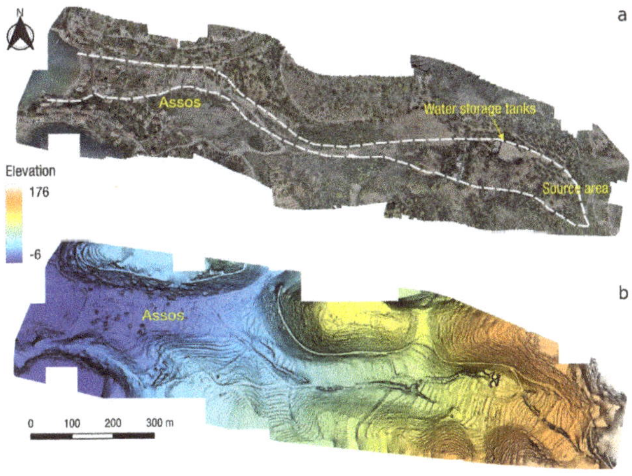

Figure 12. UAS-enabled mapping of Assos village debris flow and surrounding area (3D model can be viewed at: https://skfb.ly/6WyTT (accessed on 2 July 2022)): (**a**) orthophotomap and (**b**) digital surface model (surveyed on 30 September 2020).

Similar to Assos, just outside of the village of Fiskardo (Figure 13), which is situated at the northern part of Erissos peninsula, a debris flow was documented. The debris flow had a lesser extent than the one that occurred in Assos, but was still significant, with the flow reaching a run-out distance of approximately 400 m. Significant erosion features were observed near the source of the debris flow. Eroded zones reached, at places, a depth of 2 m. The debris flow is also visible in Sentinel-2 satellite imagery, as shown in the pre-and-post-event images (Figure 13). Moreover, several rotational landslides were identified along the steep central coastline, particularly near Myrtos beach. Sentinel-2 pre-and-post event images of landslides and debris flows near Myrtos beach, and the collapsed bridge at Agkonas village are shown in Figure 14.

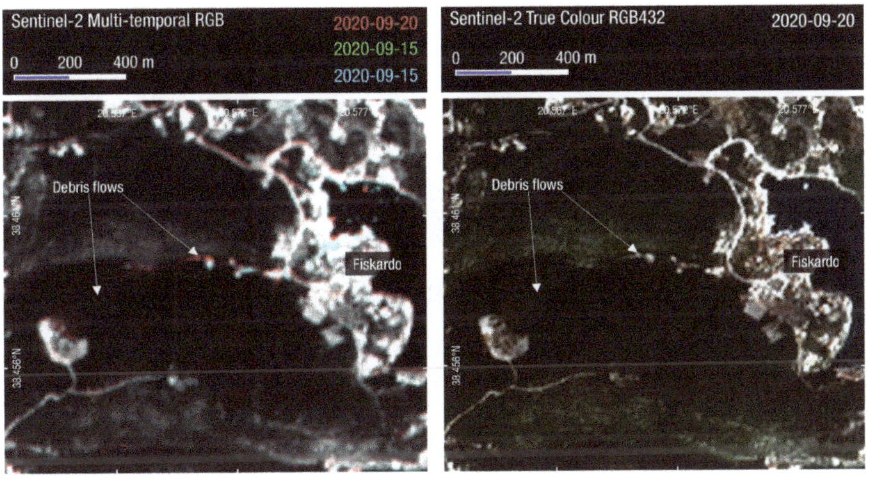

Figure 13. (**Left**): Sentinel-2 multi-temporal composite (Band 4) Fiskardo, Cephalonia. (**Right**): Sentinel-2 true color composite (Bands 4-3-2). Red colors mark the location of debris flows upstream of Fiskardo.

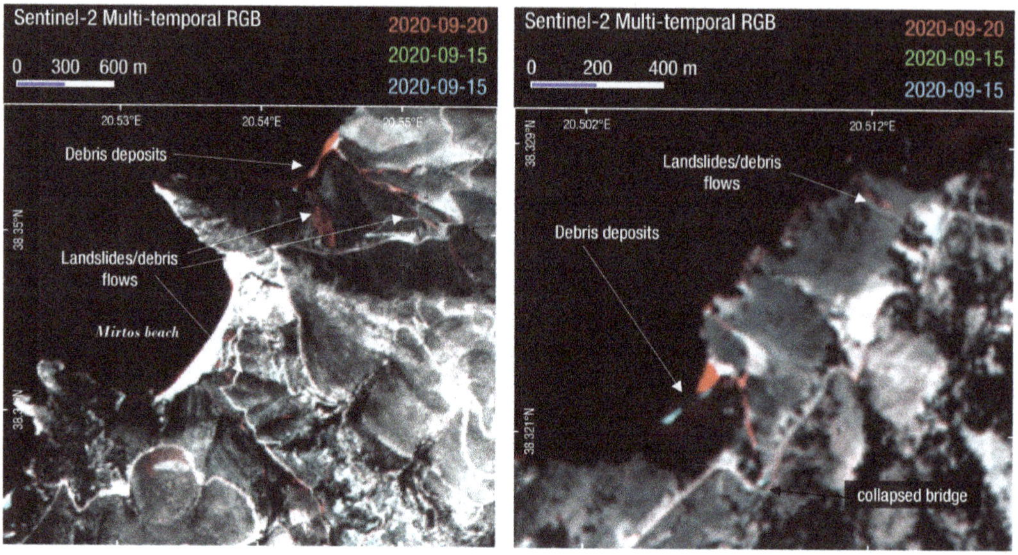

Figure 14. Sentinel-2 multi-temporal composites (Band 4) of central Cephalonia. (**Left**): Mirtos beach area; (**Right**): Site of collapsed bridge on the E.O. Argostoliou-Fiskardou (38.3195° N, 20.5081° E). Red colors mark the location of debris flows, landslides and deposition of debris along the coast.

4. Discussion

4.1. Comparison with Automated Mapping Methods

Rapid response to a major landslide event, such as hurricanes or earthquakes, is nowadays feasible due to the abundance of open remote sensing data available shortly after an event, from satellite imagery (e.g., Sentinel-1 & Sentinel-2, Landsat 8/9). The first satellite images available during or shortly after a meteorological event can provide a fast and comprehensive map of significant slides and effects, as described in the previous section. During the period that follows first response and early reconstruction, there is the need for more thorough and detailed landslide inventories. While this can be performed by manually digitizing landslides and effects using very high-resolution satellite imagery or other remote sensing data such as UAS surveys, new tools and a wealth of open satellite data can automate this workflow [42]. The ability to have access and process huge amounts of satellite imagery on the cloud through Google Earth Engine [43] and similar platforms has revolutionized remote sensing in geohazard response and analysis.

We present a comparison of our manual rapid mapping using Sentinel-2 images, with a series of recent workflows and codes that use the multi-temporal analysis of satellite imagery in Google Earth Engine [44–47]. While this comparison is not straightforward due to the different workflow, data and time frame used by either rapid manual mapping or automated multi-temporal analysis, this is an interesting case study to compare them. The wide extent and large number of landslides in the area of Karditsa, Thessaly is suitable for this use and comparison (Figure 15).

We selected four change detection techniques, that use multi-temporal analysis of open satellite data (Sentinel-1, Sentinel-2, Landsat); (a) method from [46] that calculates relative difference in the normalized difference vegetation index (rdNDVI) calculated from cloud-free composites of Sentinel-2, (b) HazMapper code by [44] that is based on the normalized difference vegetation index (dNDVI) calculated from cloud-free composites of Sentinel-2 [48], (c) ALDI, automated landslide detection index algorithm based on normalized difference vegetation index (NDVI) differencing of Landsat time series within Google Earth Engine taking into account seasonality [47] and (d) a SAR backscatter and amplitude change approach that uses multi-temporal stacks of Copernicus Sentinel-1 images [45].

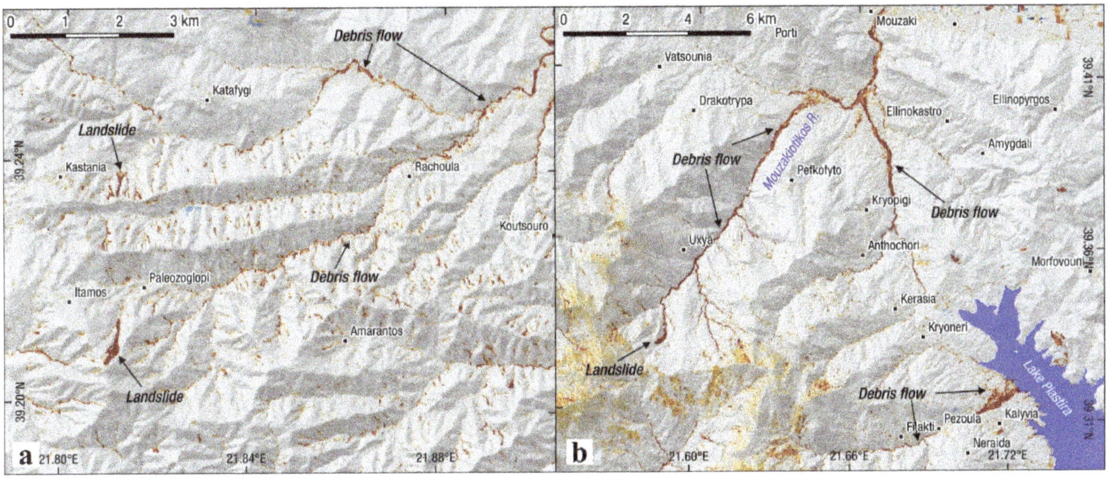

Figure 15. Change detection results from HazMapper [44,48] utilizing multi-temporal Sentinel-2 imagery. Red colors show negative rdNDVI changes that correspond to land cover changes due to landslides and debris flows. Results were in par with manual mapping in the area of high-density landslides southeast of Plastira Lake (**a**), while in most areas change detection was more successful in depicting the major debris flows such as the ones upstream of Mouzakiotikos River (**b**).

Treshold values for positive and negative landslide and debris flow identification was determined for each method results based on proposed values and also local conditions after cross comparing with mapped landslides and imagery (Table 2 and Figure S2). Examining the different change detection results we obtained from these four methods, we can observe that all methods captured the large debris flows that occurred along the mountainous watersheds such as Mouzakiotikos river (Figure 15) and Megalos river in Filakti, near Lake Plastira (Figure 16). Large landslides with significant dimensions and/or long run-out distance were also identified. As the majority of landslides triggered during Medicane Ianos were of small dimensions (less than 20–30 m), these were mostly missed or were undifferentiated from the background scatter noise.

Table 2. Positive and negative detection of mapped landslides by the different change detection methods.

Method	Positive	Negative	Success %	Threshold Value
rdNDVI [46]	872	800	**52.15**	−0.1<
HazMapper [44]	911	786	**53.68**	−10<
ALDI [47]	659	1037	**38.85**	>0.025
SAR backscatter change by [45]	837	851	**49.58**	−1 < I ratio > 1

A quick estimate of the positive or negative detection was performed by extracting pixel values on the mapped landslide inventory from the various methods. Table 2 presents the final percentage and number of positively and negatively detected landslides.

Histograms of the detection results of Table 2 are presented in Figure 17. The highest success ratio is achieved by the methods of [44,46] with almost comparable results, as expected due to the similarities in the calculated change detection data (Sentinel-2 10 m) and parameters (time-series of normalized difference vegetation index). The worst percentage was achieved by ALDI [47], probably due to the lower resolution data used (Landsat multispectral bands with 30 m resolution instead of 10 m for Sentinel-2). SAR backscatter change [45] was close to the first two method results, with the difference attributed to the coarser pixel resolution of Sentinel-1 SAR imagery (20 × 4 m, translated into >15 m ground resolution when geometrically corrected to terrain) and the challenging oblique scanning of SAR satellites that can lead to terrain shadows.

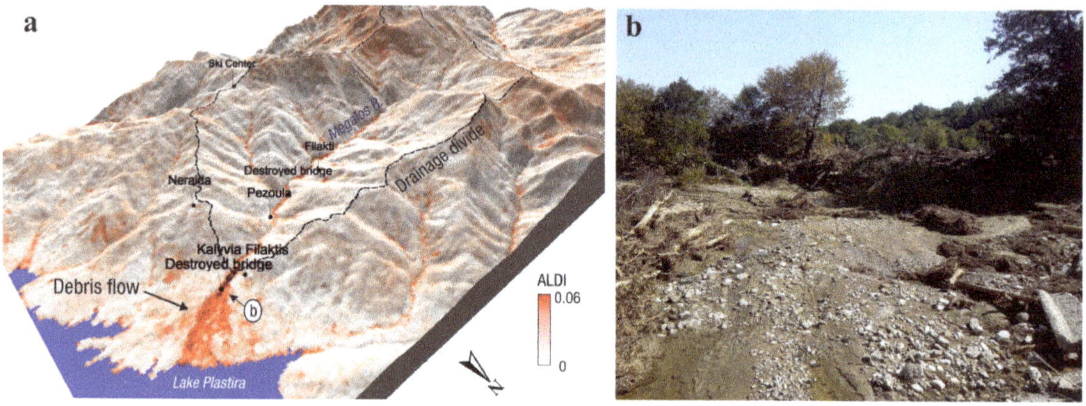

Figure 16. Debris flow and landslides along Megalos River in Filakti/Pezoula area (**a**). 3D view of the Megalos River watershed with debris flows along the main river course and its exit to Lake Plastira (red colors of ALDI change detection index values). Flooding, severe sedimentation and material transportation from upstream was mapped at its exit (photo (**b**), view towards east with location marked in (**a**)) and near Pezoula. Road bridges were destroyed in both locations, with positions labeled in (**a**).

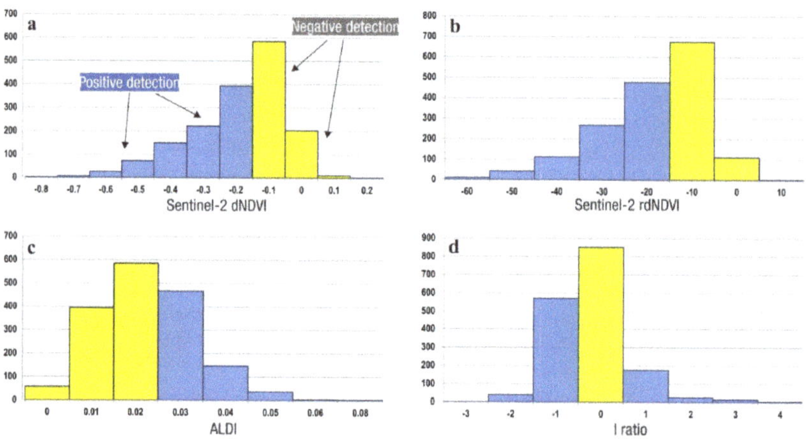

Figure 17. Histograms of pixel values by automated mapping methods corresponding to mapped landslides; (**a**) [46], (**b**) HazMapper by [44], (**c**) ALDI [47] and (**d**) SAR backscatter change by [45]. Yellow bars mark negative detections and blue bars positive detection, based on set threshold values for each method.

In conclusion, rapid manual mapping of landslides for Ianos outperformed automated mapping methods based on Google Earth Engine time-series change detection of satellite imagery. The most prominent factor was the small size of the majority of landslides, which was barely close to the ground resolution threshold of 10 m for Sentinel-2, making it harder to identify by automated change detection algorithms. A large number of false positives were visually identified in the various change detection results, a factor that could negatively affect early response and mapping (Figures S1 and S2). Debris flow and large slides were identified in most automated mapping results (Figure S1), while smaller and shallow landslides were difficult to positively identify due to variations in image reflectance and terrain shadows. Surface manifestation of small shallow landslides kept being modified in the days and weeks following the landslide event, due to erosion and vegetation regrowth, a factor that affects the signal in the long time-series used by automated change detection methods. It is worth mentioning that a number of landslides occurred in the winter months following the Ianos landslide event. These landslides would be included in most automated change detection results due to the larger time frame examined.

4.2. Older Landslides/Hazard

The mountainous area of Karditsa, Thessaly is known to be significantly affected by numerous landslides, many affecting communities and infrastructure [44]. The Agrafa mountain area, where the majority of landslides triggered by Medicane Ianos occurred, is one of the most hazardous areas in Greece for landslides. Landslide occurrence is mostly due to the factors of highly susceptible lithology (flysch and molassic sediments) and high precipitation rates of the Pindos mountain range. We compare the landslide inventory of Medicane Ianos with historically reported landslide locations in Figure 18. Historical locations were retrieved from [49,50] and official reports of the Institute of Geological and Mineral Exploration. A visual examination shows that Ianos landslides occurred along roughly the same locations of historical landslides, revealing a relation with long-term climatic and lithological/geomorphological conditions.

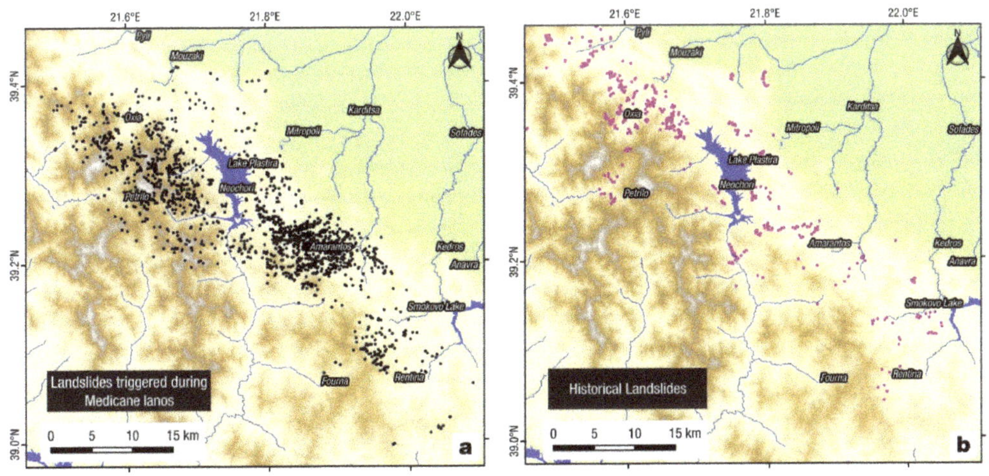

Figure 18. Comparison of (**a**) Medicane Ianos triggered landslides (black dots) and (**b**) historical occurrences of landslides in the region (purple dots).

5. Conclusions

Medicane Ianos in September 2020 caused one of the most widespread catastrophic events for Greece in the last few decades. While most of the areas affected were already susceptible to geohazard occurrences, the areal extent and scale of the almost simultaneous (within three days) occurrence of failures and damage over a large area, represent a unique challenge for future response and mitigation. The area of western Thessaly was the hardest hit by an unprecedent number of landslides and debris flows in the mountains, while the low plains were covered by 475.5 km^2 of flood waters [14,17].

We examined in detail two of the most affected areas, Cephalonia island in the Ionian Sea and Karditsa area in Thessaly, using early remote sensing data and post-event field surveys. Karditsa suffered more than 1500 landsides and numerous debris flows that caused widespread damage and disruption. Cephalonia Island did not experience the number of landsides of western Thessaly, but was severely hit by landslides and debris flows in key location and settlements, destroying bridges and disrupting transportation within the island. Assos settlement was nearly buried by a major debris flow. A dominant factor in these effects was the heavy precipitation experienced by those two areas during Medicane Ianos, in combination with local conditions highly susceptible to mass wasting.

The landside inventory acquired for Karditsa region is the first detailed inventory due to a major atmospheric event in Greece and a valuable asset for studying and mitigating future Medicane events in mainland Greece and surrounding regions. The detailed distribution of landslides can be studied along with meteorological and atmospheric observations, with possible implementation in the hurricane impact model development and forecasting.

As more data sources (satellite imagery) and tools (UAS sensors and platforms, rapid remote sensing big data processing, etc.) will be available to scientists in the near future, response and rapid mapping during future landslide events will be more thorough and faster, based on the experience that data collection from events such as Medicane Ianos provide.

Supplementary Materials: The following supporting information can be downloaded at: https://www.mdpi.com/article/10.3390/app122312443/s1, Figure S1: Comparison of field surveyed landslides with automatic satellite mapping results (a–g). Left: field photos (acquired 1–3 October 2020); right; location map with automatic satellite mapping results (HazMapper). Blue dots show mapped landslides from Sentinel-2 images and triangles with letter marking field locations of surveyed landslides. Location map for sites a–g (h) with Sentinel-2 landslides (orange), Figure S2: Automatic satellite mapping results. Left: method results; middle: method results with mapped landslides (blue dots) and field survey locations (orange triangles); right: Sentinel-2 true color image, acquired on 30 September 2020.

Author Contributions: Conceptualization, S.V. and G.P.; methodology, S.V. and G.P.; validation, S.V., G.P., V.M., C.S., V.K., E.K., I.F., G.Z., J.M. and O.-J.K.; formal analysis, S.V. and G.P.; investigation, S.V., G.P., V.M., C.S., V.K., E.K., I.F., G.Z., J.M. and O.-J.K.; data curation, S.V. and G.P.; writing—original draft preparation, S.V., G.P. and O.-J.K.; writing—review and editing, S.V., G.P., D.Z., G.Z. and O.-J.K.; visualization, S.V. and G.Z.; supervision, G.P., V.M., C.S., D.Z. and G.Z.; project administration, D.Z. and G.Z.; funding acquisition, D.Z. All authors have read and agreed to the published version of the manuscript.

Funding: Part of this research (Geotechnical Extreme Events Reconnaissance-GEER Association work) was funded by the National Science Foundation through the Geotechnical Engineering Program under Grant No. CMMI-1826118.

Institutional Review Board Statement: Not applicable.

Informed Consent Statement: Not applicable.

Data Availability Statement: A detailed description of the field surveys and reconnaissance after Medicane Ianos can be found in the Geotechnical Extreme Events Reconnaissance Report https://doi.org/10.18118/G6MT1T (accessed on 3 July 2022) and the companion web map: https://elxisgroup.com/GEER-MedicaneIanos-FieldMap/ (accessed on 3 July 2022).

Acknowledgments: We would like to thank Athanassios Ganas, Charalambos Kolovos and Vivi Diamantopoulou for discussions and assistance in the field surveys.

Conflicts of Interest: The authors declare no conflict of interest.

References

1. Gorum, T.; van Westen, C.J.; Korup, O.; van der Meijde, M.; Fan, X.; van der Meer, F.D. Complex rupture mechanism and topography control symmetry of mass-wasting pattern, 2010 Haiti earthquake. *Geomorphology* **2013**, *184*, 127–138. [CrossRef]
2. Larsen, M.C. Rainfall-triggered landslides, anthropogenic hazards, and mitigation strategies. *Adv. Geosci.* **2008**, *14*, 147–153. [CrossRef]
3. Gariano, S.L.; Guzzetti, F. Landslides in a changing climate. *Earth Sci. Rev.* **2016**, *162*, 227–252. [CrossRef]
4. Miglietta, M.M. Mediterranean tropical-like cyclones (Medicanes). *Atmosphere* **2019**, *10*, 206. [CrossRef]
5. Emanuel, K. Genesis and maintenance of "Mediterranean hurricanes". *Adv. Geosci.* **2005**, *2*, 217–220. [CrossRef]

6. Kassis, D.; Varlas, G. Hydrographic effects of an intense "medicane" over the central-eastern Mediterranean Sea in 2018. *Dyn. Atmos. Oceans* **2020**, *93*, 101185. [CrossRef]
7. Shaltout, M.; Omstedt, A. Recent sea surface temperature trends and future scenarios for the Mediterranean Sea. *Oceanologia* **2014**, *56*, 411–443. [CrossRef]
8. Koseki, S.; Mooney, P.A.; Cabos, W.; Gaertner, M.; de la Vara, A.; González-Alemán, J.J. Modelling a tropical-like cyclone in the Mediterranean Sea under present and warmer climate. *Nat. Hazards Earth Syst. Sci.* **2021**, *21*, 53–71. [CrossRef]
9. Cavicchia, L.; von Storch, H.; Gualdi, S. A long term climatology of medicanes. *Clim. Dyn.* **2014**, *43*, 1183–1195. [CrossRef]
10. Smart, D. Medicane 'Ianos' over the central Mediterranean 14–20 September 2020. *Weather* **2020**, *75*, 352–353. [CrossRef]
11. Prat, A.C.; Federico, S.; Torcasio, R.C.; D'Adderio, L.P.; Dietrich, S.; Panegrossi, G. Evaluation of the sensitivity of medicane ianos to model microphysics and initial conditions using satellite measurements. *Remote Sens.* **2021**, *13*, 4984. [CrossRef]
12. Lagouvardos, K.; Karagiannidis, A.; Dafis, S.; Kalimeris, A.; Kotroni, V. Ianos—A hurricane in the Mediterranean. *Bull. Am. Meteorol. Soc.* **2022**, *103*, E1621–E1636. [CrossRef]
13. Zimbo, F.; Ingemi, D.; Guidi, G. The tropical-like cyclone "Ianos" in September 2020. *Meteorology* **2022**, *1*, 29–44. [CrossRef]
14. Lekkas, E.; Nastos, P.; Cartalis, C.; Diakakis, M.; Gogou, M.; Mavroulis, S.; Spyrou, N.-I.; Kotsi, E.; Vassilakis, E.; Katsetsiadou, K.-N.; et al. Impact of medicane "IANOS" (September 2020) on Cephalonia and Ithaki Islands. In *Newsletter of Environmental, Disaster and Crises Management Strategies*; National Capodistrian University of Athens: Athens, Greece, 2020; ISSN 2653-9454. Available online: https://edcm.edu.gr/images/docs/newsletters/Newsletter_20_2020_Ianos.pdf (accessed on 3 July 2022).
15. Loli, M.; Mitoulis, S.A.; Tsatsis, A.; Manousakis, J.; Kourkoulis, R.; Zekkos, D. Flood characterization based on forensic analysis of bridge collapse using UAV reconnaissance and CFD simulations. *Sci. Total Environ.* **2022**, *822*, 153661. [CrossRef]
16. Tegos, A.; Ziogas, A.; Bellos, V.; Tzimas, A. Forensic hydrology: A complete reconstruction of an extreme flood event in data-scarce area. *Hydrology* **2022**, *9*, 93. [CrossRef]
17. Zekkos, D.; Zalachoris, G.; Alvertos, A.E.; Amatya, P.M.; Blunts, P.; Clark, M.; Dafis, S.; Farmakis, I.; Ganas, A.; Hille, M.; et al. *The September 18–20 2020 Medicane Iaonos Impact on Greece—Phase I Reconnaissance Report*; Geotechnical Extreme Events Report–Reconnaissance Report, GEER-068; Geer Association: San Francisco, CA, USA, 2020. [CrossRef]
18. Kirschbaum, D.; Stanley, T. Satellite-based assessment of rainfall-triggered landslide hazard for situational awareness. *Earth's Futur.* **2018**, *6*, 505–523. [CrossRef] [PubMed]
19. Khan, S.; Kirschbaum, D.B.; Stanley, T.A.; Amatya, P.M.; Emberson, R.A. Global landslide forecasting system for hazard assessment and situational awareness. *Front. Earth Sci.* **2022**, *10*, 878996. [CrossRef]
20. Kallergis, G. Hydrogeological study of Kalabaka sub-basin (western Thessaly). In *Geological and Geophysical Studies 1970, XIV (1)*; Institute of Geology and Subsurface Investigations: Athens, Greece, 1970; 197p.
21. Lekkas, E. Geological Structure and Geodynamic Evolution of Koziakas Range (Western Thessaly); Geological Monographs. PhD Thesis, Department of Geology, National and Kapodistrian University of Athens, Athens, Greece, 1988; 281p.
22. Brunn, J.H. Contribution à l' étude géologique du Pinde septentrional et d'une partie de la Macédoine occidentale. *Ann. Géol. Pays Hell.* **1956**, *7*, 1–358.
23. Aubouin, J. Contribution a l'étude géologique de la Grèce septentrionale: Les confins de l'Épire et de la Thessalie. *Ann. Geol. Pays Hellen.* **1959**, *10*, 1–484.
24. Ferrière, J. Paléogéographies et Tectoniques Superposées Dans les Hellénides Internes au Niveau de l' Othrys et du Pélion (Grèce). Thèse, Sciences, Université de Lille, Lille, France, 1982. *Soc. Géol. Nord* **1982**, *8*, 970.
25. Nielsen, A.A.; Conradsen, K.; Simpson, J.J. Multivariate Alteration Detection (MAD) and MAF postprocessing in multispectral, bitemporal image data: New approaches to change detection studies. *Remote Sens. Environ.* **1998**, *64*, 1–19. [CrossRef]
26. Brigot, G.; Colin-Koeniguer, E.; Plyer, A.; Janez, F. Adaptation and evaluation of an optical flow method applied to coregistration of forest remote sensing images. *IEEE J. Sel. Top. Appl. Earth Obs. Remote Sens.* **2016**, *9*, 2923–2939. [CrossRef]
27. Avdis, V.; Manakos, K. Geological map of Greece in scale 1:50000. In *Map Sheet Fournas*; Institute of Geological and Mineral Exploration: Athens, Greece, 1991.
28. Karfakis, I. Geological map of Greece in scale 1:50000. In *Map Sheet Mouzakion*; Institute of Geological and Mineral Exploration: Athens, Greece, 1983.
29. Lyberis, N.; Chotin, P.; Bodozis, C. Geological map of Greece in scale 1:50000. In *Map Sheet Agrafa*; Institute of Geological and Mineral Exploration: Athens, Greece, 1983.
30. Manakos, K. Geological map of Greece in scale 1:50000. In *Map Sheet Mirofillon*; Institute of Geological and Mineral Exploration: Athens, Greece, 1993.
31. Savoyat, E. Geological map of Greece in scale 1:50000. In *Map Sheet Karditsa*; Institute of Geological and Mineral Exploration: Athens, Greece, 1969.
32. Bessette-Kirton, E.K.; Coe, J.A.; Schulz, W.H.; Cerovski-Darriau, C.; Einbund, M.M. Mobility characteristics of debris slides and flows triggered by Hurricane Maria in Puerto Rico. *Landslides* **2020**, *17*, 2795–2809. [CrossRef]
33. Papathanassiou, G.; Valkaniotis, S.; Ganas, A.; Stampolidis, A.; Rapti, D.; Caputo, R. Floodplain evolution and its influence on liquefaction clustering: The case study of March 2021 Thessaly, Greece, seismic sequence. *Eng. Geol.* **2022**, *298*, 106542. [CrossRef]
34. Underhill, J.R. Late Cenozoic deformation ofifhe Hellenic foreland, Western Greece. *Geol. Soc. Am. Bull.* **1989**, *101*, 613–634. [CrossRef]

35. British Petroleum Co.; University of Munich; Migiros, G. Geological map of Greece in scale 1:50000. In *2 Map Sheets—Cephalonia Island*; Institute of Geological and Mineral Exploration: Athens, Greece, 1985.
36. Lekkas, E.; Danamos, G.; Mavrikas, G. Geological structure and evolution of Kefallonia and Ithaki islands. *Bull. Geol. Soc. Greece* **2001**, *34*, 11–17. [CrossRef]
37. Lekkas, E. (Ed.) *Neotectonic Map of Greece. Cephalonia—Ithaki Sheet, Scale 1:100.000*; University of Athens—Earthquake Planning and Protection Organization: Athens, Greece, 1996.
38. Valkaniotis, S.; Ganas, A.; Papathanassiou, G.; Papanikolaou, M. Field observations of geological effects triggered by the January–February 2014 Cephalonia (Ionian Sea, Greece) earthquakes. *Tectonophysics* **2014**, *630*, 150–157. [CrossRef]
39. Lekkas, E.L.; Mavroulis, S.D. Earthquake environmental effects and ESI 2007 seismic intensities of the early 2014 Cephalonia (Ionian Sea, western Greece) earthquakes (January 26 and February 3, Mw 6.0). *Nat. Hazards* **2015**, *78*, 1517–1544. [CrossRef]
40. Nikolaou, S.; Zekkos, D.; Asimaki, D.; Gilsanz, R. Reconnaissance highlights of the 2014 sequence of earthquakes in Cephalonia, Greece. In Proceedings of the 6th International Conference on Earthquake Geotechnical Engineering, Christchurch, New Zealand; International Society for Soil Mechanics and Geotechnical Engineering (ISSMGE): London, UK, 2015.
41. Athanasopoulos, G.; Kechagias, G.; Zekkos, D.; Batilas, A.; Karatzia, X.; Lyrantzaki, F.; Platis, A. Lateral spreading of ports in the 2014 Cephalonia, Greece, earthquakes. *Soil Dyn. Earthq. Eng.* **2020**, *128*, 105874. [CrossRef]
42. Burrows, K.; Milledge, D.; Walters, R.J.; Bellugi, D. Integrating empirical models and satellite radar can improve landslide detection for emergency response. *Nat. Hazards Earth Syst. Sci.* **2021**, *21*, 2993–3014. [CrossRef]
43. Gorelick, N.; Hancher, M.; Dixon, M.; Ilyushchenko, S.; Thau, D.; Moore, R. Google earth engine: Planetary-scale geospatial analysis for everyone. *Remote Sens. Environ.* **2017**, *202*, 18–27. [CrossRef]
44. Scheip, C.M.; Wegmann, K.W. HazMapper: A global open-source natural hazard mapping application in Google Earth Engine. *Nat. Hazards Earth Syst. Sci.* **2021**, *21*, 1495–1511. [CrossRef]
45. Handwerger, A.L.; Huang, M.-H.; Jones, S.Y.; Amatya, P.; Kerner, H.R.; Kirschbaum, D.B. Generating landslide density heatmaps for rapid detection using open-access satellite radar data in Google Earth Engine. *Nat. Hazards Earth Syst. Sci.* **2022**, *22*, 753–773. [CrossRef]
46. Lindsay, E.; Frauenfelder, R.; Rüther, D.; Nava, L.; Rubensdotter, L.; Strout, J.; Nordal, S. Multi-temporal satellite image composites in google earth engine for improved landslide visibility: A case study of a glacial landscape. *Remote Sens.* **2022**, *14*, 2301. [CrossRef]
47. Milledge, D.G.; Bellugi, D.G.; Watt, J.; Densmore, A.L. Automated determination of landslide locations after large trigger events: Advantages and disadvantages compared to manual mapping. *Nat. Hazards Earth Syst. Sci.* **2022**, *22*, 481–508. [CrossRef]
48. Scheip, C.; Wegmann, K. HazMapper v1.0 Source Code (Version 1.0). Zenodo. Available online: https://zenodo.org/record/4103348#.Y43JIX3MJPY (accessed on 2 July 2022).
49. Pyrgiotis, L. Engineering Geological Conditions in Karditsa County: Landslides Phenomena in Flysch Formations. Ph.D. Thesis, Department of Geology, University of Patras, Patras, Greece, 1997; p. 362.
50. Apostolidis, E. Engineering-Geological Conditions in the Western Thessaly Basin: Geomechanical Characteristics of the Quaternary Deposits: Analysis Using Geographic Information Systems. Ph.D. Thesis, Department of Geology, University of Patras, Patras, Greece, 2014; p. 1119.

MDPI
St. Alban-Anlage 66
4052 Basel
Switzerland
Tel. +41 61 683 77 34
Fax +41 61 302 89 18
www.mdpi.com

Applied Sciences Editorial Office
E-mail: applsci@mdpi.com
www.mdpi.com/journal/applsci

www.ingramcontent.com/pod-product-compliance
Lightning Source LLC
Chambersburg PA
CBHW040758250926
51850CB00007BA/1024
* 9 7 8 3 0 3 6 5 6 5 3 9 2 *